# Financial Mathematics and Fintech

This series addresses the emerging advances in mathematical theory related to finance and application research from all the fintech perspectives. It is a series of monographs and contributed volumes focusing on the in-depth exploration of financial mathematics such as applied mathematics, statistics, optimization, and scientific computation, and fintech applications such as artificial intelligence, block chain, cloud computing, and big data. This series is featured by the comprehensive understanding and practical application of financial mathematics and fintech. This book series involves cutting-edge applications of financial mathematics and fintech in practical programs and companies.

The Financial Mathematics and Fintech book series promotes the exchange of emerging theory and technology of financial mathematics and fintech between academia and financial practitioner. It aims to provide a timely reflection of the state of art in mathematics and computer science facing to the application of finance. As a collection, this book series provides valuable resources to a wide audience in academia, the finance community, government employees related to finance and anyone else looking to expand their knowledge in financial mathematics and fintech.

The key words in this series include but are not limited to:

a) Financial mathematics
b) Fintech
c) Computer science
d) Artificial intelligence
e) Big data

More information about this series at https://link.springer.com/bookseries/16497

Zhiyong Zheng

# Modern Cryptography Volume 1

## A Classical Introduction to Informational and Mathematical Principle

Springer

Zhiyong Zheng
School of Mathematics
Renmin University of China
Beijing, China

ISSN 2662-7167 ISSN 2662-7175 (electronic)
Financial Mathematics and Fintech
ISBN 978-981-19-0919-1 ISBN 978-981-19-0920-7 (eBook)
https://doi.org/10.1007/978-981-19-0920-7



This Springer imprint is published by the registered company Springer Nature Singapore Pte Ltd.
The registered company address is: 152 Beach Road, #21-01/04 Gateway East, Singapore 189721, Singapore

# Preface

I organized several seminars on cryptography, the students generally reflected that cryptography doesn't need much mathematics, and computer language and computer working environment are more important. Later, I reviewed several common cryptography textbooks at home and abroad. If so, these textbooks are for engineering students, and the purpose is to cultivate cryptographic engineers. It is my original intention to write a textbook of theoretical cryptography for students of mathematics department and science postgraduates, which systematically teaches the statistical characteristics of cryptographic system, the computational complexity theory of cryptographic algorithm and the mathematical principles behind various encryption and decryption algorithms.

With the rapid development of the new generation of digital technology, China has entered the era of information, network and intelligence. Cryptography is not only the cornerstone of national security in the information age, but also a sharp sword to protect people's property security, personal privacy and personal dignity. After the establishment of the first-class discipline of Cyberspace Security, China has established the first-class discipline of security. In particular, on December 19, 2019, China officially promulgated the code law to formulate a law for a discipline. This is rare all over the world. Lately, the central government explicitly requests to cultivate our own cryptography professionals. It can be seen that the discipline construction and personnel training of cryptography have been promoted to the height of national security, which has become a major national strategic demand. Writing a textbook on cryptography theory aims to cultivate our own cryptographers, which is the ultimate reason for writing this book.

Cryptosystem is an ancient art. Since the birth of human beings, there has been cryptosystem. For example, the means of communication used by human beings in war, the marks and conventions used by special groups can be classified into the category of cryptosystem art. Among them, the famous Caesar cryptosystem can be regarded as the representative work of ancient cryptosystem. For thousands of years, cryptosystem, as a technology, relies on personal intelligence and ingenuity. Occasionally, some mathematical ideas and methods were used fragmentarily. This era of cryptographers changed fundamentally only after the great American mathematician M. Shannon came out.

In 1948 and 1949, Shannon successively published two epoch-making papers in the technical bulletin of Bell laboratory. In the first paper, Shannon established the mathematical theory of communication and established the random measurement of information by using the method of probability theory, thus laid the foundation of modern information theory. In the second paper, Shannon established the informatics principle of cryptography, introduced the probability and statistics principle system of mathematics into cryptography structure and cryptanalysis, and transformed the ancient cryptography technology from art to science. Therefore, people not only call Shannon the father of modern information theory, but also the father of modern cryptography.

After Shannon's great changes from the era of cryptographer to the era of cryptoscience, the ancient cryptology technology ushered in the second historic leap in 1976, that is, the era of symmetric cryptography changed into the era of public key cryptography. In 1976, two Stanford University scholars W. Diffie and M. Hellman published a pioneering paper on asymmetric cryptography in *IEEE Transactions on Information Theory* and then entered the era of public key cryptography. Public key cryptography and mathematics are more deeply crossed and integrated, making cryptography an inseparable branch of mathematics. The era characteristic of public key cryptography is to change the cryptography from a few users to mass consumer products, which greatly improves the efficiency and social value of the cryptography. Nowadays, asymmetric cryptosystem is widely used in message authentication, identity authentication, digital signature, digital currency and blockchain architecture, which cannot be replaced by classical cryptosystem.

Based on Shannon's information theory, this book systematically introduces the information theory, statistical characteristics and computational complexity theory of public key cryptography, focusing on the three main algorithms of public key cryptography, RSA, discrete logarithm and elliptic curve cryptosystem, strives to know what it is and why it is, and lays a solid theoretical foundation for new cryptosystem design, cryptoanalysis and attack.

Lattice theory-based cryptography is a representative technology of postquantum cryptography, which is recognized by the academic community as being able to resist quantum computing attacks. At present, the theory and technology of lattice cryptography have not entered university textbooks, and various achievements and introductions have been scattered in research papers at home and abroad in the past two decades. The greatest feature of this book is that it systematically simplifies and combs the theory and technology of lattice cryptography, making it a classroom textbook for senior college students and postgraduates of cryptography, which will play an important role in accelerating the training of modern cryptography talents in China.

This book requires the reader to have a good foundation in algebra, number theory and probability statistics. It is suitable for senior students majoring in mathematics, compulsory for cryptography and science and engineering postgraduates. It can also be used as the main reference book for scientific researchers engaged in cryptography research and cryptographic engineering.

The main contents of this book have been taught in the seminar. My doctoral students Hong Ziwei, Chen Man, Xu Jie, Zhang Mingpei, Associate Professor Huang Wenlin and Dr. Tian Kun have all put forward many useful suggestions and help for the contents of this book. In particular, Chen Man has devoted a lot of time and energy to text printing and proofreading. Here, I would like to express my deep gratitude to them!

Beijing, China
November 2021

Zhiyong Zheng

# Contents

# Acronyms

1. $[x]$ denotes the largest integer not greater than the real number $x$, $\lceil x \rceil$ denotes the smallest integer not less than the real number $x$, so there are

$$[x] \leq x < [x] + 1, \lceil x \rceil - 1 < x \leq \lceil x \rceil.$$

2. $\mathbb{C}$ denotes a complex field, $\mathbb{R}$ denotes a real number field, $\mathbb{Q}$ denotes a rational number field, $\mathbb{F}_q$ denotes a finite field of $q$ elements, $q = p^r$, $p$ is prime, $\mathbb{Z}$ denotes a integer ring, $\mathbb{Z}_m$ denotes a residue class ring of mod$m$ ($m \geq 1$).
3. $(a, b)$ denotes the greatest common divisor of two integers and sometimes a two-dimensional vector.
4. $a$mod$n$ denotes the minimum nonnegative residue of the integer $a$ modulo $n$, i.e., $0 \leq a\text{mod}n < n$. Sometimes it means minimum absolute residue, i.e., $|a\text{mod}n| < \frac{1}{2}n$.
5. Let $\mathbb{F}$ be a field, $\mathbb{F}[x]$ denotes a polynomial ring of one variable over the field $\mathbb{F}$. Sometimes the variable $T$ is used, i.e., $\mathbb{F}[T]$, where $\mathbb{F} = \mathbb{C}, \mathbb{R}, \mathbb{Q}$, or $\mathbb{F} = \mathbb{F}_q$ is a finite field.
6. The base of logarithm log$N$ can be any real number $b > 1$. If $b = 2$, it is binary logarithm, and when $b = q$, it is $q$-base logarithm. Sometimes log$N$ also means natural logarithm, which is determined according to the specific situation.
7. $P\{A\}$ denotes the probability of occurrence of random event $A$.
8. If $G$ is a group, $a \in G$ is the element of the group. Then $o(a)$ denotes the order of $a$.

# Chapter 1
# Preparatory Knowledge

Modern cryptography and information theory is a branch of mathematics which develops rapidly. Almost all mathematical knowledge, such as algebra, geometry, analysis, probability and statistics, has very important applications in information theory. Especially, some modern mathematical theories, such as algebraic geometry, elliptic curve and ergodic theory, play more and more important roles in coding and cryptography. It can be said that information theory is the most dynamic branch of modern mathematics with wide application, strong intersection. This chapter requires the reader to have a preliminary knowledge of analysis, algebra, number theory and probability statistics.

## 1.1 Injective

Let $\sigma$ be a mapping of two nonempty sets $A$ to $B$, denoted as $A \xrightarrow{\sigma} B$. Generally, the mappings between sets can be divided into three categories: injective, surjective and bijective.

**Definition 1.1** Let $\sigma$ be a mapping of two nonempty sets $A \rightarrow B$, we define

(i) $a, b \in A$, if $a \neq b \Rightarrow \sigma(a) \neq \sigma(b)$, call $\sigma$ an injective of $A \rightarrow B$, it is called injective for short.

(ii) If any $b \in B$, there is a $a \in A \Rightarrow \sigma(a) = b$, call $\sigma$ a surjective of $A \rightarrow B$.

(iii) If $A \xrightarrow{\sigma} B$ is an injective and a surjective, call $\sigma$ a bijective of $A \rightarrow B$.

(iv) Let $1_A$ be the identity mapping of $A \rightarrow A$, which is defined as

$$1_A(a) = a, \forall\, a \in A.$$

(v) Suppose $A \xrightarrow{\sigma} B \xrightarrow{\tau} C$ are two mappings, define the product mapping of $\tau$ and $\sigma$, $\tau\sigma : A \rightarrow C$, and define as



Z. Zheng, *Modern Cryptography Volume 1*, Financial Mathematics and Fintech,
https://doi.org/10.1007/978-981-19-0920-7_1

$$\tau\sigma(a) = \tau(\sigma(a)), \ \forall\, a \in A.$$

Obviously, the product of two mappings has no commutativity but has the following associative law.

*Property 1* Let $A \xrightarrow{\sigma} B \xrightarrow{\tau} C \xrightarrow{\delta} D$ be three mappings, then we have

$$(\delta \cdot \tau) \cdot \sigma = \delta \cdot (\tau \cdot \sigma). \tag{1.1}$$

***Proof*** It can be verified directly by definition.

If $A \xrightarrow{\sigma} B$ is a given mappings, obviously, there is

$$\sigma 1_A = \sigma, \ 1_B \sigma = \sigma. \tag{1.2}$$

The above formula shows that identity mapping plays the role of multiplication identity in the product of mapping.

**Definition 1.2** (i) Suppose $A \xrightarrow{\sigma} B \xrightarrow{\tau} A$ are two mappings, if $\tau\sigma = 1_A$, call $\tau$ is a left inverse mapping of $\sigma$, $\sigma$ is a right inverse mapping of $\tau$.
(ii) Let $A \xrightarrow{\sigma} B \xrightarrow{\tau} A$, If $\tau\sigma = 1_A, \sigma\tau = 1_B$, call $\tau$ is an inverse mapping of $\sigma$. Denote as $\tau = \sigma^{-1}$.

The essential properties of injective, surjective and bijective between sets are described by the following lemma.

**Lemma 1.1** *(i) If $A \xrightarrow{\sigma} B$ has an inverse mapping $B \xrightarrow{\tau} A$, that is $\sigma\tau = 1_B$ and $\tau\sigma = 1_A$, then $\tau$ is unique ( denote as $\tau = \sigma^{-1}$).*

*(ii) $A \xrightarrow{\sigma} B$ is an injective if and only if $\sigma$ has a left inverse mapping $B \xrightarrow{\tau} A$, that is $\tau\sigma = 1_A$.*

*(iii) $A \xrightarrow{\sigma} B$ is an surjective if and only if $\sigma$ has a right inverse mapping $B \xrightarrow{\tau} A$, that is $\sigma\tau = 1_B$.*

*(iv) $A \xrightarrow{\sigma} B$ is an bijective if and only if $\sigma$ has an inverse mapping $\tau$, and $\tau$ is unique.*

***Proof*** First of all, prove (i). Let $B \xrightarrow{\tau_1} A$ and $B \xrightarrow{\tau_2} A$ be two inverse mappings of $\sigma$, then we have

$$\tau_1\sigma = 1_A, \tau_2\sigma = 1_A, \text{ and } \sigma\tau_1 = 1_B, \sigma\tau_2 = 1_B,$$

From (1.2), we have

$$\tau_1 = \tau_1 1_B = \tau_1(\sigma\tau_2) = (\tau_1\sigma)\tau_2 = 1_A\tau_2 = \tau_2,$$

so if $\sigma$ has an inverse mapping, then the inverse mapping is unique.

To prove (ii), we note that if $\sigma$ has a left inverse mapping $\tau$, that is $\tau\sigma = 1_A$, then $\sigma$ must be an injective, because if $a, b \in A$, $a \neq b$, then we have $\sigma(a) \neq \sigma(b)$.

If $\sigma(a) = \sigma(b) \Rightarrow \tau(\sigma(a)) = \tau(\sigma(b)) \Rightarrow a = b$, contradiction with $a \neq b$. Conversely, if $A \xrightarrow{\sigma} B$ is an injective, then for the element $\sigma(a) \in \sigma(A)$ in $\sigma(A) \subset B$, Let $\tau(\sigma(a)) = a$, For the elements in the difference set $B \backslash \sigma(A)$, arrange an image randomly, then $B \xrightarrow{\tau} A$ satisfies $\tau\sigma = 1_A$. Similarly, we can prove (iii) and (iv), we thus complete the proof.

In many books of information theory, we often confuse injective and bijective, but they are two different concepts in mathematics, which needs attention.

## 1.2 Computational Complexity

In binary computing environment, the complexity of an algorithm is measured by the number of bit operations. Bit, short for "Binary digit," is the basic amount of information, one bit represents one digit of binary system, two bits represent two digits of binary system, so what is "bit operation"?

To understand "bit operation" accurately, we start with the $b$-ary expression of real number. Let $b > 1$ be a positive integer, and any nonnegative real number $x$can be uniquely expanded into the following geometric series.

$$\begin{aligned} x &= \sum_{-\infty < i \leq k-1} d_i b^i \\ &= d_{k-1} b^{k-1} + d_{k-2} b^{k-2} + \cdots + d_1 b + d_0 + \sum_{i=1}^{+\infty} d_{-i} b^{-i}, \end{aligned} \tag{1.3}$$

where $\forall\ d_i$ satisfies $0 \leq d_i < b$. So we can express $x$ as

$$x = (d_{k-1} d_{k-2} \cdots d_0 d_{-1} d_{-2} \cdots)_b, \tag{1.4}$$

where $(d_{k-1} d_{k-2} \cdots d_0)_b$ is called a $b$-ary integer, $(0.d_{-1} d_{-2} \cdots)_b$ is called a $b$-ary decimal, and

$$x = (d_{k-1} d_{k-2} \cdots d_0)_b + (0.d_{-1} d_{-2} \cdots)_b. \tag{1.5}$$

If $b = 2$, then $x = (d_{k-1} d_{k-2} \cdots d_0 d_{-1} \cdots)_2$ is called the binary representation of $x$. If $b = 10$, then

$$\begin{aligned} x &= (d_{k-1} d_{k-2} \cdots d_1 d_0 d_{-1} d_{-2} \cdots)_{10} \\ &= d_{k-1} d_{k-2} \cdots d_1 d_0 . d_{-1} d_{-2} \cdots . \end{aligned} \tag{1.6}$$

It is our customary decimal expression. It is worth noting that in any system, integers and integers are one-to-one correspondence, and decimals and decimals are one-to-one correspondence. For example, integer in decimal system corresponds to integer in binary system, so does decimal. In other words, the real number of (0, 1) interval on the real number axis corresponds to the decimal number of (0, 1) under the binary system one by one. It should be noted that we often ignore binary decimal; in fact, it is the main technical support of various arithmetic codes, such as Shannon code.

Now let us see the $b$-ary expression of a positive integer $n$ in the decimal system. Let

$$n = (d_{k-1}d_{k-2}\cdots d_1 d_0)_b, 0 \leq d_i < b, d_{k-1} \neq 0,$$

$k$ is the number of $b$-ary digits of $n$.

**Lemma 1.2** *The number $k$ of $b$-ary digits of positive integer $n$ can be calculated according to the following formula:*

$$k = [\log_b n] + 1, \tag{1.7}$$

*$[x]$ denotes the largest integer not greater than the real number $x$.*

***Proof*** Because of $d_{k-1} \neq 0$, there is $b^{k-1} \leq n < b^k$, that is

$$k - 1 \leq \log_b n < k,$$

There's $k - 1 \leq [\log_b n]$ on the left, $[\log_b n] + 1 \leq k$ on the right, and together there's

$$k = [\log_b n] + 1.$$

We complete the proof of Lemma 1.2.

Now let us see the addition operation in $b$-ary system. For simplicity, we consider the addition of two positive integers in binary system. Let $n = (1111000)_2$, $m = (11110)_2$, then $n + m = 1111000 + 0011110 = 10010110$, that is $n + m = (10010110)_2$. The addition of numbers on the same bit actually includes the following five contents (or operations).

1. Observe the numbers in the same bit and note if there are progressions in the right bit(Every two goes into one).
2. If the upper and lower digits of the same bit are 0, and there is no progression on the right side, the sum of the two digits is 0.
3. If both the upper and lower digits of the same digit are 0, but there is a progression, or if one of the two digits is 0 and the other is 1, and there is no progression, the two digits in this digit add up to 1.
4. If two digits of the same digit have one 0, the other one is 1, and there is one progression, or two digits are 1, and there is no progression, the result of addition is 0, and one progression is put forward.
5. If two digits are 1 and have one progression, the sum result is 1 and one progression forward.

**Definition 1.3** A bit operation is an addition operation on the same bit in binary addition. Suppose $A$ is an algorithm in binary system, we use Time($A$) to represent the number of bit operations in algorithm $A$, that is, Time($A$) = completes the total number of bit operations performed by algorithm $A$.

It is easy to deduce the number of bit operations of binary about addition and subtraction by definition. Let $n, m$ be two positive integers, and their binary expression bits are $k$ and $l$ respectively, then

$$\text{Time}(n \pm m) = \max\{k, l\}. \tag{1.8}$$

In the same way, the number of bit operations required for the multiplication of B and D in binary system is satisfied

$$\text{Time}(nm) \leq (k + l) \cdot \min\{k, l\} \leq 2kl. \tag{1.9}$$

It is very convenient to estimate the number of bit operations by using the symbol "$O$" commonly used in number theory. If $f(x)$ and $g(x)$ are two real valued functions, $g(x) > 0$, suppose there are two absolute constants $B$ and $C$ such that when $|x| > B$, we have

$$|f(x)| \leq Cg(x), \quad \text{notes } f(x) = O(g(x)).$$

This sign indicates that when $x \to \infty$, the order of growth of $f(x)$ is the same as that of $g(x)$. For example, let $f(x) = a_d x^d + a_{d-1} x^{d-1} + \cdots + a_1 x + a_0 (a_d > 0)$, then

$$f(x) = O(|x|^d), \quad \text{or } f(n) = O(n^d),\ n \geq 1.$$

For any $\varepsilon > 0$, there is

$$\log n = O(n^\varepsilon),\ n \geq 1.$$

From the lemmas of 1.2, (1.8) and (1.9), we have

**Lemma 1.3** *Let $n, m$ be two positive integers, $k$ and $l$ are the bits of their binary expression, respectively, if $m \leq n$, then $l \leq k$, and*

$$\mathit{Time}(n \pm m) = O(k) = O(\log n);$$

$$\mathit{Time}(nm) = O(kl) = O(\log n \log m);$$

$$\mathit{and} \quad \mathit{Time}\left(\frac{n}{m}\right) = O(kl) = O(\log n \log m).$$

In the above lemma, division is similar to multiplication. Next, we discuss the number of bit operations required to convert a binary representation into a decimal representation, and the number of bit operations required for $n!$ to operate in binary.

**Lemma 1.4** *Let k be the number of digits in binary of n, then*

$$Time(n \text{ convert to decimal expression}) = O(k^2) = O(\log^2 n)$$

*and*

$$Time(n!) = O(n^2k^2) = O(n^2 \log^2 n).$$

***Proof*** To convert $n = (d_{k-1}d_{k-2}\cdots d_1d_0)_2$ to decimal expression. Then divide $n$ by $10 = (1010)_2$, and the remainder is 0, 1, 10, 11, 100, 101, 110, 111, 1000 or 1001, one of these binary numbers, these ten numbers correspond to one of the numbers from 0 to 9 and are denoted as $a_0 (0 \le a_0 \le 9)$, put $a_0$ as the decimal number of $n$. Similarly, divide the quotient by $10 = (1010)_2$, and the remainder is converted into a number from 0 to 9 as the ten digits of $n$ in decimal system. If we go on like this, we use division $[\log_{10} n] + 1$ times, the bit operation required for each division is $O(4k)$, so

$$\text{Time}(n \text{ convert to decimal expression}) \le k \cdot O(4k) = O(k^2).$$

In the same way, we can prove the bit operation estimation of $n!$. We complete the proof of Lemma 1.4.

Let us deduce the computational complexity of some common number theory algorithms. Let $m$ and $n$ be two positive integers, then there is a nonnegative integer $r$ such that $m \equiv r(\text{mod } n)$, where $0 \le r < n$, we call $r$ the smallest nonnegative residue of $m$ under mod $n$, and denote as $r = m \bmod n$. If $1 \le m \le n$, Euclid's division method is usually used to find the greatest common divisor $(n, m)$ of $n$ and $m$. If $(m, n) = 1$, then there is a positive integer $a$ such that $ma \equiv 1(\text{mod } n)$, $a$ is called the multiplicative inverse of $m$ under mod $n$, denote as $m^{-1} \bmod n$. By Bezout formula, if $(n, m) = 1$, then there are integers $x$ and $y$ such that $xm + yn = 1$, we usually use the extended Euclid algorithm to find $x$ and $y$. If we find x, we actually calculate $m^{-1} \bmod n$. Under the above definitions and notations, we have

**Lemma 1.5** *(i) Suppose m and n are two positive integers, then*

$$Time(\text{calculate } m \bmod n) = O(\log n \cdot \log m).$$

*(ii) Suppose m and n are two positive integers, and $m \le n$, then*

$$Time(\text{calculate } (n, m)) = O(\log^3 n).$$

*(iii) Suppose m and n are two positive integers, and $(m, n) = 1$, then*

$$Time(\text{calculate } m^{-1} \bmod n) = O(\log^3 \max(n, m)).$$

*(iv) Suppose $n, m, b$ are positive integers, $b < n$, then*

$$Time(b^m \bmod n) = O(\log m \cdot \log^2 n).$$

***Proof*** To find the minimum nonnegative residue $r$ of $m$ under $\bmod n$ is actually a division with remainder!

$$m = kn + r,\ 0 \le r < n.$$

From the lemma 1.3,

$$\text{Time}(\text{calculate } m \bmod n) = O(\log n \cdot \log m),$$

(i) holds. The Euclid algorithm used to calculate the greatest common divisor $(n, m)$ of $n$ and $m$, in fact, it is a division of $O(\log n)$ times with remainder, so

$$\text{Time}(\text{calculate } (n, m)) = O(\log^3 n).$$

In Euclid algorithm, we can get $x$ and $y$ by pushing from bottom to top, such that $xm + yn = 1$, this incremental process is called the expansion of Euclid algorithm, therefore, if $m \le n$, then

$$\text{Time}(\text{calculate } m^{-1} \bmod n) = \text{Time } (\text{calculate}(n, m)) = O(\log^3 n).$$

(iv) the computational complexity of the power of an integer under $\bmod n$. the proof method is the famous "repeated square method" . Let

$$\begin{aligned} m &= (m_{k-1}m_{k-2}\cdots m_1m_0)_2 \\ &= m_0 + 2m_1 + 4m_2 + \cdots + 2^{k-1}m_{k-1} \end{aligned}$$

be the binary representation of $m$, where $m_i = 0$ or 1. First, let $a = 1$. if $m_0 = 1$, replace $a$ with $b$, if $m_0 = 0$, then $a = 1$ remains unchanged, and let $b_1 = b^2 \bmod n$, this is the first square. If $m_1 = 1$, replace $a$ with $ab_1 \bmod n$, if $m_1 = 0$, $a$ remains unchanged, and let $b_2 = b_1^2 \bmod n$, this is the second square. So if we go on to the square of $j$, we have

$$b_j \equiv b^{2^j} (\bmod n).$$

Our calculation ends after the square of $(k - 1)$; at this time, there is

$$a \equiv b^{m_0+2m_1+4m_2+\cdots+2^{k-1}m_{k-1}} \equiv b^m (\bmod n).$$

Obviously, the number of bit operations per square is $O((\log n^2)^2) = O(\log^2 n)$. There is a total of $k$ square operations, $k = O(\log m)$. So (iv) holds. We have completed the proof.

**Definition 1.4** If an algorithm $f$ involves positive integers $n_1, n_2, \ldots, n_r$, whose binary digits are $k_1, k_2, \ldots, k_r$, and there are absolute nonnegative integers $d_1, d_2, \ldots, d_r$ such that

$$\text{Time}(f) = O(k_1^{d_1} k_2^{d_2} \cdots k_r^{d_r}), \tag{1.10}$$

The complexity of algorithm $f$ is called polynomial; otherwise, it is called nonpolynomial.

From Lemma 1.4 , we can see that addition, subtraction, multiplication and division between positive integers are polynomial algorithms, but $n!$ operation is the simplest example of nonpolynomial algorithm. If we do not need an exact value of $n!$ and only need an approximate value, we can get an approximate value of $n!$ by using a polynomial term algorithm based on Stirling formula (see Sect. 1.4 of this chapter). In the formula (1.10), if $d_1 = d_2 = \cdots = d_r = 0$, the complexity of algorithm $f$ is constant, if $d_1 = d_2 = \cdots = d_r = 1$, the complexity of the algorithm $f$ is said to be linear (the same is true for quadratic, cubic, etc.). In order to characterize nonpolynomial algorithms, we introduce two concepts: exponential and subexponential algorithms.

**Definition 1.5** Suppose that an algorithm $f$ involves a positive integer $n$, and its binary digits are $k$, if

$$\text{Time}(f) = O(t^{g(k)}), \tag{1.11}$$

where $t$ is a constant greater than 1, and $g(k)$ is a polynomial function of $k$ and $\deg g \geq 1$, then the computational complexity of $f$ is exponential. If $g(k)$ is not a polynomial function, but a function smaller than a polynomial, such as $e^{\sqrt{k \log k}}$, then the computational complexity of $f$ is subexponential.

From the above definition, we can see the computational complexity of $n!$, let $k$ be the binary number of $n$, from 1.2, then $n = O(2^k)$, and then from 1.4,

$$\text{Time}(n!) = O(n^2k^2) = O(k^2 2^{2k}) = O(2^{3k}),$$

So the computational complexity of $n!$ in binary system is exponential. This is the simplest example of exponential algorithm.

Bit algorithm cannot only define the computational complexity but also describe the running speed and time complexity of computer. The so-called computer speed refers to the total number of bits that the computer can complete in unit time (such as a second, or 1 microsecond). Therefore, there is no difference between the computational complexity and the time complexity of an algorithm. We can use the figure below to illustrate, suppose that a computer can complete $10^6$ bit operations in one second. When the binary bit of the algorithm is $k = 10^6$, the following figure lists the running time of different computational complexity algorithms on this computer (Table 1.1).

Note that 1 year $\approx 3 \times 10^7$ seconds, the age of the universe is about $10^{10}$ years; when the number of binary digits $k$ is large, the algorithm with exponential or subexponential computational complexity is actually impossible to complete on the computer; therefore, the only way to solve the problem is to improve the speed of the computer.

Computational complexity is often used to describe the complexity of a problem, because the computational complexity is also time complexity when the computer hardware conditions (such as computing speed and storage capacity) remain

**Table 1.1** Time requirements of algorithms with different computational complexity ($k = 10^6$)

| Algorithm type | Complexity | Number of bit operations | Time |
|---|---|---|---|
| Constant degree | $O(1)$ | 1 | 1 microsecond |
| Linear | $O(k)$ | $10^6$ | 1 s |
| Quadratic | $O(k^2)$ | $10^{12}$ | 11.6 days |
| Cubic | $O(k^3)$ | $10^{18}$ | 32000 years |
| Subexponential | $O(e^{\sqrt{k \log k}})$ | About $1.8 \times 10^{1618}$ | $6 \times 10^{1604}$years |
| Exponential | $O(2^k)$ | $10^{301030}$ | $3 \times 10^{301016}$years |

unchanged. At present, the complexity of algorithms is defined in a model called Turing machine. Turing machine is a kind of finite state machine with infinite read and write ability. If the result of each operation and the content of the next operation are uniquely determined, such Turing machine is called deterministic Turing machine. Therefore, the determinacy of a polynomial algorithm is accomplished on a determinate Turing machine.

**Definition 1.6** If a problem can be solved by polynomial algorithm on a certain Turing machine, it is called a $P$ class problem, and the $P$ class problem is often called an easy to handle problem. If a problem can be solved by polynomial algorithm on an uncertain Turing machine, it is called a $NP$ class problem.

According to the definition, the $P$ class problem is definitely a $NP$ class problem, because it can be solved by polynomial algorithm on deterministic Turing machine, and it can also be solved by polynomial algorithm on nondeterministic Turing machine. On the other hand, is the $NP$ problem strictly larger than the $P$ problem? This is an open problem that has not been solved in the field of theoretical computer. There is neither strict proof nor counterexample to show that a problem that can be solved by polynomial on a nondeterministic Turing machine cannot be solved by polynomial algorithm on a deterministic Turing machine. It is widely speculated that the problem of $P$ class and $NP$ class is not equivalent, which is also the cornerstone of many cryptosystems.

## 1.3 Jensen Inequality

A real valued function $f(x)$ in the interval $(a, b)$ is called a strictly convex function, if for $\forall\, x_1, x_2 \in (a, b), \lambda_1 > 0, \lambda_2 > 0, \lambda_1 + \lambda_2 = 1$, we have

$$\lambda_1 f(x_1) + \lambda_2 f(x_2) \leq f(\lambda_1 x_1 + \lambda_2 x_2),$$

and the equation holds if and only if $x_1 = x_2$. By inductive method, we can prove the Jensen inequality as follows.

**Lemma 1.6** *If $f(x)$ is a strictly convex function over $(a, b)$, then for any positive integer $n > 1$, any positive number $\lambda_i (1 \le i \le n)$, $\lambda_1 + \lambda_2 + \cdots + \lambda_n = 1$ and any $x_i \in (a, b)(1 \le i \le n)$, we have*

$$\sum_{i=1}^{n} \lambda_i f(x_i) \le f(\sum_{i=1}^{n} \lambda_i x_i), \tag{1.12}$$

*the equation holds if and only if $x_1 = x_2 = \cdots = x_n$.*

***Proof*** By inductive method, the proposition holds when $n = 1$ and $n = 2$. Suppose the proposition holds for $n - 1$. When $n > 2$, let

$$x' = \frac{\lambda_1}{\lambda_1 + \lambda_2} x_1 + \frac{\lambda_2}{\lambda_1 + \lambda_2} x_2,$$

it can be seen that $x' \in (a, b)$ and $(\lambda_1 + \lambda_2)x' = \lambda_1 x_1 + \lambda_2 x_2$, therefore,

$$\begin{aligned} \sum_{i=1}^{n} \lambda_i f(x_i) &= \lambda_1 f(x_1) + \lambda_2 f(x_2) + \sum_{i=3}^{n} \lambda_i f(x_i) \\ &\le (\lambda_1 + \lambda_2) f(x') + \sum_{i=3}^{n} \lambda_i f(x_i) \\ &\le f(\lambda_1 x_1 + \lambda_2 x_2 + \cdots + \lambda_n x_n). \end{aligned}$$

We have the proposition that holds for $n$. Thus, the inequality (1.12) holds.

From the knowledge of mathematical analysis, $f(x)$ is called a strictly convex function in the interval $(a, b)$ if and only if $f''(x) < 0$. Take $f(x) = \log x$, then $f''(x) = \frac{-1}{x^2 \ln 2}$, thus $\log x$ is a strictly convex function on the interval of $(0, +\infty)$, from Jensen inequality, we have the following inequality.

**Lemma 1.7** *Let $g(x)$ be positive function, that is $g(x) > 0$, then for any integers $\lambda_i (1 \le i \le n)$, $\lambda_1 + \lambda_2 + \cdots + \lambda_n = 1$, and any $a_1, a_2, \ldots, a_n$, we have*

$$\sum_{i=1}^{n} \lambda_i \log g(a_i) \le \log \sum_{i=1}^{n} \lambda_i g(a_i), \tag{1.13}$$

*the equation holds if and only if $g(a_1) = g(a_2) = \cdots = g(a_n)$.*

***Proof*** Because $\log x$ is strictly convex, let $x_i = g(a_i)$, then $x_i \in (0, +\infty)(1 \le i \le n)$, by Jensen inequality,

$$\sum_{i=1}^{n} \lambda_i \log g(a_i) = \sum_{i=1}^{n} \lambda_i \log x_i$$
$$\leq \log(\sum_{i=1}^{n} \lambda_i x_i)$$
$$= \log(\sum_{i=1}^{n} \lambda_i g(a_i)).$$

So the lemma holds.

A real valued function $f(x)$ is called a strictly convex function in the interval $(a, b)$, if for $\forall\, x_1, x_2 \in (a, b)$, $\lambda_1 > 0$, $\lambda_2 > 0$, $\lambda_1 + \lambda_2 = 1$, we have

$$f(\lambda_1 x_1 + \lambda_2 x_2) \leq \lambda_1 f(x_1) + \lambda_2 f(x_2),$$

and the equation holds if and only if $x_1 = x_2$. By induction, we can prove the following general inequality.

**Lemma 1.8** *If $f(x)$ is called a strictly convex function in the interval $(a, b)$, then for any positive integer $n \geq 2$, any positive numbers $\lambda_i (1 \leq i \leq n)$, $\lambda_1 + \lambda_2 + \cdots + \lambda_n = 1$ and any $x_i \in (a, b)(1 \leq i \leq n)$, then we have*

$$f(\sum_{i=1}^{n} \lambda_i x_i) \leq \sum_{i=1}^{n} \lambda_i f(x_i), \tag{1.14}$$

*the equation holds if and only if $x_1 = x_2 = \cdots = x_n$.*

We know that $f(x)$ is strictly convex in the interval $(a, b)$ if and only if $f''(x) > 0$. Let $f(x) = x \log x$, then $f''(x) = \frac{1}{x \ln 2} > 0$, when $x \in (0, +\infty)$. Then we have the following logarithmic inequality.

**Lemma 1.9** *If $a_1, a_2, \ldots, a_n$ and $b_1, b_2, \ldots, b_n$ are two groups of positive numbers, then there are*

$$\sum_{i=1}^{n} a_i \log \frac{a_i}{b_i} \geq (\sum_{i=1}^{n} a_i) \log \frac{\sum_{i=1}^{n} a_i}{\sum_{i=1}^{n} b_i}. \tag{1.15}$$

***Proof*** Because $f(x) = x \log x$ is a strictly convex function, from 1.8, we have

$$f(\sum_{i=1}^{n} \lambda_i x_i) \leq \sum_{i=1}^{n} \lambda_i f(x_i),$$

where $\sum_{i=1}^{n} \lambda_i = 1$. Take $\lambda_i = \frac{b_i}{\sum_{j=1}^{n} b_j}$, $x_i = \frac{a_i}{b_i}$, then

$$\frac{1}{\sum_{j=1}^n b_j}\sum_{i=1}^n a_i \log\frac{\sum_{i=1}^n a_i}{\sum_{i=1}^n b_i} \leq \sum_{i=1}^n \frac{a_i}{\sum_{j=1}^n b_j}\log\frac{a_i}{b_i},$$

$\sum_{j=1}^n b_j$ is deleted at the same time on both sides, then there is

$$(\sum_{i=1}^n a_i)\log\frac{\sum_{i=1}^n a_i}{\sum_{i=1}^n b_i} \leq \sum_{i=1}^n a_i\log\frac{a_i}{b_i},$$

thus (1.15) holds.

The above formula is called logarithm sum inequality, which is often used in information theory.

## 1.4 Stirling Formula

In number theory (see reference 1's Apostol 1976), we can get the average asymptotic formula of some arithmetic functions by using the *Euler* sum formula, the most important of which is the following *Stirling* formula. For all real numbers $x \geq 1$, we have

$$\sum_{1\leq m\leq x} \log m = x\log x - x + O(\log x), \tag{1.16}$$

where the $O$ constant is an absolute constant. Take $x = n \geq 1$ as a positive integer, then there is *Stirling* formula

$$\log n! = n\log n - n + O(\log n). \tag{1.17}$$

In number theory, the *Stirling* formula appears in the more precise form below,

$$n! \approx \sqrt{2\pi n}(\frac{n}{e})^n$$

or

$$\lim_{n\to\infty}\frac{n!}{\sqrt{2\pi n}(\frac{n}{e})^n} = 1.$$

**Lemma 1.10** *Let* $0 \leq m \leq n$, $n, m$ *be nonnegative integer, and* $\binom{n}{m}$ *be the combination number, then*

$$\binom{n}{m} \leq \frac{n^n}{m^m(n-m)^{n-m}}. \tag{1.18}$$

***Proof***

$$n^n = (m + (n - m))^n \geq \binom{n}{m} m^m (n - m)^{n-m},$$

The (1.18) follows at once.

We define the binary entropy function $H(x)(0 \leq x \leq 1)$ as follows.

$$H(x) = \begin{cases} 0, & \text{if } x = 0. \\ -x \log x - (1 - x) \log (1 - x), & \text{if } 0 < x \leq 1. \end{cases} \tag{1.19}$$

It is obvious that $H(x) = H(1 - x)$. So we only need to consider the case of $0 \leq x \leq \frac{1}{2}$. $H(x)$ is the information entropy of binary information space (see the example 3.5 in Sect. 1.1 of Chap. 3), the image description is as follows (Fig. 1.1):

**Lemma 1.11** *Let* $0 \leq \lambda \leq \frac{1}{2}$, *then we have*
*(i)* $\sum_{0 \leq i \leq \lambda n} \binom{n}{i} \leq 2^{nH(\lambda)}$.
*(ii)* $\log \sum_{0 \leq i \leq \lambda n} \binom{n}{i} \leq nH(\lambda)$.
*(iii)* $\lim_{n \to \infty} \frac{1}{n} \log \sum_{0 \leq i \leq \lambda n} \binom{n}{i} = H(\lambda)$.

***Proof*** We first prove that $(i)$, $(ii)$ can be obtained directly from the logarithm of $(i)$.

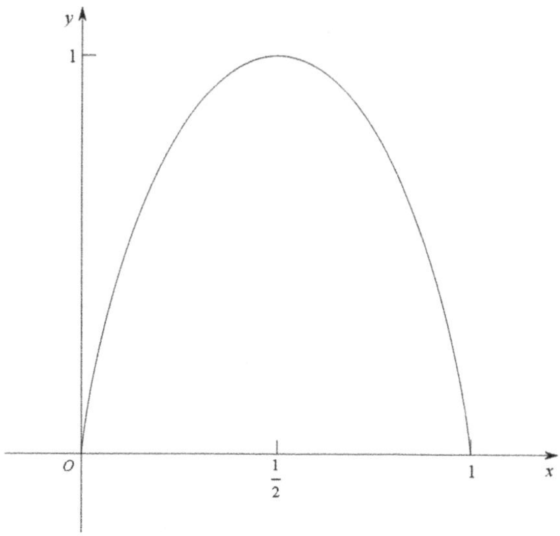


**Fig. 1.1** The information entropy of binary information space

$$
\begin{aligned}
1 = (\lambda + (1-\lambda))^n &\geq \sum_{0\leq i\leq \lambda n} \binom{n}{i} \lambda^i (1-\lambda)^{n-i} \\
&= \sum_{0\leq i\leq \lambda n} \binom{n}{i} (1-\lambda)^n (\frac{\lambda}{1-\lambda})^i \\
&\geq \sum_{0\leq i\leq \lambda n} \binom{n}{i} (1-\lambda)^n (\frac{\lambda}{1-\lambda})^{\lambda n} \\
&= 2^{-nH(\lambda)} \sum_{0\leq i\leq \lambda n} \binom{n}{i}.
\end{aligned}
$$

In order to prove that $(iii)$, we write $m = [\lambda n] = \lambda n + O(1)$, from $(ii)$, we have

$$
\frac{1}{n} \log \sum_{0\leq i\leq \lambda n} \binom{n}{i} \leq H(\lambda).
$$

on the other hand,

$$
\frac{1}{n} \log \sum_{0\leq i\leq \lambda n} \binom{n}{i} \geq \frac{1}{n} \log \binom{n}{m}
$$

$$
= \frac{1}{n} \{\log n! - \log m! - \log(n-m)!\}.
$$

From the *Stirling* formula (1.17), we have

$$
\log n! - \log m! - \log(n-m)! = n\log n - m\log m - (n-m)\log(n-m) + O(\log n).
$$

So there are

$$
\begin{aligned}
\frac{1}{n} \log \sum_{0\leq i\leq \lambda n} \binom{n}{i} &\geq \log n - \lambda \log \lambda n - (1-\lambda)\log n(1-\lambda) + O(\frac{\log n}{n}) \\
&= -\lambda \log \lambda - (1-\lambda)\log(1-\lambda) + O(\frac{\log n}{n}) \\
&= H(\lambda) + O(\frac{\log n}{n}).
\end{aligned}
$$

In the end, we have

$$
\lim_{n\to\infty} \frac{1}{n} \log \sum_{0\leq i\leq \lambda n} \binom{n}{i} = H(\lambda).
$$

Lemma 1.11 holds.

## 1.5 n-fold Bernoulli Experiment

In a given probability space, suppose that $x$ is a random event and $y$ is a random event. We denote the probability of event $x$ occurrence by $p(x)$, the probability of joint occurrence of $x$ and $y$ is denoted by $p(xy)$ and the probability of occurrence of $x$ under the condition of event $y$ is denoted by $p(x|y)$, which is called conditional probability. Obviously, there is a multiplication formula as follows:

$$p(xy) = p(y)p(x|y). \tag{1.20}$$

Two events $x$ and $y$, if $p(xy) = 0$, say $x$ and $y$ are incompatible, if $p(xy) = p(x)p(y)$, say two events are independent, or independent of each other.

A finite set of events $\{x_1, x_2, \ldots, x_n\}$ is called complete event group, if

$$\sum_{i=1}^{n} p(x_i) = 1, \text{ and } p(x_i y_j) = 0, \text{ when } i \neq j. \tag{1.21}$$

In a complete event group, we can assume that $0 < p(x_i) \leq 1 (1 \leq i \leq n)$.

Total probability formula: If $\{x_1, x_2, \ldots, x_n\}$ is a complete event group, $y$ is any random event, then we have

$$p(y) = \sum_{i=1}^{n} p(yx_i) \tag{1.22}$$

and

$$p(y) = \sum_{i=1}^{n} p(x_i)p(y|x_i). \tag{1.23}$$

**Lemma 1.12** *Let $\{x_1, x_2, \ldots, x_n\}$ is a complete event group, then the event $y$ can and can only occur simultaneously with a certain $x_i$, so for any $i$, $1 \leq i \leq n$, we have the following Bayes formula:*

$$p(x_i|y) = \frac{p(x_i)p(y|x_i)}{\sum_{j=1}^{n} p(x_j)p(y|x_j)}, \quad 1 \leq i \leq n. \tag{1.24}$$

***Proof*** From the product formula (1.20), we have

$$p(x_i y) = p(y)p(x_i|y) = p(x_i)p(y|x_i).$$

then there is

$$p(x_i|y) = \frac{p(x_i)p(y|x_i)}{p(y)}.$$

And from the total probability formula (1.23), then we can know

$$p(x_i|y) = \frac{p(x_i)p(y|x_i)}{\sum_{j=1}^{n} p(x_j)p(y|x_j)}, \quad 1 \le i \le n,$$

the Bayes formula (1.24) is proved.

Now we discuss the $n$-fold Bernoulli experiment. In statistical test, the test with only two possible results is called Bernoulli experiment, and the experiment satisfying the following agreement is called n-fold Bernoulli experiment:

(1) There are at most two possible results in each experiment: $a$ or $\bar{a}$.
(2) The probability $p$ of occurrence of $a$ in each test remains unchanged.
(3) Each experiment is statistically independent.
(4) A total of $n$ experiments were carried out.

**Lemma 1.13** (Bernoulli theorem) *In Bernoulli experiment, the probability of event $a$ is $p$, and then in the n-fold Bernoulli experiment, the probability $B(k; n, p)$ of $a$ appearing $k(0 \le k \le n)$ times is*

$$B(k; n, p) = \binom{n}{k} p^k q^{n-k}, \quad q = 1 - p. \tag{1.25}$$

***Proof*** The results of the $i$-th Bernoulli test are recorded as $x_i (x_i = a$ or $\bar{a})$, then $n$-fold Bernoulli experiment forms the following joint event $x$

$$x = x_1 x_2 \cdots x_n, \quad x_i = a \text{ or } \bar{a}.$$

Because of the independence of the experiment, when there are exactly $k$ $x_i = a$, the occurrence probability of $x$ is

$$p(x) = p(x_1)p(x_2) \cdots p(x_n) = p^k q^{n-k}.$$

Obviously, there are exactly $k$ joint events of $x_i = a$, and the total number is $x_i = a$, so

$$B(k; n, p) = \binom{n}{k} p^k q^{n-k}.$$

Lemma 1.13 holds.

In the same way, we can calculate the probability of event $a$ appearing at the $k$-th in multiple Bernoulli experiments.

**Lemma 1.14** *Suppose that $a$ and $\bar{a}$ are two possible events in Bernoulli experiments, then the probability of the first appearance of $a$ in the $k$-th Bernoulli experiment is $pq^{k-1}$.*

***Proof*** Joint event $x = x_1 x_2 \cdots x_k$ formed by $k$-fold Bernoulli experiment, where $k - 1$ $x_i = \bar{a}$, and $x_k = a$, then

$$p(x) = p(x_1) \cdots p(x_{k-1}) p(x_k) = pq^{k-1}.$$

We have completed the proof.

$n$-fold Bernoulli experiment is not only the most basic probability model in probability and statistics, but also a common tool in communication field. Next, we take the error of binary channel transmission as an example to illustrate.

***Example 1.1*** *(Error probability of binary channel)* In binary channel transmission, a codeword $x$ of length $n$ is a vector $x = (x_1, x_2, \ldots, x_n)$ in $n$-dimensional vector space $\mathbb{F}_2^n$, where $x_i = 0$ or $1 (1 \leq i \leq n)$. For convenience, let us write $x = x_1 x_2 \cdots x_n$. Due to channel interference, characters 0 and 1 may have errors in transmission, that is, 0 becomes 1, 1 becomes 0, let the error probability be $p$ ($p$ may be very small), and the error probability of each transmission is constant. Under the above assumption, the codeword $x$ with a transmission length of $n$ can be regarded as a $n$-fold Bernoulli experiment, and the error probability $B(k; n, p)$ of $k (0 \leq k \leq n)$ errors of $x$ in transmission is

$$B(k; n, p) = \binom{n}{k} p^k q^{n-k}, \ q = 1 - p.$$

## 1.6 Chebyshev Inequality

We call the variable $\xi$ defined as a real number in a probability space a random variable. For any real number $x \in (-\infty, +\infty)$, $p(x)$ is defined as the probability of the value $x$ of the random variable $\xi$, i.e.,

$$p(x) = P\{\xi = x\}, \tag{1.26}$$

Call $p(x)$ the probability function of $\xi$. If $\xi$ has only a finite number of values, or countable infinite values, that is, the value space of $\xi$ is a finite number of real numbers, or countable infinite real numbers, then $\xi$ is called discrete random variable; otherwise, $\xi$ is called continuous random variable. The distribution function $F(x)$ of a random variable $\xi$ is defined as

$$F(x) = P\{\xi \leq x\}, \ x \in (-\infty, +\infty). \tag{1.27}$$

Obviously, the distribution function $F(x)$ of $\xi$ is defined as a monotone increasing function on the whole real axis $(-\infty, +\infty)$. And it is a right continuous function, that is $F(x_0) = \lim\limits_{x \to x_0 + 0} F(x)$. The probability distribution of a random variable $\xi$ is completely determined by its distribution function $F(x)$, in fact, for any $x$,

$$p(x) = P\{\xi = x\} = F(x) - F(x - 0).$$

Let $f(x)$ be a nonnegative integrable function on the real axis. And

$$F(x) = \int_{-\infty}^{x} f(t)\,\mathrm{d}t, \tag{1.28}$$

$f(x)$ is called the density function of the random variable $\xi$. Obviously, the density function satisfies:

$$f(x) \geq 0,\ \forall x \in (-\infty, +\infty),\ \int_{-\infty}^{+\infty} f(x)\,\mathrm{d}x = 1. \tag{1.29}$$

On the other hand, the function $f(x)$ satisfying the formula (1.29) must be the density function of a random variable. Here, we introduce several common continuous random variables and their probability distribution.

1. Uniform distribution(Equal probability distribution)

A random variable $\xi$ is equal probability value in interval $[a, b]$, and $\xi$ is said to be uniformly distributed, or it is also called a random variable of uniformly distributed, and its density function is

$$f(x) = \begin{cases} \dfrac{1}{b-a}, & a \leq x \leq b. \\ 0, & \text{otherwise.} \end{cases}$$

Its distribution function $F(x)$ is

$$F(x) = \begin{cases} 0, & \text{when } x < a. \\ \dfrac{x-a}{b-a}, & \text{when } a \leq x \leq b. \\ 1, & \text{when } x > b. \end{cases}$$

2. Exponential distribution

The density function of random variable $\xi$ is

$$f(x) = \begin{cases} \lambda \mathrm{e}^{-\lambda x}, & \text{when } x \geq 0. \\ 0, & \text{when } x < 0. \end{cases}$$

where the given parameter is $\lambda \geq 0$, and its distribution function is

$$F(x) = \begin{cases} 1 - \mathrm{e}^{-\lambda x}, & \text{when } x \geq 0. \\ 0, & \text{when } x < 0. \end{cases}$$

We call $\xi$ an exponential distribution with parameter $\lambda$ or a random variable with exponential distribution.

3. Normal distribution

A continuous random variable $\xi$ whose density function $f(x)$ is defined as:

$$f(x) = \frac{1}{\sqrt{2\pi}\sigma} e^{-\frac{(x-\mu)^2}{2\sigma^2}}, \ x \in (-\infty, +\infty).$$

where $\mu$ and $\sigma$ are constants, $\sigma > 0$. We say that $\xi$ obeys the normal distribution with parameters of $\mu$ and $\sigma^2$, and denote as $\xi \sim N(\mu, \sigma^2)$. By Possion integral,

$$\int_{-\infty}^{+\infty} e^{-x^2} dx = \sqrt{\pi},$$

it is not hard to verify

$$\int_{-\infty}^{+\infty} f(x) dx = 1.$$

The distribution function $F(x)$ of normal distribution $N(\mu, \sigma^2)$ is

$$F(x) = \frac{1}{\sqrt{2\pi}\sigma} \int_{-\infty}^{x} e^{-\frac{(t-\mu)^2}{2\sigma^2}} dt.$$

When $\mu = 0, \sigma = 1$, $N(0, 1)$ is called standard normal distribution.

Let us define the mathematical expectation and variance of a random variable $\xi$. First, let us see the mathematical expectation of a discrete random variable .

(1) Let $\xi$ be a discrete random variable whose value space is $\{x_1, x_2, \ldots, x_n, \ldots\}$. And let $p(x_i) = P\{\xi = x_i\}$. If

$$\sum_{i=1}^{+\infty} |x_i| p(x_i) < \infty,$$

Then the mathematical expectation $E(\xi)$ of $\xi$ is defined as

$$E\xi = E(\xi) = \sum_{i=1}^{+\infty} x_i p(x_i). \quad (1.30)$$

(2) Let $\xi$ be a continuous random variable and $f(x)$ be its density function, if

$$\int_{-\infty}^{+\infty} |x| f(x) dx < +\infty,$$

Then the mathematical expectation $E(\xi)$ of $\xi$ is defined as

$$E\xi = E(\xi) = \int_{-\infty}^{+\infty} xf(x)\mathrm{d}x. \tag{1.31}$$

(3) Let $h(x)$ be a real valued function, then $h(\xi)$ is also a random variable, and $h(\xi)$ is called a function of the random variable $\xi$. The mathematical expectation $E(h(\xi))$ of $h(\xi)$ is $E_h(\xi)$.

**Lemma 1.15** *(1) Let $\xi$ be a discrete random variable whose value space is $\{x_1, x_2, \ldots, x_n, \ldots\}$, if $E(\xi)$ exists, then $E_h(\xi)$ also exists, and*

$$E_h(\xi) = \sum_{i=1}^{+\infty} h(x_i)\, p(x_i).$$

*(2) If $\xi$ is a continuous random variable, and $E(\xi)$ exists, then $E_h(\xi)$ also exists, and*

$$E_h(\xi) = \int_{-\infty}^{+\infty} h(x) f(x)\mathrm{d}x.$$

***Proof*** Let the value space of $\eta = h(\xi)$ be $\{y_1, y_2, \ldots, y_n, \ldots\}$, then

$$P\left\{\eta = y_j\right\} = P\Big(\bigcup_{\substack{i=1\\ h(x_i)=y_j}}^{+\infty} \{\xi = x_i\}\Big) = \sum_{\substack{i=1\\ h(x_i)=y_j}}^{+\infty} P\{\xi = x_i\}.$$

By the definition of $E(\eta)$, then

$$\begin{aligned} E_h(\xi) = E(\eta) &= \sum_{j=1}^{+\infty} y_j P\left\{\eta = y_j\right\} \\ &= \sum_{j=1}^{+\infty} y_j \sum_{\substack{i=1\\ h(x_i)=y_j}}^{+\infty} P\{\xi = x_i\} \\ &= \sum_{i=1}^{+\infty}\Big(\sum_{\substack{j=1\\ h(x_i)=y_j}}^{+\infty} y_j\Big)P\{\xi = x_i\} \\ &= \sum_{i=1}^{+\infty} h(x_i) p(x_i). \end{aligned}$$

The same can be proved (2).

The following basic properties of mathematical expectation are easy to prove.

**Lemma 1.16** *(1) If $\xi = c$ is constant, then $E(\xi) = c$.*
*(2) If $a$ and $b$ are constants, then $E(a\xi + b\xi) = aE(\xi) + bE(\xi)$.*
*(3) If $a \le \xi \le b$, then $a \le E(\xi) \le b$.*

If the mathematical expectation $E(\xi)$ of a random variable exists, then $(\xi - E\xi)^2$ is also a random variable (take $h(x) = (x - a)^2$, where $a = E(\xi)$), We define the mathematical expectation $E_h(\xi)$ of $h(\xi)$ as the variance of $\xi$, denoted as $D(\xi)$, that is

$$D(\xi) = E((\xi - E\xi)^2).$$

Denote $\sigma = \sqrt{D(\xi)}$ is the standard deviation of $\xi$. Here are some basic properties about variance.

**Lemma 1.17** *(1) $D(\xi) = E(\xi^2) - E^2(\xi)$.*
*(2) If $\xi = a$is constant, then $D(\xi) = 0$.*
*(3) $D(\xi + c) = D(\xi)$.*
*(4) $D(c\xi) = c^2 D(\xi)$.*
*(5) If $c \ne E\xi$, then $D(\xi) < E((\xi - c)^2)$.*

***Proof*** (1) can be seen from the definition,

$$\begin{aligned} D(\xi) &= E((\xi - E\xi)^2) \\ &= E(\xi^2 - 2\xi E\xi + E^2(\xi)) \\ &= E(\xi^2) - 2(E\xi)^2 + (E(\xi))^2 \\ &= E(\xi^2) - (E\xi)^2. \end{aligned}$$

(2) is trivial. Let us prove (3). By (1),

$$\begin{aligned} D(\xi + c) &= E((\xi + c)^2) - (E(\xi + c))^2 \\ &= E(\xi^2 + 2c\xi + c^2) - ((E\xi)^2 + 2cE(\xi) + c^2) \\ &= E(\xi^2) + 2cE(\xi) + c^2 - (E\xi)^2 - 2cE(\xi) - c^2 \\ &= E(\xi^2) - (E\xi)^2 = D(\xi). \end{aligned}$$

(4) can also be derived directly from (1). In fact,

$$\begin{aligned} D(c\xi) &= E(c^2\xi^2) - (E(c\xi))^2 \\ &= c^2 E(\xi^2) - c^2(E\xi)^2 \\ &= c^2 D(\xi). \end{aligned}$$

To prove (5), from Lemma 1.16, we notice that the mathematical expectation of $(\xi - E\xi)$ is 0, so if $c \ne E(\xi)$, by (3), we have

$$D(\xi) = D(\xi - c) = E((\xi - c)^2) - (E(\xi - c))^2.$$

Since the last term of the above formula is not zero, we always have

$$D(\xi) < E((\xi - c)^2).$$

(5) holds. This property indicates that $E((\xi - c)^2)$ reaches the minimum value $D(\xi)$ at $c = E\xi$. We have completed the proof.

Now we give the main results of this section; in mathematics, it is called Chebyshev type inequality, which is essentially the so-called moment inequality, because the mathematical expectation $E\xi$ of a random variable $\xi$ is the first-order origin moment and the variance is the second-order moment.

**Theorem 1.1** *Let $h(x)$ be a nonnegative real valued function of $x$, $\xi$ is a random variable, and expectation $E\xi$ exists, then for any $\varepsilon > 0$, we have*

$$P\{h(\xi) \geq \varepsilon\} \leq \frac{E_h(\xi)}{\varepsilon}, \tag{1.32}$$

*and*

$$P\{h(\xi) > \varepsilon\} < \frac{E_h(\xi)}{\varepsilon}. \tag{1.33}$$

***Proof*** We prove the theorem only for continuous random variable $\xi$. Let $f(x)$ be density function of $\xi$, then by Lemma 1.15,

$$\begin{aligned} E_h(\xi) &= \int_{-\infty}^{+\infty} h(x) f(x)\,\mathrm{d}x \\ &\geq \int_{h(x)\geq\varepsilon} h(x) f(x)\,\mathrm{d}x \\ &\geq \varepsilon \int_{h(x)\geq\varepsilon} f(x)\,\mathrm{d}x \\ &= \varepsilon P\{h(x) \geq \varepsilon\}. \end{aligned}$$

so (1.32) holds. Similarly, we can prove (1.33).

In the theorem, we can get different Chebyshev inequality by replacing $h(\xi)$ with $\xi - E\xi$.

**Corollary 1.1** (Chebyshev) *If the variance $D(\xi)$ of the random variable $\xi$ exists, then for any $\varepsilon > 0$, we have*

$$P\{|\xi - E\xi| \geq \varepsilon\} \leq \frac{D(\xi)}{\varepsilon^2}. \tag{1.34}$$

***Proof*** Take $h(\xi) = (\xi - E\xi)^2$ in Theorem 1.1, then $|\xi - E\xi| \geq \varepsilon$ if and only if $h(\xi) \geq \varepsilon^2$, from definition, $E_h(\xi) = D(\xi)$. thus

$$P\{|\xi - E\xi| \geq \varepsilon\} = P\{h(\xi) \geq \varepsilon^2\} \leq \frac{E_h(\xi)}{\varepsilon^2}.$$

The Corollary holds.

**Corollary 1.2** (Chebyshev) *Suppose that both the expected value $E\xi$ and the variance $D(\xi)$ of the random variable $\xi$ exist, then for any $k > 0$, we have*

$$P\{|\xi - E\xi| \geq k\sqrt{D(\xi)}\} \leq \frac{1}{k^2}. \tag{1.35}$$

***Proof*** Take $\varepsilon = k\sqrt{D(\xi)}$ in Corollary 1.1, then

$$P\{|\xi - E\xi| \geq k\sqrt{D(\xi)}\} \leq \frac{D(\xi)}{k^2 D(\xi)} = \frac{1}{k^2}.$$

Corollary 1.2 holds.

In mathematics, $\mu$ is often used as the expected value, $\sigma = \sqrt{D(\xi)}(\sigma \geq 0)$ as the standard deviation, that is

$$\mu = E\xi, \ \sigma = \sqrt{D(\xi)}, \ \sigma \geq 0.$$

Then the Chebyshev inequality in the Corollary 1.2 can be written as follows:

$$P\{|\xi - \mu| \geq k\sigma\} \leq \frac{1}{k^2}. \tag{1.36}$$

**Corollary 1.3** (Markov) *If the expected value of the random variable $\xi$ satisfying the positive integer $|\xi|^k$ of $k \geq 1$ exists, then*

$$P\{|\xi| \geq \varepsilon\} \leq \frac{E|\xi|^k}{\varepsilon^k}.$$

***Proof*** Take $h(\xi) = |\xi|^k$ in Theorem 1.1, Replace $\varepsilon$ with $\varepsilon^k$, then the Markov inequality is directly derived from Theorem 1.1.

Next, we introduce several common discrete random variables and their probability distribution and calculate their expected value and variance.

***Example 1.2*** *(Degenerate distribution)* A random variable $\xi$ takes a constant $a$ with probability 1, that is $\xi = a$, $P\{\xi = a\} = 1$, $\xi$ is called degenerate distribution. From Lemma 1.16, (1), $E\xi = a$, its variance is $D(\xi) = 0$.

***Example 1.3*** *(Two point distribution)* A random variable $\xi$ has only two values $\{x_1, x_2\}$, and its probability distribution is

$$P\{\xi = x_1\} = p, \ P\{\xi = x_2\} = 1 - p, \ 0 < p < 1.$$

$\xi$ is called a two-point distribution with parameter $p$, and its mathematical expectation and variance are

$$\begin{cases} E(\xi) = x_1 p + x_2(1 - p), \\ D(\xi) = p(1 - p)(x_1 - x_2)^2. \end{cases}$$

Specially, take $x_1 = 1, x_2 = 0$, then the expected value and variance of the two-point distribution are

$$E(\xi) = p, \ D(\xi) = p(1 - p).$$

***Example 1.4*** *(Equal probability distribution)* Let a random variable $\xi$ have $n$ values $\{x_1, x_2, \ldots, x_n\}$ and be equal probability distribution, that is

$$P\{\xi = x_i\} = \frac{1}{n}, \ 1 \le i \le n.$$

$\xi$ is called a equal probability distribution or uniform distribution with obeying $n$ points $x_1, x_2, \ldots, x_n$. The expected value and variance are

$$E(\xi) = \frac{1}{n}\sum_{i=1}^{n} x_i, \ D(\xi) = \frac{1}{n}\sum_{i=1}^{n}(x_i - E(\xi))^2.$$

***Example 1.5*** *(Binomial distribution)* In the $n$-fold Bernoulli experiment, the number of times $\xi$ of event $a$ is a random variable from 0 to $n$. The probability distribution is (see Bernoulli experiment)

$$P\{\xi = k\} = b(k; n, p) = \binom{n}{k} p^k q^{n-k},$$

where $0 \le k \le n$, $p$ is the probability of event $a$ occurring in each experiment. $\xi$ is called a binomial distribution with parameter $n, p$, denotes as $\xi \sim b(n, p)$. In fact, $b(k; n, p)$ is the expansion of binomial $(p + q)^n$.

**Lemma 1.18** *Let $\xi \sim b(n, p)$, then*

$$E(\xi) = np, \ D(\xi) = npq, \ q = 1 - p.$$

***Proof*** By definition,

$$\begin{aligned} E(\xi) &= \sum_{k=0}^{n} kb(k; n, p) = \sum_{k=1}^{n} k\binom{n}{k} p^k q^{n-k} \\ &= np\sum_{k=1}^{n} \binom{n-1}{k-1} p^{k-1} q^{(n-1)-(k-1)} \\ &= np\sum_{k=0}^{n-1} \binom{n-1}{k} p^k q^{n-1-k} \\ &= np\sum_{k=0}^{n-1} b(k; n-1, p) \\ &= np. \end{aligned}$$

Similarly, it can be calculated

$$E(\xi^2) = \sum_{k=0}^{n} k^2 b(k; n, p) = n^2p^2 + npq.$$

thus

$$D(\xi) = E(\xi^2) - (E(\xi))^2 = npq.$$

We have completes the proof.

**Lemma 1.19** *$p_n$ is the probability of event a in the n-fold Bernoulli experiment. If $np_n \to \lambda$, then we have*

$$\lim_{n\to\infty} b(k; n, p_n) = \frac{\lambda^k}{k!} e^{-\lambda}.$$

***Proof*** Write $\lambda_n = np_n$, then

$$\begin{aligned} b(k; n, p_n) &= \binom{n}{k} (p_n)^k (1-p_n)^{n-k} \\ &= \frac{n(n-1)\cdots(n-(k-1))}{k!} (\frac{\lambda_n}{n})^k (1-\frac{\lambda_n}{n})^{n-k} \\ &= \frac{(\lambda_n)^k}{k!} (1-\frac{1}{n})\cdots(1-\frac{k-1}{n})(1-\frac{\lambda_n}{n})^{n-k}. \end{aligned}$$

Because for fixed $k$, there is $\lim\limits_{n\to\infty} (\lambda_n)^k = \lambda^k$, and

$$\lim_{n\to\infty} (1-\frac{\lambda_n}{n})^{n-k} = e^{-\lambda},$$

also

$$\lim_{n\to\infty}(1-\frac{1}{n})(1-\frac{2}{n})\cdots(1-\frac{k-1}{n})=1.$$

So there are

$$\lim_{n\to\infty} b(k;n,p_n)=\frac{\lambda^k}{k!}e^{-\lambda}.$$

So Lemma 1.19 holds.

***Example 1.6*** *(Possion distribution)* The value of discrete random variable $\xi$ is $0, 1, \ldots, n, \ldots$, $\lambda \geq 0$ is a nonnegative real number, if the probability distribution of $\xi$ is

$$P\{\xi=k\}=p(k,\lambda)=\frac{\lambda^k}{k!}e^{-\lambda},$$

$\xi$ is called a random variable which obeys Poisson distribution. It can be proved that the expected value and variance of Poisson distribution $\xi$ are $\lambda$. When $p$ is very small, the random variable $\xi_n$ of $n$-fold Bernoulli experiment can be considered to be close to the Poisson distribution $\xi$. In this case, the probability distribution function $b(k; n, p)$ can be approximately replaced by the possion distribution, that is

$$b(k;n,p)\approx\frac{(np)^k}{k!}e^{-np}.$$

## 1.7 Stochastic Process

The so-called stochastic process is to consider the statistical characteristics of a consistent random variable $\{\xi_i\}_{i=1}^n$. We can describe it as a $n$ dimensional random vector. Let $\{\xi_i\}_{i=1}^n$ be $n$ compatible random variables of a given probability space, $\xi=(\xi_1,\xi_2,\ldots,\xi_n)$ is called an $n$-dimensional random vector with values in $\mathbb{R}^n$ in the probability space.

A stochastic process or a $n$ dimensional random vector $\xi=(\xi_1,\xi_2,\ldots,\xi_n)$ is uniquely determined by the occurrence probability of the following joint events. Let $A(\xi_i)\subset\mathbb{R}$ be the value space of random variable $\xi_i (1\leq i\leq n)$; then for any $(x_1,x_2,\ldots,x_n)\in A(\xi_1)\times A(\xi_2)\times\cdots\times A(\xi_n)\subset\mathbb{R}^n$, the probability of occurrence of the following joint event is denoted as

$$p(x_1x_2\cdots x_n)=p((x_1,x_2,\ldots,x_n))=P\{\xi_1=x_1,\xi_2=x_2,\ldots,\xi_n=x_n\}.$$

**Definition 1.7** If for any $x_i\in\mathbb{R}(1\leq i\leq n)$, we have

$$p(x_1x_2\cdots x_n)=p(x_1)p(x_2)\cdots p(x_n).$$

Called stochastic process $\{\xi_i\}_{i=1}^n$ is statistically independent.

Strictly speaking, each real number $x_i$ in the Definition 1.7 should belong to the set of *Borel* on the line to ensure the event $\{\xi_i = x_i\}$ generated by $\xi_i$ is the event in a given probability space.

Similarly, we can define a vector function $F(x_1, x_2, \ldots, x_n)$ in $\mathbb{R}^n$ as

$$F(x_1, x_2, \ldots, x_n) = P\{\xi_1 \le x_1, \xi_2 \le x_2, \ldots, \xi_n \le x_n\},$$

This is the distribution function of random vector $\xi = (\xi_1, \xi_2, \ldots, \xi_n)$. Its marginal distribution function is

$$F_i(x_i) = P\{\xi_i \le x_i\} = F(+\infty, +\infty, \ldots, x_i, +\infty, \ldots, +\infty).$$

For the following properties of stochastic process, we do not give any proof. The reader can find them in the classical probability theory textbook (see reference 1's Rényi 1970, Li 2010, Long 2020).

**Lemma 1.20** *(1) A stochastic process $\{\xi_i\}_{i=1}^n$ is statistically independent if and only if*

$$F(x_1 x_2 \cdots x_n) = F(x_1)F(x_2) \cdots F(x_n).$$

*(2) Suppose $\{\xi_i\}_{i=1}^n$ is statistically independent, for any real value function $g_i(x)$, then $\{g_i(\xi_i)\}_{i=1}^n$ is also statistically independent.*

*(3) If $\xi_i$ is n random variables, then*

$$E(\xi_1 + \xi_2 + \cdots + \xi_n) = E(\xi_1) + E(\xi_2) + \cdots + E(\xi_n).$$

*(4) If $\{\xi_i\}_{i=1}^n$ is statistically independent, the expected value $E(\xi_i)$ of each random variable existence, then the mathematical expectation of random variable $\xi = (\xi_1, \xi_2, \ldots, \xi_n)$ exists, and*

$$E(\xi) = E((\xi_1, \xi_2, \ldots, \xi_n)) = E(\xi_1)E(\xi_2) \cdots E(\xi_n).$$

**Definition 1.8** Let $\{\xi_i\}_{i=1}^{\infty}$ be a series of random variables, $\xi$ is a given random variable, if for any $\xi > 0$, we have

$$\lim_{n\to\infty} P\{|\xi_n - \xi| > \varepsilon\} = 0,$$

it is called $\{\xi_n\}$ converges to $\xi$ in probability, denoted as $\xi_n \xrightarrow{P} \xi$.

Obviously, $\xi_n \xrightarrow{P} \xi$ if and only if for any $\varepsilon > 0$, there is

$$\lim_{n\to\infty} P\{|\xi_n - \xi| \le \varepsilon\} = 1.$$

If the occurrence probability of an event is $p$, the frequency of the event in the statistical test gradually approaches its probability $p$. Strict mathematical statements and proof are attributed to the Bernoulli law of large numbers.

**Theorem 1.2** (Bernoulli) *Let $\mu_n$ be the number of occurrences of event a in the n-fold Bernoulli experiment, it is known that the probability of occurrence of a in each experiment is $p(0 < p < 1)$, then the frequency $\{\frac{\mu_n}{n}\}$ of a converges in probability to p, that is, for any $\varepsilon > 0$, there is*

$$\lim_{n\to\infty} P\{|\frac{\mu_n}{n} - p| > \varepsilon\} = 0.$$

***Proof*** Consider $\frac{\mu_n}{n}$ as a random variable, its expected value and variance are

$$E(\frac{\mu_n}{n}) = \frac{1}{n}E(\mu_n) = p$$

and

$$D(\frac{\mu_n}{n}) = \frac{1}{n^2}D(\mu_n) = \frac{pq}{n}, \quad q = 1 - p.$$

respectively. By Chebyshev inequality (1.34), we have

$$P\{|\frac{\mu_n}{n} - p| > \varepsilon\} < \frac{pq}{n\varepsilon^2}.$$

For any given $\varepsilon > 0$, we have

$$\lim_{n\to\infty} P\{|\frac{\mu_n}{n} - p| > \varepsilon\} = 0.$$

So Bernoulli's law of large numbers holds.

In order to better understand Bernoulli's law of large numbers, we can use a random process to describe it. Define

$$\xi_i = \begin{cases} 1, & \text{if event } a \text{ occurs in the } i\text{-th experiment.} \\ 0, & \text{if event } a \text{ does not occur in the } i\text{-th experiment.} \end{cases}$$

Then $\xi_i$ follows a two-point distribution with parameter $p$ (see Sect. 1.6, example 1.3), and $\{\xi_i\}_{i=1}^{+\infty}$ is an independent and identically distributed stochastic process. Obviously,

$$\mu_n = \sum_{i=1}^{n} \xi_i, \quad E(\xi_i) = p.$$

So Bernoulli's law of large numbers can be rewritten as follows:

$$\lim_{n\to\infty} P\left\{|\frac{1}{n}\sum_{i=1}^{n}\xi_i - \frac{1}{n}E(\sum_{i=1}^{n}\xi_i)| < \varepsilon\right\} = 1,$$

where $\{\xi_i\}$ is a sequence of independent random variables with the same two-point distribution of $0-1$ with parameter $p$. It is not difficult to generalize this conclusion to a more general case.

**Theorem 1.3** (Chebyshev's law of large numbers) *Let* $\{\xi_i\}_{i=1}^{+\infty}$ *be a series of independent random variables, their expected value* $E(\xi_i)$ *and variance* $D(\xi_i)$ *exist, and the variance is bounded, i.e.,* $D(\xi_i) \leq C$ *holds for any* $i \geq 1$*, then for any* $\varepsilon > 0$*, we have*

$$\lim_{n\to\infty} P\{|\frac{1}{n}\sum_{i=1}^{n}\xi_i - \frac{1}{n}\sum_{i=1}^{n}E(\xi_i)| < \varepsilon\} = 1.$$

***Proof*** By Chebyshev inequality,

$$\begin{aligned}
& P\{|\frac{1}{n}\sum_{i=1}^{n}\xi_i - \frac{1}{n}E(\sum_{i=1}^{n}\xi_i)| \geq \varepsilon\} \\
\leq & \frac{D(\frac{1}{n}\sum_{i=1}^{n}\xi_i)}{\varepsilon^2} \\
= & \frac{D(\sum_{i=1}^{n}\xi_i)}{n^2\varepsilon^2} \\
= & \frac{1}{n^2\varepsilon^2}\sum_{i=1}^{n}D(\xi_i) \\
\leq & \frac{C}{n\varepsilon^2}.
\end{aligned}$$

So there are

$$\lim_{n\to\infty} P\{|\frac{1}{n}\sum_{i=1}^{n}\xi_i - \frac{1}{n}\sum_{i=1}^{n}E(\xi_i)| \geq \varepsilon\} = 0.$$

That is, Theorem 1.3 holds.

Chebyshev's law of large numbers is more general than Bernoulli's law of large numbers, it can be understood as a sequence of independent random variables $\{\xi_i\}$, the arithmetic mean of a random variable converges to the arithmetic mean of its expected value in probability.

As a special case, we consider an independent identically distributed stochastic process $\{\xi_i\}$. Because there is the same probability distribution, there is the same expectation and variance.

**Corollary 1.4** *Let $\{\xi_i\}$ be an independent and identically distributed random process, their common expectation is $\mu$, the variance is $\sigma^2$, that is $E(\xi_i) = \mu$, $D(\xi_i) = \sigma^2 (i = 1, 2, \ldots)$, then we have*

$$\lim_{n\to\infty} P\{|\frac{1}{n}\sum_{i=1}^{n} \xi_i - \mu| < \varepsilon\} = 1,$$

*that is $\frac{1}{n}\sum_{i=1}^{n} \xi_i \xrightarrow{P} \mu$.*

In the above Corollary, the existence of variance is unnecessary, Sinchin proved an independent and identically distributed stochastic process $\{\xi_i\}$, as long as the expected value $E(\xi_i) = \mu$ exists. Then $\frac{1}{n}\sum_{i=1}^{n} \xi_i$ converges to its expected value in probability. This conclusion is called Sinchin's law of large numbers.

Finally, we state the so-called Lindbergh Levy's central limit theorem without proof.

**Theorem 1.4** (central limit theorem) *Let $\{\xi_i\}_{i=1}^{+\infty}$ is an independent and identically distributed stochastic process, the expected value is $E(\xi_i) = \mu$, the variance is $D(\xi_i) = \sigma^2 > 0 (i = 1, 2, \ldots)$, then for any $x$, we have*

$$\lim_{n\to\infty} P\{\frac{\sum_{i=1}^{n} \xi_i - n\mu}{\sigma\sqrt{n}} \leq x\} = \frac{1}{\sqrt{2\pi}} \int_{-\infty}^{x} e^{-\frac{t^2}{2}} \mathrm{d}t,$$

*That is, the sum of random variables $\sum_{i=1}^{n} \xi_i$, whose standardized variables converge to the standard normal distribution $N(0, 1)$ in probability.*

**Exercise 1** (Nie and Ding 2000)

1. Let $A, B, C$ be three nonempty sets, $A \xrightarrow{\sigma} B$ is the given mapping, $\tau_1$ and $\tau_2$ are any two mappings of $B \to C$. Prove: if $\sigma$ is surjective and $\tau_1\sigma = \tau_2\sigma$, then $\tau_1 = \tau_2$.
2. Let $\tau_1$ and $\tau_2$ be any two mappings of $A \to B$, $\sigma$ is the given mapping of $B \to C$. Prove: if $\sigma$ is injective and $\sigma\tau_1 = \sigma\tau_2$, then $\tau_1 = \tau_2$.
3. Let $A \xrightarrow{\sigma} B$ be a injective, $\tau : B \to A$ is the left inverse of $\sigma$, Is the left inverse $\tau$ of $\sigma$ unique?
4. Let $A \xrightarrow{\sigma} B$ be a surjective, Is the right inverse of $\sigma$ unique?
5. Suppose that $a, m, n$ are integers, $a \geq 0, m \geq 1, n \geq 1$, prove

$$(a^{2^m} + 1, a^{2^n} + 1) = 1 \text{ or } 2.$$

   Thus prove Polya theorem: there are infinitely many primes.
6. On the positive integer set, the Möbius function $\mu(n)$ is defined as

$$\mu(n) = \begin{cases} 1, & \text{when } n = 1, \\ 0, & \text{when } n \text{ contain square factor,} \\ (-1)^t, & \text{when } n = p_1 p_2 \cdots p_t,\ p_i \text{are different primes.} \end{cases}$$

Prove Möbius identity

$$\sum_{d|n} \mu(d) = \begin{cases} 1, & \text{when } n = 1, \\ 0, & \text{when } n > 1. \end{cases}$$

7. Suppose $\varphi(n)$ is a Euler function, prove

$$\varphi(n) = n \sum_{d|n} \frac{\mu(d)}{d}.$$

8. Let $n \geq 1$ be positive integer, prove Wilson theorem:

$$(n-1)! + 1 \equiv 0 (\bmod\, n)$$

if and only if $n$ is prime.
9. Let $n$ and $b$ be positive integers, $n > b$, prove $n$ can be uniquely expressed as the following $b$-ary number:

$$n = b_0 + b_1 b + b_2 b^2 + \cdots + b_{r-1} b^{r-1}, \text{ where } 0 \leq b_i < b, r \geq 1.$$

$n = (b_{r-1} b_{r-2} \cdots b_1 b_0)_b$ is called the $b$-ary expression of $n$ and $r$ is called the $b$-ary digit of $n$.
10. Let $f(n)$ be a complex valued function on a set of positive integers, and prove the inversion formula of Möbius:

$$F(n) = \sum_{d|n} f(d),\ \forall n \geq 1 \Leftrightarrow f(n) = \sum_{d|n} \mu(d) F(\frac{n}{d}),\ \forall n \geq 1.$$

11. Prove that the following sum formula:

$$\sum_{\substack{1 \leq r \leq n \\ (r,n)=1}} r = \frac{n\varphi(n)}{2}.$$

12. Prove: There are infinitely many primes $p$ satisfies $p \equiv -1 (\bmod\, 6)$.
13. Solve the congruence equation: $27x \equiv 25 (\bmod\, 31)$.
14. Let $p$ be a prime, $n \geq 1$ be a positive integer, find the number of solutions of quadratic congruence equation $x^2 \equiv 1 (\bmod\, p^n)$.

15. In order to reduce the number of games, 20 teams were divided into two groups, each with 10 team, find the probability that the strongest two teams will be in the same group, and the probability of the strongest two teams in different groups.
16. (Banach question). A mathematician has two boxes of matches. Each box has $N$ matches. When he uses them, he takes one match from any box and calculates the probability that one box has $k$ Matches and the other box is empty.
17. A stick of length $l$ can break at any two points, find the probability that the three pieces of the stick can form a triangle.
18. There are $k$ jars, each containing $n$ balls, numbered from 1 to $n$. Now take any ball from each jar and ask the probability of that $m$ is the largest number in the ball.
19. Take any three of the five numbers of 1, 2, 3, 4, 5 and arrange them from small to large. Let $X$ denote the number in the middle and find the probability distribution of $X$.
20. Let $F(x)$ be a distribution function of a continuous random variable, $a > 0$, prove

$$\int_{-\infty}^{+\infty} |F(x+a) - F(x)| dx = a.$$

21. (Generalization of Bernoulli's law of large numbers) Let $\mu_n$ is the number of occurrences of event $A$ in the first $n$ experiments of a series of independent Bernoulli experiments, it is known that the probability of occurrence of event $A$ in the $i$ test is $p_i$, try to write the corresponding law of large numbers and prove it.

## References

Apostol, T. M. (1976). *Introduction to analytic number theory*. Springer.
Hardy, G. H., & Wright, E. M. (1979). *An introduction to the theory of number*. Oxford University Press.
Jacobson, N. (1989). *Basic algebra (I)*. Translated by the Department of algebra, Department of mathematics, Shanghai Normal University, Beijing, Higher Education Press (in Chinese).
Leveque, W. J. (1977). *Fundamentals of number theory*. Addison-Wesley.
Li, X. (2010). *Basic probability theory*. Higher Education Press (in Chinese)
Lidl, R., & Niederreiter, H. (1983). *Finite fields*. Addison-Wesley.
Long, Y. (2020). *Probability theory and mathematical statistics*. Higher Education Press (in Chinese)
Nie, L., & Ding, S. (2000). *Introduction to algebra*. Higher Education Press (in Chinese)
Rényi, A. (1970). *Probability theory*. North-Holland.
Rosen, K. H. (1984). *Elementaty number theory and its applications*. Addison-Wesley.
Rosen, M. H. (2002). *Number theory in function fields*. Springer.
Spencer, D. (1982). *Computers in number theory*. Computer Science Press.
VanderWalden, B. L. (1963). *Algebra (I)*. Translated by Shisun Ding: Kencheng Zeng, Fuxin Hao, Beijing, Science Press (in Chinese).

VanderWalden, B. L. (1976). *Algebra (II)*. Translated by Xihua Cao: Kencheng Zeng, Fuxin Hao, Beijing, Science Press (in Chinese).
VanLint, J. H. (1991). *Introduction to coding theory*. Springer.

# Chapter 2
# The Basis of Code Theory

The channel of information transmission is called channel for short. The commonly used channels include cable, optical fiber, medium of radio wave transmission and carrier line, etc., and also include tape, optical disk, etc. The channel constitutes the physical conditions for social information to interact across space and time. In addition, a piece of social information, such as various language information, picture information, data information and so on, should be exchanged across time and space, information coding is the basic technical means. What is information coding? In short, it is the process of digitizing all kinds of social information. Digitization is not a simple digital substitution of social information, but is full of profound mathematical principles and beautiful mathematical technology. For example, the source code used for data compression and storage uses the principle of probability statistics to attach the required statistical characteristics to social information, so the source code is also called random code. The other is the so-called channel coding, which is used to overcome the channel interference. This kind of code is full of beautiful algebra, geometry and various mathematical techniques in combinatorics, in order to improve the accuracy of information transmission, so the channel coding is also called algebraic combinatorial code. The main purpose of this chapter is to introduce the basic knowledge of code theory for channel coding. Source coding will be introduced in Chap. 3.

With the hardware support of channel and the software technology of information coding, we can implement the long-distance exchange of various social information across time and space. Taking channel coding as an example, this process can be described as the following diagram (Fig. 2.1).

In 1948, American mathematician Shannon published his pioneering paper "Mathematical Principles of Communication" in the technical bulletin of Bell laboratory, marking the advent of the era of electronic information. In this paper, Shannon proved the existence of "good code" with the rate infinitely close to the channel capacity and the transmission error probability arbitrarily small by using probability theory (see Theorem in this Chap. 2.10), on the other hand, if the transmission



Z. Zheng, *Modern Cryptography Volume 1*, Financial Mathematics and Fintech,
https://doi.org/10.1007/978-981-19-0920-7_2

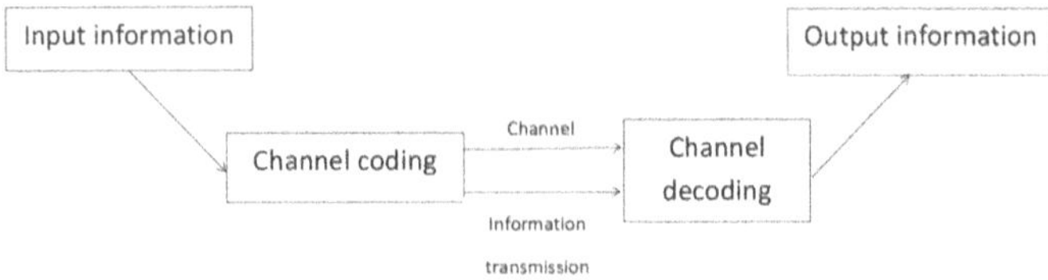


**Fig. 2.1** Channel Coding

error probability is arbitrarily small, the code rate (transmission efficiency) does not exceed an upper bound (channel capacity) (see Theorem in Chap. 3). This upper bound is called Shannon's limit, which is regarded as the golden rule in the field of electronic communication engineering technology.

Shannon's theorem is an existence proof rather than a constructive proof. How to construct the so-called good code which can not only ensure the communication efficiency (the code rate is as large as possible), but also control the transmission error rate is the unremitting goal after the advent of Shannon's theory. From Hamming and Golay to Elias, Goppa, Berrou and Turkish mathematician Arikan, from Hamming code, Golay code to convolutional code, turbo code to polar code, over the past decades, electronic communication has reached one peak after another, creating one technological miracle after another, until today's 5G era. In 1969, the U.S. Mars probe used Hadamard code to transmit image information. For the first time, mankind was lucky to witness one beautiful picture after another in outer space, in 1971, the U.S. Jupiter and Saturn probe used the famous Golay code G23 to send hundreds of frames of color photos of Jupiter and Saturn back to earth, 70 years of exploration of channel coding is a magnificent history of electronic communication.

The main purpose of this chapter is to strictly define and prove the mathematical characteristics of general codes in theory, so as to provide a solid mathematical foundation for further study of coding technology and cryptography. This chapter includes Hamming distance, Lee distance, linear code, some typical good codes, MacWilliams theorem and famous Shannon coding theorem. Master the content of this chapter, we will have a basic and comprehensive understanding of channel coding theory (error correction code).

## 2.1 Hamming Distance

In channel coding, the alphabet usually chooses a $q$-element finite field $\mathbb{F}_q$, sometimes a ring $\mathbb{Z}_m$, where $q$ is the power of a prime. Let $n \geqslant 1$ be a positive integer, $\mathbb{F}_q^n$ is an $n$-dimensional linear space over $\mathbb{F}_q$, also called codeword space.

$$\mathbb{F}_q^n = \{x = (x_1, x_2, \ldots, x_n) | \forall x_i \in \mathbb{F}_q\}.$$

A vector $x = (x_1, x_2, \ldots, x_n)$ in $\mathbb{F}_q^n$ is called a codeword of length $n$. For convenience, a codeword $x$, we write as $x = x_1x_2 \ldots x_n$, each $x_i \in \mathbb{F}_q$ is called a character, denoted by $0 = (0, 0, \ldots, 0)$.

Two codewords $x = x_1x_2 \ldots x_n$ and $y = y_1y_2 \ldots y_n$ define the number of characters whose Hamming distance is different from $x$ and $y$, that is

$$d(x, y) = \#\{i | 1 \leq i \leq n, x_i \neq y_i\}. \tag{2.1}$$

Obviously $0 \leqslant d(x, y) \leqslant n$ is a positive integer, the weight function of a codeword $x \in \mathbb{F}_q^n$ is defined as $w(x) = d(x, 0)$, that is Hamming distance between $x$ and 0. The following properties are obvious.

**Property 2.1** *If* $x, y \in \mathbb{F}_q^n$, *then*

*(i)* $d(x, y) \geq 0$, $d(x, y) = 0$ *if and only if* $x = y$.
*(ii)* $d(x, y) = d(y, x)$.
*(iii)* $w(-x) = w(x)$.
*(iv)* $d(x, y) = d(x - z, y - z)$, $\forall\, z \in \mathbb{F}_q^n$.
*(v)* $d(x, y) = w(x - y)$.

*Property (i) is called nonnegativity, property (ii) symmetry and property (iv) translation invariance. This is the basic property of distance function in mathematics, and we can analogy with the distance between two points in plane or Euclidean space.*

**Lemma 2.1** *Let* $x, y \in \mathbb{F}_q^n$ *be two codings, then*

$$w(x \pm y) \leqslant w(x) + w(y).$$

***Proof*** Because $w(-x) = w(x)$, so $w(x - y) = w(x + (-y))$. We can only prove $w(x + y) \leqslant w(x) + w(y)$. Let $x = x_1 \ldots x_n$, $y = y_1 \ldots y_n$, then

$$x + y = (x_1 + y_1)(x_2 + y_2) \ldots (x_n + y_n).$$

Obviously, if $x_i + y_i \neq 0$, then $x_i \neq 0$, or $y_i \neq 0 (1 \leqslant i \leqslant n)$. Thus $w(x + y) \leqslant w(x) + w(y)$.

$$w(x - y) = w(x + (-y)) \leqslant w(x) + w(-y) = w(x) + w(y).$$

We have completed the proof.

**Lemma 2.2** (Trigonometric inequality) *If* $x, y, z \in \mathbb{F}_q^n$ *are three codings, then*

$$d(x, y) \leqslant d(x, z) + d(z, y).$$

***Proof*** From 2.1, if $z \in \mathbb{F}_q^n$, then

$$w(x - y) \leq w(x - z) + w(z - y).$$

Then by property (v), $d(x, y) = w(x - y)$, we have

$$d(x, y) \leqslant d(x, z) + d(z, y).$$

The Lemma holds.

The nonnegativity, symmetry, translation invariance of Hamming distance and trigonometric inequality described in lemma 2 together show that Hamming distance of two codewords is equal to the distance between two points in physical space, which is a real distance function in mathematical sense. Similarly, we can define the concept of ball. A Hamming sphere with radius $\rho$ centered on codeword $x$ is defined as

$$B_\rho(x) = \{y | y \in \mathbb{F}_q^n, d(x, y) \leqslant \rho\}, \tag{2.2}$$

where $\rho$ is a nonnegative integer. Obviously, $B_0(x) = \{x\}$ contains only one codeword.

**Lemma 2.3** *For any $x \in \mathbb{F}_q^n$, $0 \leqslant \rho \leqslant n$, we have*

$$|B_\rho(x)| = \sum_{i=0}^{\rho} \binom{n}{i} (q-1)^i, \tag{2.3}$$

*where $|B_\rho(x)|$ is the number of codewords in Hamming ball $B_\rho(x)$.*

***Proof*** Let $x = x_1 x_2 \ldots x_n$, $0 \leqslant i \leqslant \rho$, $i$ given, let

$$A_i = \#\{y \in \mathbb{F}_q^n | d(y, x) = i\}.$$

Obviously,

$$A_i = \binom{n}{i} (q-1)^i,$$

so

$$|B_\rho(x)| = \sum_{i=0}^{\rho} A_i = \sum_{i=0}^{\rho} \binom{n}{i} (q-1)^i.$$

**Corollary 2.1** *For $\forall x \in \mathbb{F}_q^n$, we have*

$$|B_\rho(x)| = |B_\rho(0)|.$$

*That is to say, the number of codewords in $B_\rho(x)$ is a constant which only depends on radius $\rho$. This constant is usually denoted as $B_\rho$.*

**Definition 2.1** If $C \subsetneqq \mathbb{F}_q^n$, $C$ is called a $q$-ary code, code for short, $|C|$ is the number of codewords in code $C$. If $|C| = 1$, we call $C$ a trivial code, and all the codes we discuss are nontrivial codes.

For a code of $C$, the following five mathematical quantities are of basic importance.

**Definition 2.2** If $C$ is a code, define

Bit rate of $C$ $R = R_C = \frac{1}{n} \log_q |C|$
Minimum distance of $C$ $d = \min\{d(x, y) | x, y \in C, x \neq y\}$
Minimal weight of $C$ $w = \min\{w(x) | x \in C, x \neq 0\}$
Coverage radius of $C$ $\rho = \max\{\min\{d(x, c) | c \in C\} | x \in \mathbb{F}_q^n\}$
Disjoint radius of $C$ $\rho_1 = \max\{r | 0 \leqslant r \leqslant n, \ B_r(c_1) \cap B_r(c_2) = \phi, \ \forall c_1, \ c_2 \in C, \ c_1 \neq c_2\}$

It is important to discuss the relationship between the above five mathematical quantities for the study of codes. We begin by proving lemma 2.4.

**Lemma 2.4** *Let $d$ be minimum distance of $C$, $\rho_1$ be disjoint radius of $C$, then*

$$d = 2\rho_1 + 1, \quad d = 2\rho_1 + 2.$$

***Proof*** We can only prove $2\rho_1 + 1 \leqslant d \leqslant 2\rho_1 + 2$. If $d \leqslant 2\rho_1$, then there are codewords $c_1 \in C, c_2 \in C, c_1 \neq c_2$ such that

$$d(c_1, c_2) \leqslant 2\rho_1.$$

This means that $c_1$ and $c_2$ have at most $2\rho_1$ different characters. Without losing generality, we can make the first $2\rho_1$ characters of $c_1$ and $c_2$ different, that is

$$\begin{cases} c_1 = a_1 a_2 \ldots a_{\rho_1} a_{\rho_1+1} \ldots a_{2\rho_1} \ * * \cdots * \\ c_2 = b_1 b_2 \ldots b_{\rho_1} b_{\rho_1+1} \ldots b_{2\rho_1} \ * * \cdots * \end{cases}$$

where * represents the same character. We can put

$$x = a_1 a_2 \ldots a_{\rho_1} b_{\rho_1+1} \ldots b_{2\rho_1} \ * \cdots *,$$

this shows that

$$d(x, c_1) \leqslant \rho_1, \quad d(x, c_2) \leqslant \rho_1.$$

That is

$$x \in B_{\rho_1}(c_1) \cap B_{\rho_1}(c_2).$$

It's in contradiction with $B_{\rho_1}(c_1) \cap B_{\rho_1}(c_2) = \phi$. So we have $d \geqslant 2\rho_1 + 1$. If $d > 2\rho_1 + 2 = 2(\rho_1 + 1)$, then we can prove the following formula, which is in contradiction with the definition of disjoint radius $\rho_1$.

$$B_{\rho_1+1}(c_1) \cap B_{\rho_1+1}(c_2) = \phi, \ \forall c_1, \ c_2 \in C, \ c_1 \neq c_2.$$

Because if the above formula does not hold, then $c_1$, $c_2 \in C$, $c_1 \neq c_2$, $B_{\rho_1+1}(c_1)$ intersects with $B_{\rho_1+1}(c_2)$, we might as well make

$$x \in B_{\rho_1+1}(c_1) \cap B_{\rho_1+1}(c_2),$$

Then the trigonometric inequality of lemma 2.2 is derived

$$d(c_1, c_2) \leqslant d(c_1, x) + d(c_2, x) \leqslant 2(\rho_1 + 1),$$

It contradicts the hypothesis of $d > 2(\rho_1 + 1)$. So we have $2\rho_1 + 1 \leqslant d \leqslant 2\rho_1 + 2$. The Lemma holds.

In order to discuss the geometric meaning of covering radius $\rho$, we consider the set $\{B_\rho(c)|c \in C\}$ of balls on code $C$, if

$$\bigcup_{c \in C} B_\rho(c) = \mathbb{F}_q^n,$$

Then $\{B_\rho(c)|c \in C\}$ is called a cover of codeword space $\mathbb{F}_q^n$.

**Lemma 2.5** *Let $\rho$ be the covering radius of $C$, then $\rho$ is the smallest positive integer of $\{B_\rho(c)|c \in C\}$ covering $\mathbb{F}_q^n$.*

***Proof*** By the definition of $\rho$, for all $x \in \mathbb{F}_q^n$, there is

$$\min\{d(x, c)|c \in C\} \leqslant \rho.$$

Therefore, when $x \in \mathbb{F}_q^n$ is given, there is a codeword $c \in C \Rightarrow d(x, c) \leqslant \rho$, that is $x \in B_\rho(c)$, this shows that

$$\bigcup_{c \in C} B_\rho(c) = \mathbb{F}_q^n.$$

That is, $\{B_\rho(c)|c \in C\}$ forms a cover of $\mathbb{F}_q^n$. Obviously, $\{B_{\rho-1}(c)|c \in C\}$ can't cover $\mathbb{F}_q^n$, because if

$$\bigcup_{c \in C} B_{\rho-1}(c) = \mathbb{F}_q^n.$$

Then for any $x \in \mathbb{F}_q^n$, $\exists\, c \in C \Rightarrow x \in B_{\rho-1}(c)$, so

$$\min\{d(x, c)|c \in C\} \leqslant \rho - 1.$$

Thus

$$\rho = \max\{\min\{d(x, c)|c \in C\}|x \in \mathbb{F}_q^n\} \leqslant \rho - 1.$$

The contradiction indicates that $\rho$ is the smallest positive integer. The lemma holds.

**Lemma 2.6** *Let $d$ be the minimum distance of $C$ and $\rho$ be the covering radius of $C$, then*

$$d \leqslant 2\rho + 1.$$

***Proof*** If $d > 2\rho + 1$, Let $c_0 \in C$ be given, then we have

$$B_{\rho+1}(c_0) \cap B_\rho(c) = \phi, \ \forall c \in C, \ c \neq c_0.$$

So you can choose $x \in B_{\rho+1}(c_0)$, and $d(x, c_0) = \rho + 1$, then

$$x \notin B_\rho(c_0), \ x \notin B_\rho(c), \ \forall c \in C.$$

That is, $\{B_\rho(c)|c \in C\}$ cannot cover $\mathbb{F}_q^n$, which is contrary to lemma 2.5. So we always have $d \leqslant 2\rho + 1$. The Lemma holds.

Combining the above three lemmas, we can get the following simple but very important corollaries.

**Corollary 2.2** *Let $C \subset \mathbb{F}_q^n$ be an arbitrary $q$-ary code. $d$, $\rho$, $\rho_1$ are the minimum distance, covering radius and disjoint radius of $C$ respectively, then*

*(i)* $\rho_1 \leqslant \rho$.
*(ii) If the minimum distance of $C$ is $d = 2e + 1 \Rightarrow e = \rho_1$.*

***Proof*** (i) Directly from $2\rho_1 + 1 \leqslant d \leqslant 2\rho + 1$, if $d = 2e + 1$ is odd, then by the lemma 2.4, $d = 2\rho_1 + 1 = 2e + 1 \Rightarrow e = \rho_1$.

**Definition 2.3** A code $C$, if $\rho = \rho_1$, is called a perfect code.

**Corollary 2.3** *(i) The minimum distance of any perfect code $C$ is $d = 2\rho + 1$.*
*(ii) The minimum distance of a code $C$ is $d = 2e + 1$, Then $C$ is a perfect code if and only if $\forall\, x \in \mathbb{F}_q^n$, $\exists$ the only ball $B_e(c)$, $c \in C \Rightarrow x \in B_e(c)$.*

***Proof*** (i) can be directly launched by $2\rho_1 + 1 \leqslant d \leqslant 2\rho + 1$. To prove (ii), if $C$ is a perfect code and the minimum distance is $d = 2e + 1$, so we have $\rho_1 = \rho = e$. On the other hand, if the conditions are right, then the coverage radius of $C$ is $\rho \leqslant e = \rho_1 \leqslant \rho$, so $\rho_1 = \rho$. $C$ is a perfect code.

In order to introduce the concept of error correcting code, we discuss the so-called decoding principle in electronic information transmission. This principle is commonly known as the decoding principle of "look most like". What looks like the most? When we transmit through the channel with interference, we receive a codeword $x' \in \mathbb{F}_q^n$, and a codeword $x \in C$. If

$$d(x, x') = \min\{d(c, x')|c \in C\},$$

$x$ is the most similar codeword to $x'$ in code $C$. So we decode $x'$ to $x$. If the most similar codeword $x$ is the only one in $C$, then theoretically, $x'$ is the codeword received after $x$ transmission, so $x' \xrightarrow{x}$ is accurate.

**Definition 2.4** A code $C$ is called $e$-error correcting code ($e \geqslant 1$). If for any $x \in \mathbb{F}_q^n$, there is a $c \in C \Rightarrow x \in B_e(c)$, then $c$ is unique.

An error correcting code allows transmission errors without affecting correct decoding. For example, suppose that $C$ is a $e$-error correcting code, then for any $c \in C$, after $c$ is transmitted through the channel with interference, the codeword we receive is $x$, if an error occurs when $c$ is transmitted with no more than $e$ characters at most, that is $d(c, x) \leqslant e$, so the most similar codeword in $C$ must be $c$, so we can decode $x \xrightarrow{decode} c$ correctly.

**Corollary 2.4** *A perfect code with minimal distance $d = 2e + 1$ is e-error correcting code.*

***Proof*** Because the disjoint radius $\rho_1$ of $C$ has $\rho_1 = \rho = e$ with the covering radius $\rho$. Therefore, for any received codeword $x \in \mathbb{F}_q^n$, there exists and only exists a $c \in C \Rightarrow x \in B_e(c)$. That is, $C$ is $e$-error correction code.

Finally, we prove the main conclusion of this section.

**Theorem 2.1** *The minimum distance of a code $C$ is $d = 2e + 1$, then $C$ is a perfect code if and only if the following sphere-packing condition holds.*

$$|C| \sum_{i=0}^{e} \binom{n}{i} (q-1)^i = q^n. \tag{2.4}$$

***Proof*** If the minimum distance of $C$ is $d = 2e + 1$, and $C$ is the perfect code $\Rightarrow$ $\rho = \rho_1 = e$. So

$$\bigcup_{c \in C} B_e(c) = \mathbb{F}_q^n.$$

Then we have

$$\left| \bigcup_{c \in C} B_e(c) \right| = q^n,$$

thus

$$|C| B_e = |C| \sum_{i=0}^{e} \binom{n}{i} (q-1)^i = q^n.$$

Conversely, the sphere-packing condition (2.4) holds. Because the minimum distance of $C$ is $d = 2e + 1$, from corollary 2.2, we can see that $\rho_1 = e$, so we have

$$\bigcup_{c \in C} B_e(c) = \mathbb{F}_q^n.$$

It can be concluded that $\rho \leqslant e = \rho_1 \leqslant \rho$, thus $\rho = \rho_1$, $C$ is a perfect code. The theorem holds.

When $q = 2$, the alphabet $\mathbb{F}_2$ is a finite field of two elements $\{0, 1\}$, at this time, the coding is called binary code or binary code, and the transmission channel is called binary channel. In binary channel transmission, the most important is binary entropy function $H(\lambda)$, define as

$$H(\lambda) = \begin{cases} 0, & \text{when } \lambda = 0 \text{ or } \lambda = 1, \\ -\lambda \log \lambda - (1-\lambda)\log(1-\lambda), & \text{when } 0 < \lambda < 1. \end{cases} \tag{2.5}$$

Obviously, $H(\lambda) = H(1-\lambda)$, and $0 \leqslant H(\lambda) \leqslant 1$, $H\left(\frac{1}{2}\right) = 1$, that is $\lambda = \frac{1}{2}$ reaching the maximum. For further properties of $H(\lambda)$, please refer to Chap. 1.

**Theorem 2.2** *Let $C$ be a perfect code with minimal distance $d = 2e+1$, $R_C$ is the code rate of $C$, then*

*(i)* $1 - R_C = \frac{1}{n}\log_2 \sum_{i=0}^{e} \binom{n}{i} \leqslant H\left(\frac{e}{n}\right)$.

*(ii) When the length of codeword is $n \to \infty$, if $\lim_{n\to\infty} R_C = a$, then*

$$\lim_{n\to\infty} H\left(\frac{e}{n}\right) = 1 - a.$$

***Proof*** (i) According to the sphere-packing condition, since $C$ is the perfect code, so

$$|C| \sum_{i=0}^{e} \binom{n}{i} = 2^n.$$

We have

$$\frac{1}{n}\log_2 |C| + \frac{1}{n}\log_2 \sum_{i=0}^{e} \binom{n}{i} = 1.$$

That is

$$1 - R_C = \frac{1}{n}\log_2 \sum_{i=0}^{e} \binom{n}{i} \leqslant H\left(\frac{e}{n}\right),$$

The last inequality is derived from lemma 1.11 in the Chap. 1, so (i) holds. If there is a limit of $R_C$ when $n \to \infty$, again from lemma 1.11 in the Chap. 1, we have

$$\lim_{n\to\infty} H\left(\frac{e}{n}\right) = 1 - \lim_{n\to\infty} R_C = 1 - a.$$

The Theorem 2.2 holds.

Finally, we give an example of perfect code.

***Example 2.1*** Let $n = 2e+1$ is an odd number, then the repeated code in $\mathbb{F}_2^n$ is $A = \{0, 1\}$, where $0 = 00\ldots0$, $1 = 11\cdots1$ are Perfect codes of length n.

First, repeat code $A = \{0, 1\} \subset \mathbb{F}_2^n$ contains only two codes $0 = 0 \ldots 0 \in \mathbb{F}_2^n$, $1 = 1 \ldots 1 \in \mathbb{F}_2^n$, because $n = 2e + 1$ is an odd number, so from Corollary 2.2, Disjoint radius of $A$ is $\rho_1 = e$, let's prove that the covering radius of $A$ is $\rho = \rho_1 = e$, for any $x \in \mathbb{F}_2^n$, if $d(0, x) > e$, that is $d(0, x) \geq e + 1$, this shows that at least $e + 1$ characters are 1 in $x = x_1x_2 \ldots x_n \in \mathbb{F}_2^n$ , the maximum number of $e$ characters is 0, thus $d(1, x) \leq e$. This shows that $x \notin B_e(0)$, then $x \in B_e(1)$, that is

$$B_e(0) \cup B_e(1) = \mathbb{F}_2^n,$$

so $\rho \leq e = \rho_1 \leq \rho$, we have $\rho = \rho_1$. That is $A$ is the perfect code. Note that in this example, $e$ can take any positive integer, so the code rate of the repeat code has a limit value

$$\lim_{n\to\infty} R_A = 0, \Rightarrow \lim_{n\to\infty} H(\frac{e}{n}) = 1.$$

As the end of this chapter, we discuss and define the equivalence of two codes. Let $C \subset \mathbb{F}_q^n$ be a code of length $n$ and $S_n$ be a permutation group of $n$ elements. Any $\sigma \in S_n$ is a $n$ permutation, $x = x_1x_2 \ldots x_n \in \mathbb{F}_q^n$, We define $\sigma(x)$ as

$$\sigma(x) = x_{\sigma(1)}x_{\sigma(2)} \ldots x_{\sigma(n)} \in \mathbb{F}_q^n, \tag{2.6}$$

$$\sigma(C) = \{\sigma(c) \mid c \in C\}. \tag{2.7}$$

**Definition 2.5** Let $C$ and $C_1$ be two codes in $\mathbb{F}_q^n$, if there is $\sigma \in S_n \Rightarrow \sigma(C) = C_1$, Call $C$ and $C_1$ is equivalent, denoted as $C \sim C_1$. Obviously, equivalence is an equivalence relation between codes, because take $\sigma = 1$, then have $C \sim C$. If $C \sim C_1$, that is $C_1 = \sigma(C)$, then we have $C = \sigma^{-1}(C_1)$, that is $C \sim C_1 \Rightarrow C_1 \sim C$. Similarly, if $C \sim C_1$, $C_1 \sim C_2$, then $C \sim C_2$. Because $C_1 = \sigma(C)$, $C_2 = \tau(C_1) \Rightarrow C_2 = \tau\sigma(C)$. Another obvious property is that the function of $\sigma$ does not change the Hamming distance between two codewords, that is, we have

$$d(\sigma(x), \sigma(y)) = d(x, y), \forall \sigma \in S_n. \tag{2.8}$$

**Lemma 2.7** *Suppose $C \sim C_1$ are two equivalent codes, then they have the same code rate, the same minimum distance, the same coverage radius and the same disjoint radius. In particular, if $C$ is a perfect code, then all codes $C_1$ equivalent to $C$ are perfect codes.*

***Proof*** All the results of lemma can be easily proved by using equation (2.8).

## 2.2 Linear Code

Let $C \subset \mathbb{F}_q^n$ be a code, if $C$ is a $k$-dimensional linear subspace of $\mathbb{F}_q^n$, $C$ is called a linear code, denote as $C = [n, k]$. So for a linear code $C$, we have

$$R_C = \frac{1}{n}\log_q |C| = \frac{k}{n}, \text{ minimal distance } d = \text{minimal weight } w.$$

Let $\{\alpha_1, \alpha_2, \ldots, \alpha_k\} \subset C$ be a set of bases of linear code $C$, where

$$\alpha_i = \alpha_{i_1}\alpha_{i_2}\cdots\alpha_{i_n} \in \mathbb{F}_q^n, \ 1 \leq i \leq k.$$

**Definition 2.6** If $\{\alpha_1, \alpha_2, \ldots, \alpha_k\}$ is a set of bases of linear code $C = [n, k]$, then have $k \times n$-order matrix

$$G = \begin{bmatrix} \alpha_1 \\ \alpha_2 \\ \vdots \\ \alpha_k \end{bmatrix} = \begin{bmatrix} \alpha_{11} & \alpha_{12} & \cdots & \alpha_{1n} \\ \alpha_{21} & \alpha_{22} & \cdots & \alpha_{2n} \\ \cdots & \cdots & \cdots & \cdots \\ \alpha_{k1} & \alpha_{k2} & \cdots & \alpha_{kn} \end{bmatrix}.$$

Called generation matrix of $C$, write

$$G = [I_k, A_{k\times(n-k)}], \ I_k \text{ is } k \text{ order identity matrix.}$$

It is called the standard form of $G$.

**Lemma 2.8** *$C = [n, k]$ is a linear code, $G$ is generation matrix, then*

$$C = \{aG | a \in \mathbb{F}_q^k\}.$$

***Proof*** Because $\{\alpha_1, \alpha_2, \ldots, \alpha_k\}$ is a set of bases of linear code $C$. $\forall\, x \in C$, then

$$x = a_1\alpha_1 + a_2\alpha_2 + \cdots + a_k\alpha_k = (a_1, a_2, \ldots, a_k)\begin{bmatrix} \alpha_1 \\ \vdots \\ \alpha_k \end{bmatrix} = a \cdot G.$$

Where $a = (a_1, a_2, \ldots, a_k) \in \mathbb{F}_q^k$, the Lemma holds.

Define the inner product in $\mathbb{F}_q^n$, $x = x_1 \ldots x_n$, $y = y_1 \ldots y_n \in \mathbb{F}_q^n$, then define $< x, y >= \sum_{i=1}^n x_i y_i$, if $< x, y >= 0$, Say $x$ and $y$ orthogonal, denote as $x \perp y$.

**Definition 2.7** Let $C = [n, k]$ be a linear code whose orthogonal complement space $C^\perp$ is

$$C^\perp = \{y \in \mathbb{F}_q^n | < x, y >= 0, \forall\, x \in C\}.$$

Obviously, $C^\perp$ is an $[n, n-k]$-linear code, and $C^\perp$ is the dual code of $C$. The generating matrix $H$ of $C^\perp$ is called the check matrix of $C$.

**Lemma 2.9** *$C = [n, k]$ is a linear code, $H$ is a check matrix, then we have*

$$xH' = 0 \Leftrightarrow x \in C.$$

*Where $H'$ is the transpose matrix of $H$.*

***Proof*** We only prove the conclusion by taking the standard form of the generating matrix $G$ of $C$. Let

$$G = [I_k, A_{k\times(n-k)}] = [I_k, A], \quad A = A_{k\times(n-k)}.$$

Then the check matrix of $C$, that is, the generating matrix of dual code $C^{\perp}$ is

$$H = [-A', I_{n-k}],\ H' = \begin{bmatrix} -A \\ I_{n-k} \end{bmatrix}.$$

By Lemma 2.8, if $x \in C$, then $\exists\, a \in \mathbb{F}_q^k \Rightarrow x = aG$, thus

$$xH' = aGH' = a[I_k, A]\begin{bmatrix} -A \\ I_{n-k} \end{bmatrix} = 0.$$

Conversely, if $xH' = 0$, because $H$ is the generating matrix of $C^{\perp}$, again by Lemma 2.8, for $\forall\, y \in C^{\perp}, \exists\, b \in \mathbb{F}_q^{n-k} \Rightarrow y = bH$, thus

$$< x, y >= xy' = xH'b' = 0 \Rightarrow x \in (C^{\perp})^{\perp} = C.$$

The Lemma holds.

By Lemma 2.9, $\forall\, x, y \in \mathbb{F}_q^n$, then

$$xH' = yH' \Leftrightarrow x - y \in C.$$

Because $C$ is an additive subgroup of $\mathbb{F}_q^n$, $xH'$ is called the check value of codeword $x$. Then the check values of the two codewords are equal $\Leftrightarrow$. These two codewords are in the same additive coset of $C$. The following decoding principle of linear code is produced.

**Decoding principle:** If the $C = [n, k]$ linear code is used for coding, through an interference channel, when the received codeword is $x \in \mathbb{F}_q^n$, then find a codeword $x_0$ with the least weight in the additive coset $x + C$ of $x$, that is, $x_0$ satisfies

$$x_0 \in x + C, \text{ and } w(x_0) = \min\{w(\alpha) | \alpha \in x + C\}.$$

$x_0$ is called the leader codeword in coset $x + C$. We're going to decode $x$ into $x - x_0$.

**Lemma 2.10** *If the minimum distance of linear code $C = [n, k]$ is $d = 2e + 1$, then there is at most one codeword $x_0 \Rightarrow w(x_0) \le e$ in any additive coset $x + C$ of $C$.*

***Proof*** If $\alpha, \beta \in x + C$, and $w(\alpha) \leq e, w(\beta) \leq e$. Then $\alpha - \beta \in C$. And $w(\alpha - \beta) \leq w(\alpha) + w(\beta) = 2e$, but minimal weight of $C$ =Minimal distance of $C = 2e + 1$, so there are contradictions, thus $\alpha = \beta$. The Lemma holds.

**Corollary 2.5** *For a perfect linear code $C = [n, k]$ with minimal distance $d = 2e + 1$, then there exists and only exists a codeword with weight $\leq e$ in any additive coset $x + C$ of $C$. In other words, the leader code in any addition set is unique.*

***Proof*** $x \in \mathbb{F}_q^n \Rightarrow \exists\, c \in C$ such that $x \in B_e(c)$, that is $d(c, x) \leq e$. So $w(x - c) \leq e$.But $x - c \in x + C$. The Lemma holds.

**Definition 2.8** If any two column vectors of the generator matrix $G$ of a linear code $C = [n, k]$ are linearly independent, $C$ is called a projective code.

In order to discuss the true meaning of projective codes, we consider the $(k-1)$-dimensional projective space $PG(k-1, q)$ over $\mathbb{F}_q$.

In $\mathbb{F}_q^k$, any two vectors $a = (a_1, a_2, \ldots, a_k)$, $b = (b_1, b_2, \ldots, b_k)$, say $a \sim b$, if $\exists\, \lambda \in \mathbb{F}_q^* \Rightarrow a = \lambda b$. This is an equivalent relation on $\mathbb{F}_q^k$. Obviously $b \sim 0 \Leftrightarrow b = 0$, any $a \in \mathbb{F}_q^k, \overline{a} = \{\lambda a | \lambda \in \mathbb{F}_q^*\}$, the quotient set $\mathbb{F}_q^k/_\sim$ is called a $(k-1)$-dimensional projective space over $\mathbb{F}_q$. Denote as $PG(k-1, q)$, therefore

$$PG(k-1, q) = \mathbb{F}_q^k/_\sim = \{\overline{a} | a \in \mathbb{F}_q^k\}.$$

The number of nonzero points in $(k-1)$-dimensional projective space $PG(k-1, q)$ is

$$|PG(k-1, q)| = \frac{q^k - 1}{q - 1} = 1 + q + \cdots + q^{k-1}.$$

A linear code $[n, n-k]$, its check matrix $H$ is a $k \times n$-order matrix, and any two column vectors are linearly independent, that is $H = [a_1, a_2, \ldots, a_n]$, then $\{a_1, a_2, \ldots, a_n\} \subset PG(k-1, q)$ are $n$ with different nonzeros. So the generating matrix of an $[n, k]$ projective code consists of $n$ different nonzero points in projective space $PG(k-1, q)$. Because $n \leq |PG(k-1, q)|$, when the maximum value is reached, i.e.

$$n = |PG(k-1, q)| = \frac{q^k - 1}{q - 1}.$$

This leads to a perfect example of linear codes, called Hamming codes.

**Definition 2.9** Let $k > 1, n = \frac{q^k-1}{q-1}$, a linear code $C = [n, n-k]$ is called a Hamming code if any two column vectors of the check matrix $H$ of $C$ are linearly independent.

Since $C$ is a $n-k$-dimensional linear subspace and $C^\perp$ is a $k$-dimensional linear subspace, its generating matrix $H$ is a $k \times n$-order matrix. Therefore, if any two column vectors of $H$ are linearly independent, they represent $n$ different points in projective space $PG(k-1, q)$. Because $n = \frac{q^k-1}{q-1}$, then the construction of Hamming codes is the most possible.

**Theorem 2.3** *Any Hamming code $C = [n, n-k]$ is a perfect code, its minimum distance is $d = 3$; therefore, Hamming codes are perfect 1−error correcting codes.*

***Proof*** We first prove that the minimum distance of Hamming code $C$ is $d \geq 3$. If $d \leq 2$, there is $x = x_1x_2 \ldots x_n \Rightarrow w(x) \leq 2$, that is, there are at most two characters $x_i$ and $x_j$ are not 0. Because the minimum distance $d$ = minimum weight $w$ of a linear code.

Let $H = (\alpha_1, \alpha_2, \ldots, \alpha_n)$ be the check matrix of $C$. if $xH' = 0$, then

$$(x_1, x_2, \ldots, x_n)\begin{bmatrix}\alpha_1\\ \alpha_2\\ \vdots\\ \alpha_n\end{bmatrix} = 0.$$

We have $\alpha_i x_i + \alpha_j x_j = 0$, thus $\alpha_i$ and $\alpha_j$ are linearly related, contradiction. So $d \geq 3$, by Lemma 2.4, then the disjoint radius of $C$ is $\rho_1 \geq 1$.

On the other hand, $c \in C$, by Lemma 2.3, the number of elements in ball $B_1(c)$ is

$$|B_1(c)| = 1 + n(q-1) = q^k.$$

Because $C$ is a $(n-k)$-dimensional linear subspace, that is $|C| = q^{n-k}$, so

$$|\bigcup_{c \in C} B_1(c)| = |C|q^k = q^n = |\mathbb{F}_q^n|,$$

$$\Rightarrow \bigcup_{c \in C} B_1(c) = \mathbb{F}_q^n.$$

We have $1 \leq \rho_1 \leq \rho \leq 1 \Rightarrow \rho_1 = \rho = 1$. $C$ is a perfect code. Its minimal distance is $d = 2\rho + 1 = 3$, the Lemma holds.

Next, we discuss the weight polynomial of a linear code $C$ and prove the famous MacWilliams theorem.

$x \in C = [n, k]$, then the value of weight function $w(x)$ is from 0 to $n$, actually $w(x) = 0 \Leftrightarrow x = 0 \in C$, $w(x) = n \Leftrightarrow x = x_1 \ldots x_n, \forall\ x_i \neq 0$. So for each $i, 0 \leq i \leq n$, define

$$A(i) = \#\{x \in C | w(x) = i\}$$

and weighted polynomials of $C$.

$$A(z) = \sum_{i=0}^{n} A_i z^i, \ z \text{ is a variable.}$$

Obviously, for any given $c \in C$, then the number of codewords in $C$ whose Hamming distance to $c$ is exactly equal to $i$ is $A_i$, that is

$$A_i = \#\{x \in C | d(x, c) = i\}.$$

The codes with the above properties are called distance invariant codes; obviously, linear codes are distance invariant codes.

The following result was proved by MacWilliams in 1963; he established the relationship between the weight polynomials of a linear code $C$ and its dual code $C^{\perp}$, which is the most basic achievement in code theory.

**Theorem 2.4** (MacWilliams) *Let $C = [n, k]$ be a linear code over $\mathbb{F}_q$ and the weight polynomial be $A(z)$, $C^{\perp}$ is the dual code of $C$, the weight polynomial is $B(z)$, then*

$$B(z) = q^{-k}(1 + (q - 1)z)^n A(\frac{1 - z}{1 + (q - 1)z}).$$

*Specially, when $q = 2$,*

$$2^k B(z) = (1 + z)^n A(\frac{1 - z}{1 + z}).$$

***Proof*** Let $\psi(a)$ be an additive feature on $\mathbb{F}_q$. $\psi(a)$ can be constructed as follows:

$$\psi(a) = \exp(\frac{2\pi i tr(a)}{p}), tr(a) : \mathbb{F}_q \to \mathbb{F}_p.$$

For any $c \in C$, we define the polynomial $g_c(z)$ as

$$g_c(z) = \sum_{x \in \mathbb{F}_q^n} z^{w(x)} \psi(< x, c >), \tag{2.9}$$

therefore,

$$\sum_{c \in C} g_c(z) = \sum_{x \in \mathbb{F}_q^n} z^{w(x)} \sum_{c \in C} \psi(< x, c >), \tag{2.10}$$

Let's calculate the inner sum of (2.10). If $x \in C^{\perp}$, then

$$\sum_{c \in C} \psi(< x, c >) = |C|.$$

If $x \notin C^{\perp}$, let's prove

$$\sum_{c \in C} \psi(< x, c >) = 0. \tag{2.11}$$

If $x \in \mathbb{F}_q^n$, $x \notin C^{\perp}$, let

$$T(x) = \{y \in C \,|< y, x >= 0\} \subsetneq C,$$

so $T(x)$ is a linear subspace of $C$. $c \in C$, we consider additive cosets any two codewords $c + y_1$ and $c + y_2$ in this set , we have

$$< c + y_1, x >=< c + y_2, x >=< c, x > .$$

On the contrary, any two additive cosets $c_1 + T(x)$, $c_2 + T(x)$, if $< c_1, x >=< c_2, x >$, then $< c_1 - c_2, x >= 0$, that is $c_1 - c_2 \in T(x)$, so $c_1 + T(x) = c_2 + T(x)$. Therefore, the inner product of any two codewords in $c + T(x) \subset C$ is the same with $x$. Conversely, different additive cosets and the inner product of $x$ are not equal. Because $x \notin C^{\perp}$, $\exists\ c_0 \in C$, such that $< c_0, x >\neq 0$, let $< c_0, x >= a \neq 0$, then $< a^{-1}c_0, x >= 1$, let $c_1 = a^{-1}c_0 \in C$, then $< c_1, x >= 1$. Therefore, $\forall a \in \mathbb{F}_q \Rightarrow< ac_1, x >= a$. $< c, x >$ takes every element of $\mathbb{F}_q$, so

$$\sum_{c\in C} \psi(< x, c >) = [C : T(x)] \sum_{a\in\mathbb{F}_q} \psi(a) = 0.$$

That is, (2.11) holds. From (2.10), we can get that

$$\sum_{c\in C} g_c(z) = |C| \sum_{\substack{x\in\mathbb{F}_q^n \\ x\in C^{\perp}}} z^{w(x)} = |C|B(z). \tag{2.12}$$

Define the weight function $w(a) = 1$ for $a \in \mathbb{F}_q$, if $a \neq 0$, $w(0) = 0$. For any $x \in \mathbb{F}_q^n$, $c \in C$, write $x = x_1x_2 \ldots x_n$, $c = c_1c_2 \ldots c_n$, then it is defined by G, we have

$$\begin{aligned} g_c(z) &= \sum_{\substack{1\leq i\leq n \\ x_i\in\mathbb{F}_q}} z^{w(x_1)+w(x_2)+\cdots+w(x_n)} \psi(c_1x_1 + \cdots + c_nx_n) \\ &= \prod_{i=1}^{n} \sum_{x\in\mathbb{F}_q} z^{w(x)} \psi(c_ix). \end{aligned} \tag{2.13}$$

The inner layer sum of the above formula can be calculated as

$$\sum_{x\in\mathbb{F}_q} z^{w(x)} \psi(c_ix) = \begin{cases} 1 - z, & if c_i \neq 0, \\ 1 + (q - 1)z, & if c_i = 0. \end{cases}$$

From (2.13), then we have

$$g_c(z) = (1 - z)^{w(c)}(1 + (q - 1)z)^{n-w(c)}.$$

Thus

$$\sum_{c\in C} g_c(z) = (1+(q-1)z)^n \sum_{c\in C} (\frac{1-z}{1+(q-1)z})^{w(c)}$$
$$= (1+(q-1)z)^n A(\frac{1-z}{1+(q-1)z}).$$

Finally, from (2.12), we have

$$B(z) = \frac{1}{|C|}(1+(q-1)z)^n A(\frac{1-z}{1+(q-1)z})$$
$$= q^{-k}(1+(q-1)z)^n A(\frac{1-z}{1+(q-1)z}).$$

We have completed the proof of the theorem.

## 2.3 Lee Distance

$m > 1$ is a positive integer, $\mathbb{Z}_m$ a is residue class rings of mod $m$, if $\mathbb{Z}_m$ is the alphabet and $C \subset \mathbb{Z}_m^n$ is the proper subset, then $C$ is called an $m$-ary code. In this case, Hamming distance is not the best tool to measure error, we substitute Lee distance and Lee weight. Let $i \in \mathbb{Z}_m$, define Lee weight as

$$W_L(i) = \min\{i, m-i\}. \tag{2.14}$$

Obviously,

$$W_L(0) = 0,\ W_L(-i) = W_L(m-i) = W_L(i). \tag{2.15}$$

Suppose $a = (a_1, a_2, \ldots, a_n) = a_1a_2\ldots a_n \in \mathbb{Z}_m^n$, $b = b_1b_2\ldots b_n \in \mathbb{Z}_m^n$, define Lee weight and Lee distance on $m$-ary code $C$ as follows

$$\begin{cases} W_L(a) = \sum_{i=1}^{n} W_L(a_i) \\ d_L(a,b) = W_L(a-b). \end{cases}$$

From (2.15), we have

$$W_L(-a) = W_L(a), d_L(a,b) = d_L(b,a), \forall\, a, b \in \mathbb{Z}_m^n.$$

**Lemma 2.11** *For $\forall\, a, b, c \in \mathbb{Z}_m^n$, we have the following trigonometric inequalities*

$$d_L(a,b) \le d_L(a,c) + d_L(c,b).$$

***Proof*** Suppose $0 \le i < m, 0 \le j < m$, we have

$$W_L(i+j) \le W_L(i) + W_L(j). \tag{2.16}$$

Because $0 \le i + j \le \frac{m}{2}$, then

$$W_L(i+j) = i + j = W_L(i) + W_L(j).$$

If $\frac{m}{2} < i + j < m$, we discuss it in three ways,
(1) $i \le \frac{m}{2}, j \le \frac{m}{2}$, there is

$$W_L(i+j) = m - i - j < i + j = W_L(i) + W_L(j).$$

(2) $i \le \frac{m}{2}, j > \frac{m}{2}$, there is

$$W_L(i+j) = m - i - j \le m - j = W_L(j) \le W_L(i) + W_L(j).$$

(3) $i > \frac{m}{2}, j \le \frac{m}{2}$, there is

$$W_L(i+j) = m - i - j \le m - i = W_L(i) \le W_L(i) + W_L(j).$$

So we always have (2.16), in $\mathbb{Z}_m^n$, (2.16) can be extended to

$$W_L(a+b) \le W_L(a) + W_L(b), \forall\, a, b \in \mathbb{Z}_m^n.$$

So

$$\begin{aligned} d_L(a,b) &= W_L(a-b) = W_L((a-c)+(c-b)) \\ &\le W_L(a-c) + W_L(c-b) = d_L(a,c) + d_L(c,b). \end{aligned}$$

The Lemma holds.

Next let's make $m = 4$, the alphabet is $\mathbb{Z}_4$, On a 4-ary code, we discuss Lee weight and Lee distance. Suppose $a \in \mathbb{Z}_4^n$, $0 \le i \le 3$, let

$$n_i(a) = \#\{j | 1 \le j \le n, a = a_1a_2 \ldots a_n, a_j = i\}. \tag{2.17}$$

$n_i(a)$ is the number of characters equal to $i$ in codeword $a$. $C \subset \mathbb{Z}_4^n$ is a 4-ary code, the symmetric polynomial and Lee weight polynomial of $C$ are defined as

$$swe_C(w,x,y) = \sum_{c \in C} w^{n_0(c)} x^{n_1(c)+n_3(c)} y^{n_2(c)} \tag{2.18}$$

and

$$Lee_C(x,y) = \sum_{c \in C} x^{2n - W_L(c)} y^{W_L(c)}. \tag{2.19}$$

**Lemma 2.12** *Let $C \subset \mathbb{Z}_4^n$ is a 4-ary code with codeword length of $n$, then the symmetric polynomials and Lee weight polynomials have the following relation on $C$,*

$$Lee_C(x, y) = swe_C(x^2, xy, y^2).$$

***Proof*** $a \in \mathbb{Z}_4^n$, by definition

$$n_0(a) + n_1(a) + n_2(a) + n_3(a) = n.$$

Let $a = a_1 a_2 \dots a_n$, then

$$W_L(a) = \sum_{i=1}^{n} W_L(a_i) = n_1(a) + 2n_2(a) + n_3(a).$$

So

$$\begin{aligned} Lee_C(x, y) &= \sum_{c \in C} x^{2n_0(c)} \cdot (xy)^{n_1(c)+n_3(c)} y^{2n_2(c)} \\ &= swe_C(x^2, xy, y^2). \end{aligned}$$

The Lemma holds.

By using Lee weight and Lee distance, we can extend the MacWilliams theorem to $\mathbb{Z}_4$ codes, we have

**Theorem 2.5** *Let $C \subset \mathbb{Z}_4^n$ be a linear code and $C^\perp$ be its dual code, $Lee_C(x, y)$ be a Lee weighted polynomial of $C$, then*

$$Lee_{C^\perp}(x, y) = \frac{1}{|C|} Lee_C(x + y, x - y).$$

***Proof*** Let $\psi$ be a nontrivial characteristic of $\mathbb{Z}_4$, and let $\psi$ be

$$\psi(i) = (\sqrt{-1})^i, i = 0, 1, 2, 3.$$

Let $f(u)$ be a function defined on $\mathbb{Z}_4^n$, we let

$$g(c) = \sum_{u \in \mathbb{Z}_4^n} f(u)\psi(< c, u >). \tag{2.20}$$

As in Theorem 2.4, there are

$$\sum_{c \in C} g(c) = |C| \sum_{u \in C^\perp} f(u). \tag{2.21}$$

Take

$$f(u) = w^{n_0(u)} x^{n_1(u)+n_3(u)} y^{n_2(u)}, \quad u \in \mathbb{Z}_4^n.$$

Write $u = u_1 u_2 \dots u_n \in \mathbb{Z}_4^n$, then for each $i,\ 0 \le i \le 3$, we have

$$n_i(u) = n_i(u_1) + n_i(u_2) + \cdots + n_i(u_n).$$

Thus

$$f(u) = \prod_{i=1}^{n} f(u_i).$$

Let $c = c_1 c_2 \dots c_n \in \mathbb{Z}_4^n$, by (2.20),

$$g(c) = \prod_{i=1}^{n} (\sum_{u \in \mathbb{Z}_4} f(u)\psi(< c_i, u >)). \tag{2.22}$$

Now we calculate the inner sum on the right side of equation (2.22).

$$\sum_{u \in \mathbb{Z}_4} f(u)\psi(< c_i, u >) = \begin{cases} w + 2x + y, & \text{if } c_i = 0 \\ w - y, & \text{if } c_i = 1 \text{ or } 3\ . \\ w - 2x + y, & \text{or } c_i = 2. \end{cases}$$

by (2.22),

$$g(c) = (w + 2x + y)^{n_0(c)} (w - y)^{n_1(c) + n_3(c)} (w - 2x + y)^{n_2(c)}.$$

So

$$\sum_{c \in C} g(c) = swe_C(w + 2x + y, w - y, w - 2x + y).$$

by (2.21),

$$|C| swe_{C^\perp}(w, x, y) = swe_C(w + 2x + y, w - y, w - 2x + y).$$

By Lemma 2.12, and replace the variable, there are

$$Lee_{C^\perp}(x, y) = \frac{1}{|C|} Lee_C(x + y, x - y).$$

We have completed the proof.

## 2.4 Some Typical Codes

### 2.4.1 *Hadamard Codes*

In order to introduce Hadamard codes, we first define a Hadamard matrix of order $n$. Let $H = (a_{ij})$, if $a_{ij} = \pm 1$, and

$$HH' = nI_n = \begin{bmatrix} n & 0 & \cdots & 0 \\ 0 & n & \cdots & 0 \\ \vdots & \vdots & \cdots & \vdots \\ 0 & 0 & \cdots & n \end{bmatrix},$$

$H$ is called a Hadamard matrix of order $n$. It is easy to verify that the following $H_2$ is a Hadamard matrix of second order

$$H_2 = \begin{bmatrix} 1 & 1 \\ 1 & -1 \end{bmatrix}.$$

In order to obtain higher-order Hadamard matrices, a useful tool is the so-called Kronecker product. Let $A = (a_{ij})_{m\times m}$, $B = (b_{ij})_{n\times n}$, then $A$ and $B$'s Kronecker product $A \otimes B$ define as

$$A \otimes B = \begin{bmatrix} a_{11}B & a_{12}B & \cdots & a_{1m}B \\ a_{21}B & a_{22}B & \cdots & a_{2m}B \\ \vdots & \vdots & \cdots & \vdots \\ a_{m1}B & a_{m2}B & \cdots & a_{mm}B \end{bmatrix}.$$

Obviously, $A \otimes B$ is a square matrix of order $nm \times nm$. The following results are easy to prove.

**Lemma 2.13** *Let $A$ be a Hadamard matrix of order $m$, $B$ be a Hadamard matrix of order $n$, then $A \otimes B$ be a Hadamard matrix of order $nm \times nm$.*

***Proof*** Let $A = (a_{ij})_{m\times m}$, $B = (b_{ij})_{n\times n}$, $H = A \otimes B$, then

$$
\begin{aligned}
HH' &= \begin{bmatrix} a_{11}B & a_{12}B & \cdots & a_{1m}B \\ a_{21}B & a_{22}B & \cdots & a_{2m}B \\ \vdots & \vdots & \cdots & \vdots \\ a_{m1}B & a_{m2}B & \cdots & a_{mm}B \end{bmatrix} \cdot \begin{bmatrix} a_{11}B' & a_{21}B' & \cdots & a_{m1}B' \\ a_{12}B' & a_{22}B' & \cdots & a_{m2}B' \\ \vdots & \vdots & \cdots & \vdots \\ a_{1m}B' & a_{2m}B' & \cdots & a_{mm}B' \end{bmatrix} \\
&= \begin{bmatrix} c_{11}BB' & c_{12}BB' & \cdots & c_{1m}BB' \\ \vdots & \vdots & \cdots & \vdots \\ c_{m1}BB' & c_{m2}BB' & \cdots & c_{mm}BB' \end{bmatrix} \\
&= \begin{bmatrix} mnI_n & 0 & \cdots & 0 \\ 0 & mnI_n & \cdots & 0 \\ \vdots & \vdots & \cdots & \vdots \\ 0 & 0 & \cdots & mnI_n \end{bmatrix} \\
&= mnI_{nm}.
\end{aligned}
$$

The Lemma holds.

Since $H_2$ is a Hadamard matrix of order 2, then

$$H_2 \otimes H_2 = H_2^{\otimes 2},\ H_2 \otimes H_2 \otimes \cdots \otimes H_2 = H_2^{\otimes n}$$

are Hadamard matrix of order 4 and order $2^n$ respectively.

Let $n$ be an even number and $H_n$ be a Hadamard matrix of order $n$, take $\alpha_1, \alpha_2, \ldots, \alpha_n$ as $n$ row vectors, i.e.,

$$H_n = \begin{bmatrix} \alpha_1 \\ \alpha_2 \\ \vdots \\ \alpha_n \end{bmatrix},\ -H_n = \begin{bmatrix} -\alpha_1 \\ -\alpha_2 \\ \vdots \\ -\alpha_n \end{bmatrix}.$$

We get $2n$ row vectors $\{\pm\alpha_1, \pm\alpha_2, \ldots, \pm\alpha_n\}$, for each row vectors $\pm\alpha_i$, we replace the component $-1$ with 0, the row vector $\alpha_i$ so permuted is denoted as $\overline{\alpha_i}$, $-\alpha_i$ denote as $\overline{-\alpha_i}$, so $\overline{\pm\alpha_i}$ forms a vector of $\mathbb{F}_2^n$, denote as

$$C = \{\overline{\pm\alpha_1}, \overline{\pm\alpha_2}, \ldots, \overline{\pm\alpha_n}\} \subset \mathbb{F}_2^n.$$

$C$ is called a Hadamard code.

**Theorem 2.6** *The minimum distance of Hadamard code $C$ of length $n$ ($n$ is an even number) is $d = \frac{n}{2}$.*

***Proof*** Let $H_n$ be a Hadamard matrix of order $n$, $H_n = \begin{bmatrix} \alpha_1 \\ \alpha_2 \\ \vdots \\ \alpha_n \end{bmatrix}$, Each $\alpha_i$ is a row vector of $H_n$, substitute $\alpha_i \xrightarrow{\sigma} \overline{\alpha_i}$, such that each $\overline{\alpha_i} \subset \mathbb{F}_2^n$ become a binary codeword. We see that this kind of permutation does not change the corresponding Hamming distance, that is

$$\begin{cases} d(\alpha_i, \alpha_j) = d(\overline{\alpha}_i, \overline{\alpha}_j) \\ d(-\alpha_i, -\alpha_j) = d(\overline{\alpha}_i, \overline{\alpha}_j), \end{cases}$$

where $i \neq j$. Let us prove that the minimum distance of $C$ is $\frac{n}{2}$, let $a = a_1 a_2 \ldots a_n$, $b = b_1 b_2 \ldots b_n$ are two different row vectors of Hadamard matrix $H_n$, because of

$$ab' = 0 \Rightarrow \sum_{i=1}^{n} a_i b_i = 0.$$

And $a_i = \pm 1$, $b_i = \pm 1$. Let the number of the same character be $d_1$ and the number of different characters be $d = d(a, b)$, so there are $d_1 - d = 0$, that is $d_1 = d$, but $d_1 + d = n$, so $d = \frac{n}{2}$. The Lemma holds.

**Corollary 2.6** *$C = \{\overline{\pm\alpha_1}, \overline{\pm\alpha_2}, \ldots, \overline{\pm\alpha_n}\}$ is Hadamard code, then the Hamming distance of any two different codewords on $C$ is $\frac{n}{2}$.*

***Proof*** $\{\pm\alpha_1, \pm\alpha_2, \ldots, \pm\alpha_n\}$ s the row vector of Hadamard matrix, let $a = \pm\alpha_i$, $b = \pm\alpha_j (i \neq j)$, then

$$ab' = \pm \sum_{i=1}^{n} a_i b_i = 0 \Rightarrow d(a, b) = \frac{n}{2}.$$

A code of length $n$, number of codewords $M$, minimum distance $d$, denoted as $(n, M, d)$, different from linear code $[n, k]$ or $[n, k, d]$, Hadamard code is

$$C = (n, 2n, \frac{n}{2}).$$

When $n = 8, d = 4$, this is an extension of Hamming code. When $n = 32$, $(32, 64, 16)$ is the code used by the U.S. Mars probe in 1969 to transmit pictures taken on Mars.

## 2.4.2 *Binary Golay Codes*

In the theory and application of channel coding, binary Golay code is the most famous one. In order to introduce Golay code $G_{23}$ completely, we first introduce the concept of $t - (m, k, \lambda)$ design.

Let $S$ be a set of $m$ elements, that is $|S| = m$. The elements in $S$ are called points. Let $\mathfrak{R}$ be the set of subsets with $k$ elements in $S$, $|\mathfrak{R}| = M$, i.e.,

$$\mathfrak{R} = \{B_1, B_2, \ldots, B_M\}, B_i \subset S, |B_i| = k, 1 \leq i \leq M.$$

Element $B_i$ in $\mathfrak{R}$ is called block.

**Definition 2.10** $(S, \mathfrak{R})$ is called $t-(m, k, \lambda)$ design, if for any $T \subset S, |T| = t$, then there are exactly $\lambda$ blocks $B$ in $\mathfrak{R}$ such that $T \subset B$. If $(S, \mathfrak{R})$ is a $t-(m, k, \lambda)$ design, denote as $(S, \mathfrak{R}) = t-(m, k, \lambda)$. If $\lambda = 1$, then $t-(m, k, 1)$ is called a Steiner system.

In a $t-(m, k, \lambda)$ design $(S, \mathfrak{R})$, we introduce its occurrence matrix. For any $a \in S$, the characteristic function $\chi_i(a)$ is defined as

$$\chi_i(a) = \begin{cases} 1, & \text{if } a \in B_i, \\ 0, & \text{if } a \notin B_i, \end{cases}$$

write $S = \{a_1, a_2, \ldots, a_m\}$, $\mathfrak{R} = \{B_1, B_2, \ldots, B_M\}$, $|\mathfrak{R}| = M$. Matrix

$$A = (\chi_j(a_i))_{m\times M} = \begin{bmatrix} \chi_1(a_1) & \chi_2(a_1) & \cdots & \chi_M(a_1) \\ \chi_1(a_2) & \chi_2(a_2) & \cdots & \chi_M(a_2) \\ \vdots & \vdots & \cdots & \vdots \\ \chi_1(a_m) & \chi_2(a_m) & \cdots & \chi_M(a_m) \end{bmatrix},$$

$A$ is called the occurrence matrix of $t-(m, k, \lambda)$ design.

Let's now consider a concrete example, $2-(11, 6, 3)$ design. Where there are 11 points in $S$ and 6 points in $\mathfrak{R}$, and any two points in $S$ have exactly three blocks containing it.

**Lemma 2.14** $2-(11, 6, 3)$ *design is the only definite one, that is to say, let* $S = \{a_1, a_2, \ldots, a_{11}\}$, *then there are 11 blocks in* $\mathfrak{R}$,

$$\mathfrak{R} = \{B_1, B_2, \ldots, B_{11}\}.$$

*And for any* $a \in S$, *exactly* 6 *blocks* $B_j$ *in* $\mathfrak{R}$ *contain* $a$.

***Proof*** Suppose $\forall\, a \in S$, there is exactly $l$ $B_j$ containing it, because there are exactly 3 blocks in any 2 points, so there are $6l - l = 10 \times 3$. Then $l = 6$. In addition, suppose $|\mathfrak{R}| = M$, because each point has exactly six blocks containing it, there is $6 \times M = 11 \times 6$, we can get $M = 11$.

By Lemma 2.14, the generating matrix $N$ of $2-(11, 6, 3)$ design is an 11-order square matrix

$$N = \begin{bmatrix} \chi_1(a_1) & \chi_2(a_1) & \cdots & \chi_{11}(a_1) \\ \chi_1(a_2) & \chi_2(a_2) & \cdots & \chi_{11}(a_2) \\ \vdots & \vdots & \cdots & \vdots \\ \chi_1(a_{11}) & \chi_2(a_{11}) & \cdots & \chi_{11}(a_{11}) \end{bmatrix}.$$

And every row of $N$ has exactly six 1's and five 0's, and every column of $N$ has exactly six 1's and five 0's.

**Lemma 2.15** *Let N be the occurrence matrix of* $2-(11,6,3)$ *design, then*

$$NN' = 3I_{11} + 3J_{11},\ J_{11} = \begin{bmatrix} 1 & 1 & \cdots & 1 \\ \vdots & \vdots & \cdots & \vdots \\ 1 & 1 & \cdots & 1 \end{bmatrix}.$$

*If N is regarded as a square matrix of order 11 over* $\mathbb{F}_2$, *then*

$$NN' = I_{11} + J_{11}.$$

*Further* $rank(N) = 10$, *and the solution of linear equation system* $XN = 0$ *is exactly two repeated codewords* $0$ *and* $1 (0 = (0, 0, \ldots, 0), 1 = (1, 1, \ldots, 1))$ *in* $\mathbb{F}_2^{11}$.

***Proof*** Let $NN' = (b_{ij})_{11\times 11}$, defined by

$$b_{ij} = \sum_{k=1}^{11} \chi_k(a_i)\chi_k(a_j).$$

When $i \neq j$, $b_{ij} = 3$, when $i = j$, $b_{ij} = 6$, so we have

$$NN' = 3I_{11} + 3J_{11} \equiv I_{11} + J_{11} (\text{mod } 2).$$

Let $N (\text{mod } 2)$ still be $N$, which is a square matrix of order 11 over $\mathbb{F}_2$. we have

$$rank(N) = rank(I_{11}) - rank(J_{11}) = 10.$$

So the solution space of $XN = 0$ is a one-dimensional linear subspace of $\mathbb{F}_2^{11}$. Since each column vector of $N$ has exactly six 1's and five 0's, then

$$(1, 1, \ldots, 1)N = (0, 0, \ldots, 0) \in \mathbb{F}_2^{11}.$$

So there are exactly two solutions for $XN = 0$:

$$x = (0, 0, \ldots, 0),\ x = (1, 1, \ldots, 1).$$

The Lemma holds.

Next, let's construct a matrix $G$ of order $12 \times 24$, $G = (I_{12}, P)$, where

$$P = \begin{pmatrix} 0 & 1 \cdots 1 \\ 1 & \\ \vdots & \text{N} \\ 1 & \end{pmatrix}, \text{ and } G = \begin{bmatrix} \alpha_1 \\ \alpha_2 \\ \vdots \\ \alpha_{12} \end{bmatrix}.$$

where $\alpha_i \in \mathbb{F}_2^{24}$ is the 12 row vector of $G$. Obviously we have a weight function

$$w(\alpha_1) = 12, \; w(\alpha_i) = 8, \; 2 \leq i \leq 12. \tag{2.23}$$

**Lemma 2.16** *Let* $G = \begin{bmatrix} \alpha_1 \\ \alpha_2 \\ \vdots \\ \alpha_{12} \end{bmatrix}$, *then* $\{\alpha_1, \alpha_2, \ldots, \alpha_{12}\} \subset \mathbb{F}_2^{24}$ *is a linear independent group, and the weight of any nonzero linear combination is at least* 8, *that is*

$$w(a_1\alpha_1 + a_2\alpha_2 + \cdots + a_{12}\alpha_{12}) \geq 8, \; a_i \textit{ not all zero}. \tag{2.24}$$

***Proof*** Let's prove that $\{\alpha_i\}_{i=1}^{12}$ is a set of vectors orthogonal to each other, that is, the inner product is $<\alpha_i, \alpha_j>= \alpha_i \alpha_j' = 0$. Obviously we have

$$<\alpha_1, \alpha_j>= \alpha_1 \alpha_j' = 6 \equiv 0 (\text{mod } 2), \; j \neq 1.$$

If $i \neq 1$, $j \neq 1$, $i \neq j$, then

$$<\alpha_i, \alpha_j>= 1 + \sum_{k=1}^{11} \chi_k(a_i)\chi_k(a_j) = 4 \equiv 0 (\text{mod } 2).$$

So $<\alpha_i, \alpha_j>= 0$, when $i \neq j$, that is $\{\alpha_1, \alpha_2, \ldots, \alpha_{12}\}$ is a linear independent group of $\mathbb{F}_2^{24}$. If $a_i \in \mathbb{F}_2$, not all zero, take $a = a_1 a_2 \ldots a_{12}$, let's prove (2.24) by induction of $w(a)$. If $w(a) = 1$, the proposition holds by (2.23). When $w(a) \geq 8$, the proposition is ordinary, for $2 \leq w(a) \leq 7$, we can still prove

$$w(a_1\alpha_1 + a_2\alpha_2 + \cdots + a_{12}\alpha_{12}) \geq 8.$$

So the Lemma holds.

**Definition 2.11** The linear code [24, 12] generated by row vector group $\{\alpha_1, \alpha_2, \ldots, \alpha_{12}\}$ of $G$ in $\mathbb{F}_2^{24}$ is called Golay code, denoted as $G_{24}$. Remove the last component of $\alpha_i$, $\alpha_i \rightarrow \overline{\alpha}_i$, then $\overline{\alpha}_i \in \mathbb{F}_2^{23}$. The linear code [23, 12] generated by $\{\overline{\alpha}_1, \overline{\alpha}_2, \ldots, \overline{\alpha}_{12}\}$ in $\mathbb{F}_2^{23}$ is called Golay code, denote as $G_{23}$.

**Theorem 2.7** *Golay code $G_{23}$ is a perfect code* [23, 12] *with minimal distance of $d = 7$.*

***Proof*** Because the minimal distance of linear codes is minimal weight, by Lemma 2.16,

$$w(a_1\overline{\alpha}_1 + a_2\overline{\alpha}_2 + \cdots + a_{12}\overline{\alpha}_{12}) \geq w(a_1\alpha_1 + a_2\alpha_2 + \cdots + a_{12}\alpha_{12}) - 1 \geq 7.$$

On the one hand, $w(\alpha_i) = 8$ for $\forall\ \alpha_i, i \neq 1$, so there is $\overline{\alpha}_i \Rightarrow w(\overline{\alpha}_i) = w(\alpha_i) - 1 = 7$. So the minimum distance of $G_{23}$ is $d = 7$. On the other hand, we note that

$$|G_{23}| \sum_{i=0}^{3} \binom{23}{i} = 2^{12} \sum_{i=0}^{3} \binom{23}{i} = 2^{23}.$$

By the sphere-packing condition of Theorem 2.1 $\Rightarrow G_{23}$ is a perfect code, the Lemma holds.

### *2.4.3 3-Ary Golay Code*

In order to introduce 3-ary Golay codes, we first define a Paley matrix of order $q$. Let $q \geq 3$ be an odd number, and define a second-order real-valued multiplication characteristic $\chi(a)$ in the finite field $\mathbb{F}_q$ as

$$\chi(a) = \begin{cases} 0, & \text{if } a = 0; \\ 1, & \text{if } a \in (\mathbb{F}_q^*)^2; \\ -1, & \text{if } a \notin (\mathbb{F}_q^*)^2. \end{cases}$$

Obviously, $\chi$ is a character in $\mathbb{F}_q^*$. Because $\mathbb{F}_q^*$ is a $(q-1)$-order cyclic multiplicative group, so we have

$$\chi(-1) = (-1)^{\frac{q-1}{2}} = \begin{cases} 1, & \text{if } q \equiv 1 (\bmod 4); \\ -1, & \text{if } q \equiv 2 \text{ or } 3 (\bmod 4). \end{cases}$$

Write $\mathbb{F}_q = \{a_0, a_1, \ldots, a_{q-1}\}$ , where $a_0 = 0$, then Paley matrix $S_q$ of order $q$ is defined as

$$S_q = (\chi(a_i - a_j))_{q \times q} = \begin{bmatrix} 0 & \chi(-a_1) & \chi(-a_2) & \cdots & \chi(-a_{q-1}) \\ \chi(a_1) & 0 & \chi(a_1 - a_2) & \cdots & \chi(a_1 - a_{q-1}) \\ \chi(a_2) & \chi(a_2 - a_1) & 0 & \cdots & \chi(a_2 - a_{q-1}) \\ \cdots & \cdots & \cdots & \cdots & \cdots \\ \chi(a_{q-1}) & \chi(a_{q-1} - a_1) & \cdots & \cdots & 0 \end{bmatrix}.$$

**Lemma 2.17** *The Paley matrix $S_q$ of order $q$ has the following properties:*

(i) $S_q J_q = J_q S_q = 0$.
(ii) $S_q S_q^{'} = qI_q - J_q$.
(iii) $S_q^{'} = (-1)^{\frac{q-1}{2}} S_q$.

*Here, $I_q$ is the unit matrix of order $q$ and $J_q$ is the square matrix of order $q$ with all elements of 1.*

***Proof*** Let $S_q J_q = (b_{ij})_{q\times q}$, then for $\forall\, 0 \le i \le q-1,\ 0 \le j \le q-1$, there is

$$b_{ij} = \sum_{k=0}^{q-1} \chi(a_i - a_k) = \sum_{c\in\mathbb{F}_q} \chi(c) = 0.$$

So (i) holds. To prove (ii), let $S_q S_q' = (c_{ij})_{q\times q}$, then

$$c_{ij} = \sum_{k=0}^{q-1} \chi(a_i - a_k)\chi(a_j - a_k).$$

Obviously, we have

$$c_{ij} = \begin{cases} q-1, & \text{if } i = j; \\ -1, & \text{if } i \ne j. \end{cases}$$

So (ii) holds. To prove (iii), noticed that $\chi(-1) = (-1)^{\frac{q-1}{2}}$, so

$$S_q = \chi(-1)S_q^{'} = (-1)^{\frac{q-1}{2}} S_q^{'},$$

the Lemma holds.

Let $q = 5$, we consider the Paley matrix $S_5$ of order 5, it has been calculated that

$$S_5 = \begin{bmatrix} 0 & 1 & -1 & -1 & 1 \\ 1 & 0 & 1 & -1 & -1 \\ -1 & 1 & 0 & 1 & -1 \\ -1 & -1 & 1 & 0 & 1 \\ 1 & -1 & -1 & 1 & 0 \end{bmatrix}.$$

In $\mathbb{F}_3^{11}$, we consider a linear code $C$ whose generator matrix is

$$G = \begin{pmatrix} I_6 & \begin{matrix} 1\ 1\ 1\ 1\ 1 \\ S_5 \end{matrix} \end{pmatrix},$$

So $C$ is a six-dimensional linear subspace in $\mathbb{F}_3^{11}$, that is $C = [11, 6]$. This code is called 3-ary Golay code. In order to further discuss 3-ary Golay codes $[11, 6]$, we discuss the concept of extended codes of linear codes.

If $C \subset \mathbb{F}_q^n$ is a $q$-ary linear code of length $n$, the extension code $\overline{C}$ of $C$ is defined as

$$\overline{C} = \{(c_1, c_2, \ldots, c_{n+1}) | (c_1, c_2, \ldots, c_n) \in C, \text{ and} \sum_{i=1}^{n+1} c_i = 0\}.$$

Obviously, $\overline{C} \subset \mathbb{F}_q^{n+1}$ is a linear code.

**Lemma 2.18** *If $C \subset \mathbb{F}_q^n$ is a linear code, the generation matrix is $G$ and the test matrix is $H$, then the length of extension code $\overline{C} \subset \mathbb{F}_q^{n+1}$ is $n + 1$, its generation matrix $\overline{G}$ and test matrix $\overline{H}$ are*

$$\overline{G} = [G, \beta], \text{ and } \overline{H} = \begin{pmatrix} 1\ 1 \cdots 1 & 1 \\ & 0 \\ H & \vdots \\ & 0 \end{pmatrix},$$

*respectively. Where $\beta$ is a column vector and satisfies that the sum of all column vectors of $\beta$ and $G$ is 0. Further, let $q = 2$, if the minimum distance $d$ of $C$ is odd, then the minimum distance of $\overline{C}$ is $d + 1$.*

***Proof*** The generation matrix and check matrix of $\overline{C}$ can be given directly by definition. The minimal weight $w = w(c)$ of $C$ can be obtained by $c = c_1 c_2 \ldots c_n \in C$, because $q = 2$, so there are $w$ $c_i = 1$, and $w$ is an odd number, then $w \neq 0$, let $c_{n+1} = 1$, then

$$c^* = c_1 c_2 \ldots c_{n+1} \in \overline{C} \text{ and } w(c^*) = d + 1.$$

This is the minimal weight in $\overline{C}$. The lemma is proved.

Consider the extension codes $\overline{C} = [12, 6]$ of 3-ary Golay code $C = [11, 6]$, its generating matrix is

$$\overline{G} = \begin{pmatrix} & 1\ 1\ 1\ 1\ 1 & 0 \\ & & -1 \\ I_6 & S_5 & \vdots \\ & & -1 \end{pmatrix}, \tag{2.25}$$

Note that the sum of the components of each row vector of $S_5$ is 0, and the inner product of the different row vectors is - 1, and the inner product of the same row vector is 1, so

$$\overline{G} \cdot \overline{G}' = 0.$$

Therefore, the extended code $\overline{C}$ is a self-dual code, that is $(\overline{C})^{\perp} = \overline{C}$.

**Theorem 2.8** *3-ary Golay code $C$ is a perfect linear code* $[11, 6]$, *its minimum distance is 5, so it is a 2-error correcting code.*

***Proof*** The weight of each row vector of $\overline{G}$ is 6, according to the calculation, the weight of the linear combination of row vectors of $\overline{G}$ is 6, so the minimum distance of extension code $\overline{C}$ is 6 $\Rightarrow$ the minimum distance of $C$ is 5. So the disjoint radius of $C$ is $\rho_1 = 2$. And because

$$|C| = 3^6, \ \sum_{i=0}^{2} \binom{11}{i} 2^i = 3^5,$$

then the condition of sphere packing satisfy

$$|C| \sum_{i=0}^{2} \binom{11}{i} 2^i = 3^{11}.$$

Thus by Theorem 2.1, $C$ is a perfect code, the Theorem holds.

***Remark 2.1*** It is worth noting that J.H.VanLint in 1971 (See reference 2 [24]), $A.Tiet\ddot{a}v\ddot{a}inen$ in 1973(See reference 2 [43]) independently proved that perfect codes (nontrivial) with minimal distance greater than 3 have only 2-ary Golay codes $G_{23}$ and 3-ary Golay codes over any finite field.

### 2.4.4 Reed–Muller Codes

Reed and Muller proposed a class of 2-ary linear codes based on finite geometry in 1954. In order to discuss the structure and properties of these codes, we first prove some results in number theory.

**Lemma 2.19** *Let $p$ be a prime, $k, n$ be two nonnegative integers whose $p$-ary is expressed as*

$$n = \sum_{i=0}^{l} n_i p^i, \ k = \sum_{i=0}^{l} k_i p^i.$$

*Then*

$$\binom{n}{k} \equiv \prod_{i=0}^{l} \binom{n_i}{k_i} (\text{mod } p), \ where \binom{n_i}{k_i} = 0, \ if \, k_i > n_i.$$

***Proof*** If $k = 0$, then $k_i = 0$, so the above formula holds. If $n = k$, then $n_i = k_i$, the above formula also holds. We might as well make $1 \leq k < n$, note the polynomial congruence

$$(1 + x)^p \equiv 1 + x^p (\text{mod } p),$$

so we have

$$\begin{aligned}(1+x)^n &= (1+x)^{\sum_{i=0}^{l} n_i p^i}\\ &\equiv \prod_{i=0}^{l}(1+x^{p^i})^{n_i} \pmod p,\end{aligned}$$

Comparing the coefficients of the $x^k$ terms on both sides of the above formula, if there is a $k_j > n_j$, then the $x^k$ terms do not appear on the right side of the above formula, which means that the coefficients of the $x^k$ terms on the left side are

$$\binom{n}{k} \equiv 0 (\text{mod } p).$$

If $k_i \leq n_i$, $\forall\, 0 \leq i \leq l$, then

$$\binom{n}{k} \equiv \prod_{i=0}^{l} \binom{n_i}{k_i} (\text{mod } p).$$

We complete the proof of Lemma.

Massey defined the concept of polynomial weight for the first time in 1973, on a finite field with characteristic 2 ($q = 2^r$), a polynomial $f(x) \in \mathbb{F}_q[x]$, whose Hamming weight is defined as

$$w(f(x)) = \text{The number of nonzero coefficients of } f(x).$$

**Lemma 2.20** (Massey, 1973) *Let* $f(x) = \sum_{i=0}^{l} b_i (x+c)^i \in \mathbb{F}_q[x]$ *and* $b_l \neq 0$, *let* $i_0$ *be the smallest subscript* $i$ *of* $b_i \neq 0$, *then*

$$w(f(x)) \geq w((x+c)^{i_0}).$$

***Proof*** $l = 0$, then $i_0 = 0$, the lemma holds. Let $l < 2^n$ be lemma, we consider $2^n \leq l < 2^{n+1}$, write $f(x)$ as

$$\begin{aligned}f(x) &= \sum_{i=0}^{2^n-1} b_i (x+c)^i + \sum_{i=2^n}^{l} b_i (x+c)^i\\ &= f_1(x) + (x+c)^{2^n} f_2(x)\\ &= f_1(x) + c^{2^n} f_2(x) + x^{2^n} f_2(x),\end{aligned}$$

where $\deg f_1(x) < 2^n$, $\deg f_2(x) < 2^n$. There are two situations to discuss:

(i) If $f_1(x) = 0$, then $w(f(x)) = 2w(f_2(x))$. Because $i_0 \geq 2^n$, so

$$\begin{aligned} w((x+c)^{i_0}) &= w((x^{2^n} + c^{2^n})(x+c)^{i_0-2^n}) \\ &= 2w((x+c)^{i_0-2^n}). \end{aligned}$$

From inductive hypothesis

$$w(f_2(x)) \geq w((x+c)^{i_0-2^n}).$$

So there are

$$w(f(x)) = 2w(f_2(x)) > 2w((x+c)^{i_0-2^n}) = w((x+c)^{i_0}).$$

(ii) $f_1(x) \neq 0$, $i_1$ is the subscript of $f_1(x)$, $i_2$ is the subscript of $f_2(x)$. If the term not 0 in $f_1(x)$ plus the corresponding term of $c^{2^n} f_2(x)$ becomes 0, then $x^{2^n} f_2(x)$ will have corresponding terms that are not zero, so we always have

$$w(f(x)) \geq w(f_1(x)), \ w(f(x)) \geq w(f_2(x)).$$

If $i_1 < i_2$, then $i_0 = i_1$, from inductive hypothesis,

$$w(f(x)) \geq w(f_1(x)) \geq w((x-c)^{i_1}) = w((x-c)^{i_0}).$$

Similarly, if $i_2 < i_1$, then $i_0 = i_2$, there is

$$w(f(x)) \geq w(f_2(x)) \geq w((x-c)^{i_2}) = w((x-c)^{i_0}).$$

If $i_1 = i_2$ , then it can always be changed into the case of $i_1 \neq i_0$, so we always have Lemma holds.

Next, we use Massey's method to construct Reed–Muller codes. Let $m \geq 1$, $\mathbb{F}_2^m$ be an $m$-dimensional affine space, denote as $AG(m, 2)$, $\alpha \in AG(m, 2)$ is a point in affine space, write $\alpha$ as an $m$-dimensional column vector, let $\{u_0, u_1, \ldots, u_{m-1}\}$ be the standard base of $\mathbb{F}_2^m$, that is

$$\alpha = \begin{bmatrix} a_0 \\ a_1 \\ \vdots \\ a_{m-1} \end{bmatrix}, u_0 = \begin{bmatrix} 1 \\ 0 \\ 0 \\ \vdots \\ 0 \end{bmatrix}, \ldots, u_{m-1} = \begin{bmatrix} 0 \\ 0 \\ 0 \\ \vdots \\ 1 \end{bmatrix},$$

where $a_i = 0$ or 1. Let's establish a $1-1$ correspondence between the points in the integer set $\{0 \leq j < 2^m\}$ and $AG(m, 2)$. Let $0 \leq j < 2^m$, then

$$j = \sum_{i=0}^{m-1} a_{ij} 2^i, \, a_{ij} \in \mathbb{F}_2.$$

We define

$$x_j = \sum_{i=0}^{m-1} a_{ij} u_i = \begin{bmatrix} a_{0j} \\ a_{1j} \\ \vdots \\ a_{(m-1)j} \end{bmatrix} \in \mathbb{F}_2^m,$$

Because when $j_1 \neq j_2$, there is $x_{j_1} \neq x_{j_2}$, So $\{x_j | 0 \leq j < 2^m\}$ gives all the points in $\mathbb{F}_2^m$. Write $n = 2^m$ and consider the matrix

$$E = [x_0, x_1, \ldots, x_{n-1}] = \begin{bmatrix} a_{00} & a_{01} & \cdots & a_{0(n-1)} \\ a_{10} & a_{11} & \cdots & a_{1(n-1)} \\ \vdots & \vdots & \cdots & \vdots \\ a_{(m-1)0} & a_{(m-1)1} & \cdots & a_{(m-1)(n-1)} \end{bmatrix}_{m \times n},$$

Each row vector $\alpha_i = (a_{i0}, a_{i1}, \ldots, a_{i(n-1)})(0 \leq i \leq m-1)$ of $E$ is a vector of $\mathbb{F}_2^n$, which is written as

$$E = \begin{bmatrix} \alpha_0 \\ \alpha_1 \\ \vdots \\ \alpha_{m-1} \end{bmatrix} = (a_{ij})_{m \times n} (0 \leq i < m, 0 \leq j < 2^m = n).$$

For each $i, 0 \leq i < m$, define a linear subspace in $\mathbb{F}_2^m$,

$$B_i = \{x_j \in \mathbb{F}_2^m | a_{ij} = 0\}.$$

Obviously, $B_i$ is a linear subspace, and the additive coset of $B_i$ is called an $m-1$-dimensional plat in $\mathbb{F}_2^m$. We consider $A_i = B_i + u_i$,

$$A_i = \{x_j \in \mathbb{F}_2^m | a_{ij} = 1, 0 \leq j < n\} \Rightarrow |A_i| = 2^{m-1}.$$

We define the characteristic function $\chi_i(\alpha)$ in $\mathbb{F}_2^m$ according to $A_i$,

$$\chi_i(\alpha) = \begin{cases} 1, & \text{if } \alpha \in A_i; \\ 0, & \text{if } \alpha \notin A_i. \end{cases}$$

where $\alpha \in \mathbb{F}_2^m$. So each row vector $\alpha_i (0 \leq i < m)$ in $E$ can be expressed as

$$\alpha_i = (\chi_i(x_0), \chi_i(x_1), \ldots, \chi_i(x_{n-1})).$$

For any two vectors $\alpha = (b_0, b_1, \ldots, b_{n-1})$, $\beta = (c_0, c_1, \ldots, c_{n-1})$ in $\mathbb{F}_2^n$, define the product vector

$$\alpha\beta = (b_0c_0, b_1c_1, \ldots, b_{n-1}c_{n-1}) \in \mathbb{F}_2^n.$$

So for $0 \le i_1, i_2 < m$, we have the product of row vectors of $E$

$$\alpha_{i_1}\alpha_{i_2} = (\chi_{i_1}(x_0)\chi_{i_2}(x_0), \chi_{i_1}(x_1)\chi_{i_2}(x_1), \ldots, \chi_{i_1}(x_{n-1})\chi_{i_2}(x_{n-1})).$$

So the $j$-th ($0 \le j < 2^m$) component of $\alpha_{i_1}\alpha_{i_2}$ is

$$\chi_{i_1}(x_j)\chi_{i_2}(x_j) = \begin{cases} 1, & \text{if } x_j \in A_{i_1} \cap A_{i_2}; \\ 0, & \text{if } x_j \notin A_{i_1} \cap A_{i_2}. \end{cases}$$

From the definition of $A_i$, obviously,

$$|A_{i_1} \cap A_{i_2}| = 2^{m-2}.$$

**Lemma 2.21** *Let $i_1, i_2, \ldots, i_s$ be the number of $s (0 \le s < m)$ different indexes from 0 to $m - 1$, then*

$$|A_{i_1} \cap A_{i_2} \cap \cdots \cap A_{i_s}| = 2^{m-s},$$

*And $\alpha_{i_1}\alpha_{i_2} \cdots \alpha_{i_s} \in \mathbb{F}_2^n$ has a weight function*

$$w(\alpha_{i_1}\alpha_{i_2} \cdots \alpha_{i_s}) = 2^{m-s}.$$

***Proof*** The first conclusion is obvious. Let's just prove the second conclusion,

$$\alpha = \alpha_{i_1}\alpha_{i_2} \cdots \alpha_{i_s} = (x_{i_1}(x_0) \cdots x_{i_s}(x_0), x_{i_1}(x_1) \cdots x_{i_s}(x_1), \ldots, x_{i_1}(x_{n-1}) \cdots x_{i_s}(x_{n-1}))$$

has $2^{m-s}$ $x_j \in A_{i_1} \cap A_{i_2} \cap \cdots \cap A_{i_s}$, so there are $2^{m-s}$ components in $\alpha$ that are 1 and the others are 0, so

$$w(\alpha) = w(\alpha_{i_1}\alpha_{i_2} \cdots \alpha_{i_s}) = 2^{m-s},$$

the Lemma holds.

For $0 \le l < 2^m$, $I(l)$ is defined as an indicator set,

$$I(l) = \{i_1, i_2, \ldots, i_s \mid l = \sum_{i=0}^{m-1} a_{il}2^i \text{ satisfy } a_{il} = 0\}.$$

The following properties of the indicator set $I(l)$ are obvious:

(i) If $l_1 \neq l_2 \Rightarrow I(l_1) \neq I(l_2)$.
(ii) $\bigcup_{0\leq l<n} I(l) = \{0, 1, 2, \ldots, m-1\}$.
(iii) If $l = n-1 \Rightarrow I(n-1)$ is an empty set.

The above properties are easy to verify, such as $(iii)$, because $l = n-1 = 2^m - 1 = 1 + 2 + \cdots + 2^{m-1}$, so the subscripts $i$ of $a_{il} = 0$ don't exist, that is $I(n-1) = \varnothing$.

Sometimes we can write indicator sets $I(l) = \{i_1, i_2, \ldots, i_s\}_l$.

**Lemma 2.22** *Let* $0 \leq l < n = 2^m$, $I(l) = \{i_1, i_2, \ldots, i_s\}$, *re hypothesis*

$$\alpha_{i_1}\alpha_{i_2}\cdots\alpha_{i_s} = (b_{l0}, b_{l1}, \ldots, b_{l(n-1)}) \in \mathbb{F}_2^n,$$

*then in the ring* $\mathbb{F}_2[x]$, *there is*

$$(1+x)^l = \sum_{j=0}^{n-1} b_{lj} x^{n-1-j}. \tag{2.26}$$

***Proof*** For $0 \leq j < n$, write $j = \sum_{i=0}^{m-1} a_{ij} 2^i$, then

$$n - 1 - j = \sum_{i=0}^{m-1} c_{ij} 2^i, \text{ where } c_{ij} = 1 - a_{ij}.$$

By Lemma 2.19,

$$\binom{l}{n-1-j} \equiv \prod_{i=0}^{m-1} \binom{a_{il}}{c_{ij}} (\bmod 2).$$

If

$$\binom{l}{n-1-j} \equiv 1 (\bmod 2),$$

then when $a_{il} = 0, \Rightarrow c_{ij} = 0 \Rightarrow a_{ij} = 1$, that is to say

$$\binom{l}{n-1-j} \equiv 1 (\bmod 2) \Leftrightarrow a_{ij} = 1, \text{ for } \forall\, i \in I(l).$$

on the other hand, from Lemma 2.21,

$$b_{lj} = 1 \Leftrightarrow x_j \in A_{i_1} \bigcap A_{i_2} \bigcap \cdots \bigcap A_{i_s} \Leftrightarrow a_{ij} = 1, \text{ when } i \in I(l).$$

Compare the $x^{n-1-j}$ terms on both sides of formula (2.26), so we have

$$(1+x)^l = \sum_{j=0}^{n-1} b_{lj} x^{n-1-j}.$$

The Lemma holds.

For any $0 \leq l < n = 2^m$, we define the index set $I(l) = \{i_1, i_2, \ldots, i_s\}$ and the vector in $\mathbb{F}_2^n$.

$$N_l = \alpha_{i_1}\alpha_{i_2}\cdots\alpha_{i_s}.$$

The index set $I(l)$ corresponding to different $l$ is different, so the corresponding vector $N_l$ is different; since the index set corresponding to $l = n-1$ is an empty set, the corresponding vector $N_{n-1}$ is defined as

$$N_{n-1} = (1, 1, \ldots, 1) = e.$$

Let $e_0 = (1, 0, \ldots, 0), \ldots, e_{n-1} = (0, 0, \ldots, 1)$ be a set of standard bases of $\mathbb{F}_2^n$.

**Lemma 2.23** *For $0 \leq j < n$, we have*

$$e_j = \prod_{i=0}^{m-1} (\alpha_i + (1 + a_{ij})e),$$

*where $\alpha_i$ is the $i$-th row of matrix $E$.*

***Proof*** For vector $\alpha$ in $\mathbb{F}_2^n$, its complement vector $\overline{\alpha}$ is defined to replace the component of 1 in $\alpha$ with 0, and the component of 0 in $\alpha$ with 1. So there are

$$\alpha + \overline{\alpha} = e = (1, 1, \ldots, 1), \forall\, \alpha \in \mathbb{F}_2^n.$$

When $0 \leq j < n$ is given, we define the $j$-th complement of row vector $\alpha_i (0 \leq i < m)$ of matrix $E$ as

$$\overline{\alpha_i(j)} = \begin{cases} \alpha_i, & \text{if } a_{ij} = 1; \\ \overline{\alpha_i}, & \text{if } a_{ij} = 0. \end{cases}$$

Obviously, there is

$$\alpha_i + (1 + a_{ij})e = \overline{\alpha_i(j)},$$

from the definition of index set $I(l)$, we have

$$\overline{\alpha_i(j)} = \begin{cases} \alpha_i, & \text{if } i \notin I(j); \\ e - \alpha_i, & \text{if } i \in I(j). \end{cases}$$

Now let's calculate

$$\prod_{i=0}^{m-1}(\alpha_i + (1 + a_{ij})e) = \prod_{i=0}^{m-1}\overline{\alpha_i(j)}$$
$$= \prod_{i\in I(j)}(e - \alpha_i)\prod_{i\notin I(j)}\alpha_i = b.$$

where $b \in \mathbb{F}_2^n$, let $b = (b_0, b_1, \ldots, b_{n-1})$. Obviously, $b_j = 1$. If $k \neq j$, then

$$b_k = \prod_{i\notin I(j)} a_{ik} \cdot \prod_{i\in I(j)}(1 - a_{ik}) = 0.$$

Thus $b = e_j$. We have completed the proof of Lemma.

**Lemma 2.24** $\{N_l\}_{0\le l<n}$ *constitutes a group of bases of* $\mathbb{F}_2^n$, *where* $N_{n-1} = e = (1, 1, \ldots, 1)$.

***Proof*** $\{N_l\}_{0\le l<n}$ has exactly $n$ different vectors, let's prove that they are linearly independent. Let

$$N_l = \alpha_{i_1}\alpha_{i_2}\cdots\alpha_{i_s} = (b_{l0}, b_{l1}, \ldots, b_{l(n-1)}),$$

$$\sum_{l=0}^{n-1} c_l N_l = (\sum_{l=0}^{n-1} c_l b_{l0}, \sum_{l=0}^{n-1} c_l b_{l1}, \ldots, \sum_{l=0}^{n-1} c_l b_{l(n-1)})$$

be a linear combination. Where $c = (c_0, c_1, \ldots, c_{n-1}) \neq 0$. Because

$$f(x) = \sum_{l=0}^{n-1}(1 + x)^l \in \mathbb{F}_2[x],\ f(x) \neq 0.$$

By Lemma 2.22, we have

$$f(x) = \sum_{j=0}^{n-1}(\sum_{l=0}^{n-1} c_l b_{lj})x^{n-1-j}.$$

So if there's a component $\sum_{l=0}^{n-1} c_l b_{lj} \neq 0$, that is $\{N_l\}_{0\le l<n}$ is a group of bases. The Lemma holds.

**Definition 2.12** Let $0 \le r < m$, a linear code of order $r$Ⅱ Reed–Muller code $R(r, m)$ be

$$R(r, m) = L(\{\alpha_{i_1}\alpha_{i_2}\ldots\alpha_{i_s} | 0 \le s \le r\}) \subset \mathbb{F}_2^n,$$

the vector corresponding to $s = 0$ is $e$.

Obviously, when $r = 0$, $R(0, m)$ corresponds to the repeated code in $\mathbb{F}_2^n$:

$$R(0, m) = \{(0, 0, \ldots, 0), (1, 1, \ldots, 1)\}.$$

For general $r$, $0 \leq r < m$, $R(r, m)$ is a $t$-dimensional linear subspace in $\mathbb{F}_2^n$, where

$$t = \sum_{s=0}^{r} \binom{m}{s}.$$

**Lemma 2.25** *The dual code of Reed–Muller code $R(r, m)$ of order $r$ is $R(m - r - 1, m)$.*

***Proof*** The dimensions of $R(r, m)$ and $R(m - r - 1, m)$ are

$$dim(R(r, m)) = \sum_{s=0}^{r} \binom{m}{s}$$

and

$$dim(R(m - r - 1, m)) = \sum_{s=0}^{m-r-1} \binom{m}{s}.$$

Because

$$\begin{aligned}
&\sum_{s=0}^{r} \binom{m}{s} + \sum_{s=0}^{m-r-1} \binom{m}{m-s} \\
&= \sum_{s=0}^{r} \binom{m}{s} + \sum_{s=r+1}^{m} \binom{m}{s} \\
&= \sum_{s=0}^{m} \binom{m}{s} = (1+1)^m \\
&= 2^m = n.
\end{aligned}$$

That is

$$dim(R(r, m)) + dim(R(m - r - 1, m)) = n.$$

Let $\alpha_{i_1}\alpha_{i_2}\cdots\alpha_{i_s}, \alpha_{j_1}\alpha_{j_2}\cdots\alpha_{j_t}$ be the basis vectors of R(r,m) and R(m-r-1,m), respectively. Let

$$\alpha = \alpha_{i_1}\alpha_{i_2}\cdots\alpha_{i_s},\ \beta = \alpha_{j_1}\alpha_{j_2}\cdots\alpha_{j_t},$$

by Lemma 2.21,

$$w(\alpha) = 2^{m-s}, w(\beta) = 2^{m-t},\ s \leq r < m,\ t \leq m - r - 1,$$

because $s + t < m$, the product $\alpha\beta = \alpha_{i_1}\alpha_{i_2}\cdots\alpha_{i_s} \cdot \alpha_{j_1}\alpha_{j_2}\cdots\alpha_{j_t}$ has weight

$$w(\alpha\beta) = w(\alpha_{i_1}\alpha_{i_2}\cdots\alpha_{i_s}\cdot\alpha_{j_1}\alpha_{j_2}\cdots\alpha_{j_t}) = 2^{m-(s+t)},$$

so

$$<\alpha, \beta> = 0,$$

That is, the dual code of $R(r, m)$ is $R(m - r - 1, m)$. The Lemma holds.

**Theorem 2.9** *Reed–Muller code $R(r, m)$ of order $r$ have minimal distance $d = 2^{m-r}$, specially, when $r = m - 2$, $R(m - 2, m)$ is a linear code $[n, n - m - 1]$.*

***Proof*** From Lemma 2.21, we have

$$w(\alpha_{i_1}\alpha_{i_2}\cdots\alpha_{i_s}) = 2^{m-s},$$

so the minimum distance of R(r,m) is $d \le 2^{m-r}$, on the other hand, let $I_1(r)$ be the value of all $l$ of corresponding $\{i_1, i_2, \ldots, i_s\}$ under the condition of $s \le r$, let

$$\alpha_{i_1}\alpha_{i_2}\cdots\alpha_{i_s} = (b_{l0}, b_{l1}, \ldots, b_{l(n-1)}),$$

then

$$f(x) = \sum_{l\in I_1(r)} (1 + x)^l = \sum_{j=0}^{n-1}\Big(\sum_{l\in I_1(r)} c_l b_{lj}\Big)x^{n-1-j}.$$

Therefore, the weight function of linear combination has the following relationship:

$$w\Big(\sum_{l\in I_1(r)} c_l\alpha_{i_1}\alpha_{i_2}\cdots\alpha_{i_s}\Big) = w(f(x)).$$

Define $i_0$ as

$$i_0 = \min\{l | l \in I_1(r)\}.$$

Obviously,

$$i_0 = 1 + 2 + \cdots + 2^{m-r-1} = 2^{m-r} - 1,$$

from Lemma 2.20, then there is

$$w(f(x)) \ge w((x + 1)^{i_0}) = i_0 + 1 = 2^{m-r}.$$

Because the combination numbers

$$\binom{i_0}{k} = \binom{2^{m-r} - 1}{k} \quad (0 \le k \le 2^{m-r} - 1)$$

are all odd, this is because

$$i_0 = 1 + 2 + \cdots + 2^{m-r-1},\ k = k_0 + k_1\cdot 2 + \cdots + k_{m-r-1}2^{m-r-1},$$

$\forall\, k_i \le 1$, so as to deduce

$$\binom{i_0}{k} \equiv \prod \binom{l}{k_i} (\bmod 2).$$

So there is

$$\binom{i_0}{k} \equiv 1 (\bmod 2).$$

In the end, we have $d = 2^{m-r}$. If let $r = m - 2$, then the minimum distance is 4. The dimension of $R(m-2, m)$ is

$$\begin{aligned} t &= \sum_{s=0}^{m-2} \binom{m}{s} = \sum_{s=0}^{m} \binom{m}{s} - \binom{m}{m-1} - \binom{m}{m} \\ &= 2^m - m - 1 \\ &= n - m - 1. \end{aligned}$$

So $R(m-2, m)$ is a linear code $[n, n-m-1]$. The theorem is proved.

Because $R(m-2, m)$ is in the form of linear code $[n, n-k]$, and the minimum distance is 4, so we consider $R(m-2, m)$ as a class of extended Hamming codes. Although it is not perfect, Hamming codes are perfect linear codes.

## 2.5 Shannon Theorem

In the channel transmission, due to the interference of the channel, a codeword $x \in C$ cannot be decoded correctly after it is sent, the probability of this error is recorded as $p(x)$, which is called the error probability of codeword $x$. According to Hamming distance, after code $C$ is selected, according to the decoding principle of "look most like", the error probability $p(x)$ of a codeword $x \overset{\text{sending}}{\longrightarrow} x'$ satisfies

$$\begin{cases} p(x) = 0, & \text{if } d(x, x') \le \rho_1 < \frac{1}{2}n; \\ p(x) > 0, & \text{if } d(x, x') > \rho_1. \end{cases}$$

where $\rho_1$ is the disjoint radius of code $C$. Therefore, the error probability $p(x)$ of code word $x$ is related to code $C$. The error probability of code $C$ is

$$p(C) = \frac{1}{|C|} \sum_{x \in C} p(x).$$

It is difficult to calculate the error probability of a codeword mathematically, we take the binary channel as an example, $C \subset \mathbb{F}_2^n$ is a binary code of length $n$, to calculate the error probability $p(x)$ of $x \in C$, we agree that the transmission error probability of character 0 is $p$, $p < \frac{1}{2}$, that is the probability of receiving 0 as 1 after transmission, and the probability of character 1 transmission error is also $p$, although the probability of error is very low, that is, the value of $p$ is very small, the probability of error exists due to the interference of channel. We further agree that the error probability of each transmission of character 0 or 1 is $p$, which is called memoryless channel. In the memoryless binary channel, the transmission of a codeword $x = x_1x_2 \ldots x_n \in C$ just constitutes the $n$-fold Bernoulli test, this probability model provides a theoretical basis for calculating the error probability of codeword $x$, let's take 2-tuple code as an example.

**Lemma 2.26** *Let $A_n$ be a binary repeated code of length n, that is $A_n = \{0, 1\} \subset \mathbb{F}_2^n$, $p(A_n)$ is the probability of error, then*

$$\lim_{n\to\infty} p(A_n) = 0.$$

***Proof*** The transmission of codeword $0 = (0, 0, \ldots, 0)$ is regarded as $n$-fold Bernoulli test, the character 0 has only two results of 0 and 1 after each transmission, the probability of occurrence of 0 is $q = 1 - p$, and the probability of occurrence of 1 is $p < \frac{1}{2}$. Let $0 \le k \le n$, then the probability of 0 appearing $k$ times is

$$\binom{n}{k} q^k p^{n-k}.$$

If $k > \frac{1}{2}n$, then there are $k > \frac{1}{2}n$ 0 characters in the received codeword after the codeword 0 is transmitted, suppose $0 \to \overline{0}$, then $d(0, \overline{0}) \le n - k < \frac{1}{2}n$. Because the disjoint radius of repeat code is $\frac{1}{2}n$, according to the decoding principle, we can always decode $\overline{0} \to 0$ correctly; therefore, the error of codeword $0 = (0, 0, \ldots, 0) \in \mathbb{F}_2^n$ occurs if and only if when $k \le \frac{1}{2}n$, the error probability is

$$p(0) = \sum_{0 \le k \le \frac{n}{2}} \binom{n}{k} q^k p^{n-k}.$$

Similarly, the error probability of codeword $1 = (1, 1, \ldots, 1) \in \mathbb{F}_2^n$ is

$$p(1) = \sum_{0 \le k \le \frac{n}{2}} \binom{n}{k} q^k p^{n-k}.$$

Therefore, the error probability of repeat code $A_n$ is

$$p(A_n) = \sum_{0\leq k\leq \frac{n}{2}} \binom{n}{k} q^k p^{n-k}.$$

To calculate the limit value $n \to \infty$ of the above equation, let's see

$$\sum_{0\leq k\leq \frac{n}{2}} \binom{n}{k} < \sum_{0\leq k\leq n} \binom{n}{k} = 2^n.$$

Because $p < \frac{1}{2}$, so $p < q$, and when $k \leq \frac{n}{2}$, we have

$$k \log \frac{q}{p} \leq \frac{n}{2} \log \frac{q}{p}.$$

It can be directly proved by the above formula

$$q^k p^{n-k} \leq (qp)^{\frac{n}{2}}.$$

Thus

$$p(A_n) \leq 2^n (qp)^{\frac{n}{2}} = (4qp)^{\frac{n}{2}}.$$

Because when $p < \frac{1}{2}$,

$$p^2 - p + \frac{1}{4} = (p - \frac{1}{2})^2 > 0,$$

so

$$p(1-p) = pq < \frac{1}{4}, \text{ that is } 4pq < 1.$$

Therefore,

$$0 \leq \lim_{n\to\infty} p(A_n) \leq \lim_{n\to\infty} (4qp)^{\frac{n}{2}} = 0.$$

The Lemma holds.

Below, we assume that the channel transmission is binary memoryless symmetric channel. Each code is binary code. The error probability of each transmission of characters 0 and 1 is $p$, $q = 1 - p$, $p < \frac{1}{2}$. For given codeword length $n$ and the number of codewords $M = M_n$, we define Shannon's probability $P^*(n, M_n, p)$ as

$$P^*(n, M_n, p) = \min\{P(C) | C \subset \mathbb{F}_2^n,\ |C| = M_n\}.$$

Shannon proved the following famous theorem in 1948.

**Theorem 2.10** (Shannon) *In a memoryless symmetric binary channel, let* $0 < \lambda < 1 + p \log p + q \log q$ *be a given real number,* $M_n = 2^{[\lambda n]}$*, then we have*

$$\lim_{n\to\infty} P^*(n, M_n, p) = 0.$$

In order to understand the meaning of Shannon's theorem and prove it, we need some auxiliary conclusions.

**Lemma 2.27** $0 < \lambda < 1 + p\log p + q\log q$ *is a given real number, any binary code* $C \subset \mathbb{F}_2^n$, *if* $|C| = 2^{[\lambda n]}$, *then the code rate* $R_C$ *of* $C$ *satisfies*

$$\lambda - \frac{1}{n} < R_C \leq \lambda.$$

*Specially, When* $n \to \infty$, *the rate of* $C$ *approaches* $\lambda$.

***Proof***

$$|C| = 2^{[\lambda n]} \Rightarrow \log_2 |C| = [\lambda n] \leq \lambda n.$$

Therefore,

$$R_C = \frac{1}{n}\log_2 |C| \leq \lambda.$$

From the properties of square bracket function,

$$\lambda n < [\lambda n] + 1,$$

so

$$\lambda n - 1 < [\lambda n] = \log_2 |C|.$$

There are

$$\lambda - \frac{1}{n} < \frac{1}{n}\log_2 |C| = R_C.$$

The Lemma 2.27 holds.

Combining Lemma 2.27, the significance of Shannon's theorem is that the code rate tends to the capacity $1 - H(p)$ of a channel when the code length $n$ increases and tends to infinity, and there exists a code $C$ whose error probability is arbitrarily small, according to Shannon's understanding, this kind of code is called "good code". Shannon first proved the existence of "good codes" under more general conditions by probability method. Theorem 2.10 is only a special case of Shannon's channel coding theorem. To prove Shannon theorem, we must accurately estimate the error probability of a given number of codewords under the principle of decoding.

**Lemma 2.28** *In the memoryless binary channel, let the probability of each transmission error of characters* 0 *and* 1 *be* $p$, $q = 1 - p$, *a codeword* $x = x_1x_2\ldots x_n \in \mathbb{F}_2^n$ *has exactly* $\omega$ *characters error during transmission, then for any* $\varepsilon > 0$, *let* $b = \sqrt{\frac{npq}{\varepsilon}}$, *we have*

$$P\{\omega > np + b\} \leq \varepsilon.$$

***Proof*** For any a codeword $x = x_1x_2\ldots x_n \in \mathbb{F}_2^n$, when transmitted in a memoryless binary channel, it can be regarded as an $n$-fold Bernoulli test, $\omega$ with exactly $\omega$ errors in $x$ can be regarded as a discrete random variable with a value of $0, 1, 2, \dots, n$, the probability of occurrence of $\omega$ is (i.e., the probability of the value $\omega$ of the random variable $\omega$)

$$b(\omega, n, p) = \binom{n}{\omega} p^{\omega} q^{n-\omega}.$$

Therefore, the probability distribution of $\omega$ obeys the discrete random variable of binomial distribution. From Lemma 1.18 of the first chapter, the expected value $E(\omega)$ and variance $D(\omega)$ of $\omega$ are as follows:

$$E(\omega) = np, \ \ D(\omega) = npq.$$

From the Chebyshev inequality of corollary 1.2, for any $k > 0$,

$$P\{|\omega - E(\omega)| \geq k\sqrt{D(\omega)}\} \leq \frac{1}{k^2}.$$

Take $k = \frac{1}{\sqrt{\varepsilon}}$, then we have

$$P\{w > np + b\} \leq P\{|\omega - np| > b\} \leq \varepsilon.$$

That is

$$P\{w > np + b\} \leq \varepsilon.$$

The Lemma 2.28 holds.

**Lemma 2.29** *Take* $\rho = [np + b]$, *where* $b = \sqrt{\frac{np(1-p)}{\varepsilon}}$, *then*

$$\frac{\rho}{n} \log \frac{\rho}{n} = p \log p + O(\frac{1}{\sqrt{n}}),$$
$$(1 - \frac{\rho}{n}) \log(1 - \frac{\rho}{n}) = q \log q + O(\frac{1}{\sqrt{n}}).$$

***Proof*** When $\varepsilon > 0$ is given, $b = O(\sqrt{n})$, so $\rho$ can be rewritten as

$$\rho = np + O(\sqrt{n}), \ \ \frac{\rho}{n} = p + O(\frac{1}{\sqrt{n}}).$$

Thus

$$\begin{aligned}\frac{\rho}{n} \log \frac{\rho}{n} &= (p + O(\frac{1}{\sqrt{n}})) \log(p + O(\frac{1}{\sqrt{n}})) \\ &= (p + O(\frac{1}{\sqrt{n}}))(\log p + \log(1 + O(\frac{1}{\sqrt{n}}))).\end{aligned}$$

For the real number $x$ of $|x| < 1$, we have the following Taylor expansion

$$\log(1+x) = x - \frac{1}{2}x^2 + \frac{1}{3}x^3 - \frac{1}{4}x^4 \cdots.$$

So when $|x| < 1$, we have

$$\log(1+x) = O(|x|),$$

thus

$$\log(1 + O(\frac{1}{\sqrt{n}})) = O(\frac{1}{\sqrt{n}}),$$

we have

$$\begin{aligned}\frac{\rho}{n}\log\frac{\rho}{n} &= (p + O(\frac{1}{\sqrt{n}}))(\log p + O(\frac{1}{\sqrt{n}}))\\ &= p\log p + O(\frac{1}{\sqrt{n}}).\end{aligned}$$

Similarly, for the second asymptotic formula,

$$(1-\frac{\rho}{n})\log(1-\frac{\rho}{n}) = q\log q + O(\frac{1}{\sqrt{n}}),$$

the Lemma 2.29 holds.

To prove Shannon theorem, we define the following auxiliary functions, and for any two codewords $x, y \in \mathbb{F}_2^n$, $\rho \geq 0$, define

$$f_\rho(x, y) = \begin{cases} 0, & \text{if } d(x, y) > \rho;\\ 1, & \text{if } d(x, y) \leq \rho.\end{cases}$$

Let $C = \{x_1, x_2, \ldots, x_M\} \subset \mathbb{F}_2^n$ be a binary code of $|C| = M$, define

$$g_i(y) = 1 - f_\rho(y, x_i) + \sum_{j\neq i} f_\rho(y, x_j).$$

**Lemma 2.30** *Assuming $y \in \mathbb{F}_2^n$ is a given codeword, then*

$$\begin{cases} g_i(y) = 0, & \textit{if } x_i \in C \textit{ is the only codeword so that } d(y, x_i) \leq \rho,\\ g_i(y) \geq 1, & \textit{otherwise}.\end{cases}$$

***Proof*** If there is a unique $x_i \in C$ such that $d(y, x_i) \leq \rho$, then $f_\rho(y, x_i) = 1$, but $f_\rho(y, x_j) = 0 (i \neq j)$, therefore

$$g_i(y) = 1 - f_\rho(y, x_i) + \sum_{j\neq i} f_\rho(y, x_j) = 0.$$

If $d(y, x_i) > \rho$, then $f_\rho(y, x_i) = 0$, so

$$g_i(y) = 1 - f_\rho(y, x_i) + \sum_{j \neq i} f_\rho(y, x_j) = 1 + \sum_{j \neq i} f_\rho(y, x_j) \geq 1.$$

If $d(y, x_i) \leq \rho$, but there is at least one $x_k \neq x_i$ such that $d(y, x_k) \leq \rho$, then

$$\begin{aligned} g_i(y) &= 1 - f_\rho(y, x_i) + \sum_{j \neq i} f_\rho(y, x_j) \\ &= 1 + \sum_{j \neq i, j \neq k} f_\rho(y, x_j) \geq 1. \end{aligned}$$

The Lemma 2.30 holds.

With the above preparation, we give the proof of Shannon's theorem.

***Proof*** (The proof of Theorem 2.10) According to the assumptions of the theorem, we assume that $0 < \lambda < 1 + p \log p + q \log q$ is a given positive real number ($p < \frac{1}{2}$).

$$M = M_n = 2^{[\lambda n]}, \ |C| = M.$$

Let

$$|C| = \{x_1, x_2, \ldots, x_M\} \subset \mathbb{F}_2^n,$$

$\varepsilon > 0$ is any given positive number,

$$b = \sqrt{\frac{npq}{\varepsilon}}, \ \rho = [pn + b].$$

Because of $p < \frac{1}{2}$, when $n$ is sufficiently large, we have $\rho = pn + O(\sqrt{n}) < \frac{1}{2}n$. In order to calculate the error probability of codeword $x_i \in C$, suppose $x_i \xrightarrow{\text{transmit}} y$, if $d(x_i, y) \leq \rho$, and there is a unique codeword $x_i \in C$ such that $d(y, x_i) \leq \rho$, so according to the decoding principle of "look the most like", $x_i$ is the most similar codeword in $C$, so we can decode it correctly as $y \rightarrow^{transmit} x_i$, in this case, the error probability of $x_i$ is 0. Otherwise, there will be real decoding error. On the other hand, $y$ becomes $x_i$, and the occurrence probability of the received codeword after transmission is the conditional probability $p = (y|x_i)$, so the error probability of $x_i$ is estimated as

$$\begin{aligned} P_i = p(x_i) &\leq \sum_{y \in \mathbb{F}_2^n} p(y|x_i) g_i(y) \\ &= \sum_{y \in \mathbb{F}_2^n} p(y|x_i)(1 - f_\rho(y, x_i)) + \sum_{y \in \mathbb{F}_2^n} \sum_{\substack{j=1 \\ j \neq i}}^{M} p(y|x_i) f_\rho(y, x_j). \end{aligned} \quad (2.27)$$

According to the definition of $f_\rho(y, x_i)$, the first term of the above formula is the probability that the received codeword $y$ sent by $x_i$ is not in ball $B_\rho(x_i)$, i.e.

$$\sum_{y\in\mathbb{F}_2^n} p(y|x_i)(1 - f_\rho(y, x_i)) = P\{\text{received codewords } y|y \notin B_\rho(x_i)\}.$$

Because $\omega = d(y, x_i)$ is exactly the number of $\omega$ error characters in $x_i \to y$, from the Chebyshev inequality of Lemma 2.28, we have

$$P\{\text{received codewords}|y \notin B_\rho(x_i)\} = P\{\omega > \rho\} \leq P\{\omega \geq np + b\} < \varepsilon,$$

from (2.27), we have

$$P_i = p(x_i) \leq \varepsilon + \sum_{y\in\mathbb{F}_2^n} \sum_{\substack{j=1\\ j\neq i}}^{M} p(y|x_i) f_\rho(y, x_j). \tag{2.28}$$

Because the definition of the error probability $p(C)$ of code $C$, so there is

$$p(C) = \frac{1}{M}\sum_{i=1}^{M} p(x_i) \leq \varepsilon + M^{-1} \sum_{i=1}^{M} \sum_{y\in\mathbb{F}_2^n} p(y|x_i) \sum_{\substack{j=1\\ j\neq i}}^{M} f_\rho(y, x_j).$$

Since $C$ is randomly selected, we can regard $p(C)$ as a random variable, so Shannon's probability $P * (n, M_n, p)$ is the minimum value of $p(C)$, so it is less than the expected value of $p(C)$, i.e.

$$\begin{aligned} P^*(n, M_n, p) &\leq E(P(C)) \\ &\leq \varepsilon + M^{-1} \sum_{i=1}^{M} \sum_{y\in\mathbb{F}_2^n} \sum_{\substack{j=1\\ j\neq i}}^{M} E(p(y|x_i) \cdot f_\rho(y, x_j)). \end{aligned}$$

When $i$ is given, the random variables $p(y|x_i)$ and $f_\rho(y, x_j)(j \neq i)$ are statistically independent, so

$$E(p(y|x_i) \cdot f_\rho(y, x_j)) = E(p(y|x_i))E(f_\rho(y, x_j)).$$

So there is

$$P^*(n, M_n, p) \leq \varepsilon + M^{-1} \sum_{i=1}^{M} \sum_{y\in\mathbb{F}_2^n} \sum_{\substack{j=1\\ j\neq i}}^{M} E(p(y|x_i))E(f_\rho(y, x_j)). \tag{2.29}$$

Let's calculate the expected value of $f_\rho(y, x_j)$, because $y$ is selected in $\mathbb{F}_2^n$ with equal probability, so

$$\begin{aligned}E(f_\rho(y, x_j)) &= \sum_{y\in\mathbb{F}_2^n} p(y) f_\rho(y, x_j)\\ &= \frac{1}{2^n}|B_\rho(x_j)|\\ &= \frac{1}{2^n}|B_\rho(0)|.\end{aligned}$$

So there is

$$\begin{aligned}P^*(n, M_n, p) &= \varepsilon + M^{-1}\sum_{i=1}^{M}\sum_{y\in\mathbb{F}_2^n} E(p(y|x_i))\sum_{\substack{j=1\\ j\neq i}}^{M} E(f_\rho(y, x_j))\\ &= \varepsilon + M^{-1}\sum_{i=1}^{M}\sum_{y\in\mathbb{F}_2^n} E(p(y|x_i))\frac{(M-1)|B_\rho(0)|}{2^n}.\end{aligned} \tag{2.30}$$

Now let's calculate the expected value of $p(y|x_i)(y$ fixed, $x_i$ randomly selected in $C)$

$$E(p(y|x_i)) = \sum_{i=1}^{M} p(x_i)p(y|x_i) = p(y),$$

thus

$$\sum_{i=1}^{M}\sum_{y\in\mathbb{F}_2^n} E(p(y|x_i)) = \sum_{i=1}^{M}\sum_{y\in\mathbb{F}_2^n} p(y) = M.$$

From (2.30),

$$P^*(n, M_n, p) \leq \varepsilon + \frac{M-1}{2^n}|B_\rho(0)|,$$

$$\log_2(P^*(n, M_n, p) - \varepsilon) \leq \log_2 M + \log_2 |B_\rho(0)| - n.$$

That is

$$\frac{1}{n}\log_2(P^*(n, M_n, p) - \varepsilon) \leq \frac{1}{n}\log_2 M + \frac{1}{n}\log_2 |B_\rho(0)| - 1.$$

From Lemma 1.11 of Chap. 1,

$$\frac{1}{n}\log_2 B_\rho(0) = \frac{1}{n}\log_2\sum_{i=0}^{\rho}\binom{n}{i} \leq H(\frac{\rho}{n}),$$

where $H(x) = -x \log x - (1-x)\log(1-x)(0 < x < \frac{1}{2})$ is the binary entropy function, so there is

$$\frac{1}{n}\log_2(P^*(n, M_n, p) - \varepsilon) \le \frac{1}{n}\log_2 M + H(\frac{\rho}{n}) - 1.$$

By hypothesis $M = 2^{[\lambda n]}$, $\rho = [pn + b]$, $b = O(\sqrt{n})$, we have

$$\begin{aligned}\frac{1}{n}\log_2(P^*(n, M_n, p) - \varepsilon) &\le \frac{[\lambda n]}{n} + H(\frac{\rho}{n}) - 1\\ &= \lambda + H(\frac{\rho}{n}) - 1 + O(\frac{1}{n}).\end{aligned}$$

By Lemma 2.29,

$$\begin{aligned}H(\frac{\rho}{n}) &= -(\frac{\rho}{n}\log\frac{\rho}{n} + (1-\frac{\rho}{n})\log(1-\frac{\rho}{n}))\\ &= -(p\log p + q\log q + O(\frac{1}{\sqrt{n}})).\end{aligned}$$

So

$$\frac{1}{n}\log_2(P^*(n, M_n, p) - \varepsilon) \le \lambda - (1 + p\log p + q\log q) + O(\frac{1}{\sqrt{n}}).$$

By hypothesis $\lambda < 1 + p\log p + q\log q$, when $n$ is sufficiently large, we have

$$\frac{1}{n}\log_2(P^*(n, M_n, p) - \varepsilon) \le -\beta(\beta > 0).$$

Therefore, $0 \le P^*(n, M_n, p) \le \varepsilon + 2^{-\beta n}$, take the limit $n \to \infty$ on both sides, finally,

$$\lim_{n\to\infty} P^*(n, M_n, p) = 0.$$

We completed the proof of the theorem.

According to Shannon, the code rate is close to a given normal number $\lambda$,

$$0 < \lambda < 1 + p\log p + q\log q = 1 - H(p),$$

the code with arbitrarily small error probability is called "good code", we further analyze the construction of this kind of "good code". (Shannon only proved the existence of "good code" in probability).

**Theorem 2.11** *For given $\lambda$, $0 < \lambda < 1 + p\log p + q\log q(p < \frac{1}{2})$, $M_n = 2^{[\lambda n]}$, if there is a perfect code $C_n$, and $|C_n| = M_n$, then we have*

$$\lim_{n\to\infty} p(C_n) = 0.$$

***Proof*** If perfect code $C_n$ exists, by Lemma 2.27,

$$\lambda - \frac{1}{n} \le R_{C_n} \le \lambda.$$

Therefore, the code rate of $C_n$ can be arbitrarily close to $\lambda$, the error probability of $C_n$ can be arbitrarily small, so $C_n$ is a "good code" in the mathematical sense. To prove Theorem 2.11, because $C_n$ is a perfect code, the minimum distance $d_n$ is defined as

$$d_n = 2e_n + 1,\ e_n < \frac{n}{2}.$$

Because of $\lim\limits_{n\to\infty} R_{C_n} = \lambda$, by Theorem 2.2, we have

$$\lim_{n\to\infty} H(\frac{e_n}{n}) = 1 - \lambda > H(p).$$

Because the binary entropy function $H(x)$ is a monotone continuous rising function $(0 < x < \frac{1}{2})$. So we have the limit $\lim\limits_{n\to\infty} \frac{e_n}{n}$, and

$$\lim_{n\to\infty} \frac{e_n}{n} > p,\ \text{that is}\ \frac{e_n}{n} > p,\ \text{When}\ n\ \text{is sufficiently large.}$$

Now consider the error probability $p(x)$ of codeword $x = x_1x_2\dots x_n \in C_n$, since $C_n$ is $e_n$ error correction code, so $x \to x'$, when $d(x, x') \le e_n$, we can always decode correctly, at this time, the error probability of $x$ is 0. Therefore, $x$ transmission error, that is, the case where $x'$ cannot be decoded correctly occurs only in case $d(x', x) = w_n > e_n$. At this point we have(When $n$ is sufficiently large)

$$\frac{w_n}{n} > \frac{e_n}{n} > p + \varepsilon,\ (\text{exist a}\ \varepsilon > 0)$$

So the error probability $p(x)$ of $x \in C_n$ is estimated

$$\begin{aligned} p(x) &\le P\{\frac{w_n}{n} > p + \varepsilon\} \\ &\le P\{|\frac{w_n}{n} - p| > \varepsilon\}. \end{aligned}$$

Because when $n \to \infty$, the random variable sequence $\{w_n\}$ is a Bernoulli random process (i.e., for each $n$, it is $n$-folds Bernoulli test). From theorem 1.2 in Chap. 1, we have

$$\lim_{n\to\infty} p(x) \le \lim_{n\to\infty} P\{|\frac{w_n}{n} - p| > \varepsilon\} = 0.$$

For $\forall\ x \in C_n$ holds, so

$$\lim_{n\to\infty} p(C_n) = 0.$$

The Theorem 2.11 holds.

From the proof of Theorems 2.10 and 2.11, it can be seen that Shannon randomly selects a code and randomly selects a codeword, which essentially regards the input information as a random event in a given probability space, and the transmission process of information is essentially a random process. The fundamental difference between Shannon and other mathematicians at the same time is that he regards information or a code as a random variable. The mathematical model of information transmission is a dynamic probability model rather than a static algebraic model. The most important method to study a code naturally is probability statistics rather than the algebraic combination method of traditional mathematics. From the perspective of probability theory, Theorems 2.10 and 2.11 regard a code as a random variable, but they have great particularity. The probability distribution of this random variable obeys Bernoulli binomial distribution, especially the statistical characteristics of code rate, which are not clearly expressed. It is the core content of Shannon's information theory to study the relationship between random variables with general probability distribution and codes. One of the most basic concepts is information entropy, or code entropy. Using the concept of code entropy, the statistical characteristics of a code are clearly displayed. Therefore, we see a basic framework and prototype of modern information theory. In the next chapter, we explain and prove these basic ideas and results of Shannon information theory in detail. One of the most important results is Shannon channel coding theorem (see Theorem 3.12 in Chap. 3). Shannon uses the probability method to prove that the so-called good code with a code rate up to the transmission capacity and an arbitrarily small error probability exists for the general memoryless channel (whether symmetrical or not). On the contrary, the code rate of a code with an arbitrarily small error probability must not be greater than the capacity of the channel. This channel capacity is called Shannon's limit, which has been pursued for a long time in the field of electronic communication engineering technology. People want to find a channel coding scheme with error probability in a controllable range (e.g., less than $\varepsilon$) and transmission efficiency (i.e., code rate) reaching Shannon's limit. In today's 5G era, this engineering technical problem seems to have been overcome. Returning to theorem 2.10, we see that the upper limit $1 - H(p)$ of the code rate is the channel capacity of the memoryless symmetric binary channel (see example 2 in Sect. 8 of Chap. 3). From this example, we can get a glimpse of Shannon's channel coding theory.

**Exercise 2**

1. Please design a code of length 7, which contains 8 codewords, where the Hamming distance of any two codewords is $\geq 4$. The code is transmitted through symmetric binary channel, assuming the error probability of characters 0 and 1 is $p$, calculate the success probability of codeword transmission.
2. Let $C$ be a binary code of length 16, satisfy

   $(i)$ Each codeword has a weight of 6.
   $(ii)$ Any two codewords have Hamming distance of 8.

Prove: $|C| \leq 16$. Does the binary code $C$ of $|C| = 16$ exist?

3. Let $C$ be a binary code of length $n$ and an error correcting code of one character, prove

$$|C| \leq \frac{2^n}{n+2} \ (n\ is\ even).$$

4. Let $C$ be a binary perfect code of length $n$, and the minimum distance is 7. Prove: $n = 7$ or $n = 23$.
5. Let $C \subset \mathbb{F}_q^n$ be a linear code, $C = [n, k]$ and any $k$ coordinates be symmetric, prove: the minimum distance of $C$ is $d = n - k + 1$.
6. Suppose $C = [2k + 1, k] \subset \mathbb{F}_2^{2k+1}$, and $C \subset C^{\perp}$, write the difference set $C^{\perp} \backslash C$.
7. Let $x = x_1 x_2 \ldots x_6 \in \mathbb{F}_2^6$, Decide Hamming ball $|B_1(x)|$. We can find a code $C \subset \mathbb{F}_2^6$? Where $|C| = 9$, satisfy the Hamming distance of any two different codewords in $C$ is $\geq 3$?
8. Let $C = [n, k] \subset \mathbb{F}_q^n$ be a linear code, the generating matrix is $G$, if every column of $G$ is not all zero, prove

$$\sum_{x \in C} w(x) = n(q-1)q^{k-1}.$$

Where $w(x)$ is the weight of codeword $x$.

9. Let $C = [n, k]$ be a linear binary code, and there is a codeword with odd weight in $C$, prove that the codewords with even weight in $C$ form a linear code $[n, k - 1]$.
10. Let $C$ be a linear binary code, the generating matrix $G$ is

$$\begin{pmatrix} 1\,0\,0\,0\,1\,0\,1 \\ 0\,1\,0\,0\,1\,0\,1 \\ 0\,0\,1\,0\,0\,1\,1 \\ 0\,0\,0\,1\,0\,1\,1 \end{pmatrix},$$

Please decode the received codewords as follows: $y_1 = (1101011)$, $y_2 = (0110111)$, $y_3 = (0111000)$.

11. Let $p$ be a prime, is there a self-dual linear code $C = [8, 4]$ over $\mathbb{F}_p$?
12. Let $R_k$ be the rate of binary Hamming codes, find $\lim_{k \to \infty} R_k = ?$
13. Let $C$ be a linear binary code, the weight distribution polynomial is $A(z)$, finding the weight distribution polynomial $B(z)$ of dual code $C^{\perp}$.
14. Let $C = [n, k] \subset \mathbb{F}_2^n$, weight distribution polynomial be $A(z)$, we use binary symmetric channel to transmit codewords, and the error probability is $p$ (the error probability of characters 0 and 1), we hope that a codeword transmission error can be detected, and calculate the probability that a codeword transmission error will not be detected.
15. There is no linear code $C = [15, 8]$ with minimum distance 5 over any finite field $\mathbb{F}_q$.
16. Let $n = 2^m$, proved that Reed–Muller code $R(1, m)$ is Hadamard code of length $n$ .

17. Proved that ternary Golay has 132 codewords and its weight is 5. Let $x$ be the codeword of weight 5, consider all pairs $(x, 2x)$, where $w(x) = 5$, take the component whose coordinate component is not zero as a subset. Proved that there are 66 such subsets and form $4 - (11, 5, 1)$ designs.
18. If the minimum distance $d$ of a binary code $C = (n, M, d)$ is even, prove that there exists a binary code such that all its codewords have even weights.
19. Let $H$ be a Hadamard matrix $H_{12}$, define

$$A = H - I, G = (I, A), I \text{ is the unit matrix.}$$

Proved that $G$ is the generating matrix of ternary code [24, 12] and the minimum distance is 9.
20. Let $C = [4, 2]$ be a ternary Hamming code. $H$ is the check matrix of $C$, let $I$ be the unit matrix of order 4, $J$ is a square matrix of order 4 with all elements of 1, define

$$G = \begin{bmatrix} J + I & I & I \\ 0 & H & -H \end{bmatrix},$$

prove that $G$ generates a ternary code $C = [12, 6]$ and the minimum distance is 6.

## References


Barg, A. M., Katsman, S. L., & Tsfasman, M. A. (1987). Algebraic Geometric Codes from Curves of Small Genus. *Probl. of Information Transmission,23*, 34–38.

Berlekamp, E. R. (1972). Decoding the Golay Code, JPL Technical Report 32-1256 (Vol. IX, pp. 81–85). Jet Propulsion Laboratory.

Berlekamp, E. R. (1968). *Algebraic Coding Theory*. NewYork: McGraw-Hill.

Best, M. R. (1980). Binary cades with a minimum distance of four. *IEEE Transactions of Information Theory, 26*, 738–742.

Best, M. R. (1978). On the Existence of Perfect Codes, Report ZN 82/78. Amsterdam: Mathematical Centre.

Bussey, W. H. (1905). Galois field tables for $p^n$ 169. *Bull Amer Math Soc, 12*, 22–38.

Bussey, W. H. (1910). Tables of Galois fields of order less than 1000. *Bull Amer Math Soc, 16*, 188–206.

Cameron, P. J., & van Lint, j. H. (1991). *Designs, graphs, codes and their links*. London Math Soc Student Texts (Vol. 22). Cambridge University Press.

Conway, J. H., & Sloane, N. J. A. (1994). Quaternary constructions for the binary single- error-correcting codes of Julin, Best, and others. *Designs, Codes and Cryptography,41*, 31–42.

Curtis, C. W., & Reiner, I. (1962). *Representation Theory of Finite Groups and Associative Algebras*. New York-London: Interscience.

Delsarte, P., & Goethals, J. M. (1975). Unrestricted codes with the Golay parameters are unique. *Discrete Math.,12*, 211–224.

Elias, P. *Coding for noisy channels*. IRE Conv, Record, part 4 (pp. 37–46).

Feller, W. (1950). *An introduction to probability theory and its applications* (Vol. I). Wiley.

Feng, G.-L., & Rao, T. R. N. (1994). A simple approach for construction of algebraic-geometric codes from affine plane curves. *IEEE Trans. Info. Theory,40*, 1003–1012.

Forney, G. D. (1970). Convolutional codes I: Algebraic structure. *IEEE Trans Info Theory, 16*, 720–738: Ibid, 17, 360 (1971).
Gallagher, R. G. (1968). Information Theory and Reliable Communication. New York: Wiley.
Goethals, J. M. (1977). The extended Nadler code is unique. *IEEE Trans Info, 23*, 132–135.
Goppa, V. D. (1970). A new class of linear error- correcting codes. *Problems of Info Transmission, 6*, 207–212.
Goto, M. (1975). A note on perfect decimal AN codes. *Info Control, 29*, 385–387.
Goto, M., & Fukumara, T. (1975). Perfect nonbinary AN codes with distance three. *Info Control, 27*, 336–348.
Graham, R. L., & Sloane, N. J. A. (1980). Lower bounds for constant weight codes. *IEEE Trans Info Theory, 26*, 37–40.
Gritsenko, V. M. (1969). Nonbinary arithmetic correcting codes. *Problems of Info Transmission, 5*, 15–22.
Helgert, H. J., & Stinaff, R. D. (1973). Minimum distance bounds for binary linear codes. *IEEE Trans Info Theory, 19*, 344–356.
Høholdt, T., & Pellikaan, R. (1995). On the decoding of algebraic-geometric codes. *IEEE Transactions of Info Theory, 41*, 1589–1614.
Høholdt, T., van Lint, J. H., & Pellikaan, R. (1998). Algebraic geometry codes. In V. S. Pless, W. C. Huffman & R. A. Brualdi (Eds.),*Hand-book of coding theory*. Elsevier Science Publishers.
Hong, Y. (1984). On the nonexistence of unknown perfect 6- and 8-codes in Hamming schemes $H(n, q)$ with $q$ arbitrary. Osaka J. Math., 21, 687–700.
Justesen, J. (1975). An algebraic construction of rate $\frac{1}{v}$ convolutional codes. *IEEE Trans Info Theory, 21*, 577–580.
Justesen, J., Larsen, K. J., Jensen, E. H., Havemose, A., & Høholdt, T. (1989). Construction and decoding of a class of algebraic geometry codes. *IEEE Transactions of Info Theory, 35*, 811–821.
Kasami, T. (1969). An upper bound on k/n for affine invariant codes with fixed d/n. *IEEE Trans Info Theory, 15*, 171–176.
Kerdock, A. M. (1972). A class of low-rate nonlinear codes. Info and Control, 20, 182–187.
Levenshtein, V. I. (1975). Minimum redundancy of binary error-correcting codes. *Info Control, 28*, 268–291.
Macwilliams, F. J., & Sloane, N. J. A. (1977). The Theory of Error-correcting Codes. Amsterdam-New York-Oxford: North Holland.
Massey, J. L., & Garcia, O. N. (1972). Error-correcting codes in computer arithmetic. In J. T. Ton (Eds.), *Advances in information systems science* (Vol. 4, Ch. 5). Plenum Press.
Massey, J. L., Costello, D. J., & Justesen, J. (1973). Polynomial weights and code construction. *IEEE Trans. Info. Theory,19*, 101–110.
McEliece, R. J. (1977). The theory of information and coding. In *Encyclopedia of mathematics and its applications* (Vol. 3). Addison-Wesley.
McEliece, R. J. (1979). The bounds of Delsarte and Lovasz and their applications to coding theory. In G. Longo (Eds.), *Algebraic coding theory and applications*CISM Courses and Lectures (Vol. 258). Springer.
McEliece, R. J., Rodemich, E. R., Rumsey, H. C., & Welch, L. R. (1977). New upper bounds on the rate of a code via the Delsarte- MacWilliams inequalities. *IEEE Trans. Info. Theory,23*, 157–166.
Peek, J. H. (1985). Communications Aspects of the Compact Disc Digital Audio System. IEEE Communications Magazine, 23(2), 7–15.
Peterson, W. W., & Weldon, E. J. (1972). Error-correcting codes (2nd Edn). MIT Press.
Piret, P. (1977). Algebraic properties of convolutional codes with automorphisms, Ph.D. Dissertation. University of Catholique de Louvain.
Piret, P. (1988). *Convolutional codes, an algebraic approach*. The MIT Press.
Posner, E. C. (1968). Combinatorial structures in planetary reconnaissance. In E. B. Mann (Eds.), *Error correcting codes* (pp. 15–46). Wiley.
Rao, T. R. N. (1974). Error Coding for Arithmetic Processors. New York-London: Academic Press.

Roos, C. (1979). On the structure of convolutional and cyclic convolutional codes. IEEE Trans. Info. Theory, 25, 676–683.
Shannon, C. E. (1948). A mathematical theory of communication. *Bell System Technology Journal, 27*, 379–423, 623–656.
Sloane, N. J. A., Reddy, S. M., & Chen, C. L. (1972). New binary codes. *IEEE Trans. Info. Theory, 18*, 503–510.
Solomon, G., & van Tilborg, H. C. A. A connection between block and convolutional codes. *SIAM Journal of Applied Mathematics, 37*, 358–369.
Stichtenoth, H. (1993). *Algebraic function fields and codes*. Springer, Universitext.
Tietäväinen, T. (1973). On the nonexistence of perfect codes over finite fields. *SIAM Journal of Applied Mathematics, 24*, 88–96.
van der Geer, G., & van Lint, J. H. (1988). *Introduction to coding theory and algebraic geometry*. Birkhäuser.
van Lint, J. H. (1971). Nonexistence theorems for perfect error-correcting codes. In *Computers in Algebra and Theory* (Vol. IV) (SIAM-AMS Proceedings).
van Lint, J. H. (1972). A new description of the Nadler code. *IEEE Trans Info Theory, 18*, 825–826.
van Lint, J. H. (1975). A survey of perfect codes. *Rocky Mountain Journal of Math, 5*, 199–224.
van Lint, J. H. (1990). Algebraic geometric codes. In D. Ray-Chaudhuri (Eds.), *Coding theory and design theory I*, The IMA Volumes in Math and Appl 20. Springer.
van Lint, J. H. (1999). *Introduction to coding theory*, GTM86, Springer.
van Lint, J. H., & Macwilliams, F. J. (1978). Generalized quadratic residue codes. *IEEE Trans Info Theory, 24*, 730–737.
van Lint, J. H., & Wilson, R. M. (1992). *A course in combinatorics*. Cambridge University Press.
van Oorschot, P. C., & Vanstone, S. A. (1989). *An introduction to error correcting codes with applications*. Kluwer.

# Chapter 3
# Shannon Theory

## 3.1 Information Space

According to Shannon, a message $x$ is a random event. Let $p(x)$ be the probability of occurrence of event $x$. If $p(x) = 0$, this event does not occur; If $p(x) = 1$, this event must occur. When $p(x) = 0$ or $p(x) = 1$, information $x$ can be called trivial information or spam information. Therefore, the real mathematical significance of information $x$ lies in its uncertainty, that is $0 < p(x) < 1$. Quantitative research on the uncertainty of nontrivial information constitutes all the starting point of Shannon's theory; this starting point is now called information quantity or information entropy, or entropy for short. Shannon and his colleagues at Bell laboratory considered "bit" as the basic quantitative unit of information. What is "bit"? We can simply understand it as the number of bits in the binary system. However, according to Shannon, the binary system with $n$ digits can express up to $2^n$ numbers. From the point of view of probability and statistics, the probability of occurrence of these $2^n$ numbers is $\frac{1}{2^n}$. Therefore, a bit is the amount of information contained in event $x$ with probability $\frac{1}{2}$. Taking this as the starting point, Shannon defined the self-information $I(x)$ contained in an information $x$ as

$$I(x) = -\log_2 p(x). \tag{3.1}$$

Therefore, one piece of information $x$ contains $I(x)$-bit information, when $p(x) = \frac{1}{2}$, then $I(x) = 1$. Equation (3.1) is Shannon's first extraordinary progress in information quantification. On the other hand, with the emergence of Telegraph and telephone, binary is widely used in the conversion and transmission of information. Therefore, we can assert that without binary, there would be no Shannon's theory, let alone the current informatics and information age. The purpose of this section is to strictly mathematically deduce and simplify the most basic and important conclusions in Shannon's theory. First, we start with the rationality of the definition of formula (3.1).

If $I(x)$ is used to represent the self-information of a random event $x$, the greater the probability of occurrence $p(x)$, the smaller the uncertainty. Therefore, $I(x)$ should be a monotonic decreasing function of probability $p(x)$. If $xy$ is a joint event and

 
Z. Zheng, *Modern Cryptography Volume 1*, Financial Mathematics and Fintech,
https://doi.org/10.1007/978-981-19-0920-7_3

is statistically independent, that is, $p(xy) = p(x)p(y)$, then the self-information amount is $I(xy) = I(x) + I(y)$. Of course, the self-information amount $I(x)$ is nonnegative, that is $I(x) \geq 0$. Shannon prove, the self-information $I(x)$ satisfying the above three assumptions must be

$$I(x) = -c \log p(x),$$

where $c$ is a constant. This conclusion can be derived directly from the following mathematical theorems.

**Lemma 3.1** *If the real function $f(x)$ satisfies the following conditions in interval $[1, +\infty)$:*

*(i)* $f(x) \geq 0$,
*(ii) If* $x < y \Rightarrow f(x) < f(y)$,
*(iii)* $f(xy) = f(x) + f(y)$.

*Then $f(x) = c \log x$, where $c$ is a constant.*

***Proof*** Repeated use condition $(iii)$, then there is

$$f(x^k) = kf(x), \ k \geq 1$$

for any positive integer $k$. Take $x = 1$, then the above formula holds if and only if $f(1) = 0$. It can be seen from $(ii)$ that $f(x) > 0$ when $x > 1$. Let $x > 1$, $y > 1$ and $k \geq 1$ given, you can always find a nonnegative integer $n$ to satisfy

$$y^n \leq x^k < y^{n+1},$$

Take logarithms on both sides to get

$$\frac{n}{k} \leq \frac{\log x}{\log y} < \frac{n+1}{k},$$

On the other hand, we have

$$nf(y) \leq kf(x) < (n+1)f(y),$$

thus

$$|\frac{f(x)}{f(y)} - \frac{\log x}{\log y}| \leq \frac{1}{k},$$

when $k \to \infty$, we have

$$\frac{f(x)}{f(y)} = \frac{\log x}{\log y}, \ \forall x, y \in (1, +\infty).$$

Therefore,

$$\frac{f(x)}{\log x} = \frac{f(y)}{\log y} = c, \ \forall\, x, y \in (1, +\infty).$$

That is $f(x) = c \log x$. The Lemma holds.

In Lemma 3.1, let $I(x) = f(\frac{1}{p(x)})$, then $f(x)$ satisfies the condition $(i)$, $(ii)$ and $(iii)$, thus $I(x) = -c \log p(x)$. That is (3.1) holds.

In order to introduce the definition of information space, we use $X$ to represent a finite set of original information, or a countable and additive information set, which is called source state set. It can be an alphabet, a finite number of symbols or a set of numbers. For example, 26 letters in English and 2-element finite field $\mathbb{F}_2$ are commonly used source state sets. Elements in $X$ can be called messages, events, etc., or characters. We often use English capital letters such as $X$, $Y$, $Z$ to represent a source state set, and lowercase Greek letters $\xi$, $\eta$, ... to represent a random variable in a given probability space.

**Definition 3.1** The value space of a random variable $\xi$ is a source state set $X$; the probability distribution of characters on $X$ as events is defined as

$$p(x) = P\{\xi = x\}, \ \forall\, x \in X. \tag{3.2}$$

We call $(X, \xi)$ an information space in a given probability space, when the random variable $\xi$ is clear, we usually record the information space $(X, \xi)$ as $X$. If $\eta$ is another random variable valued on $X$, and $\xi$ and $\eta$ obey the same probability distribution, that is

$$P\{\xi = x\} = P\{\eta = x\}, \ \forall\, x \in X.$$

Call two information spaces $(X, \xi) = (X, \eta)$, usually recorded as $X$.

As can be seen from Definition 3.1, an information space $X$ constitutes a finite complete event group, that is, we have

$$\sum_{x \in X} p(x) = 1, \ 0 \le p(x) \le 1, \ x \in X. \tag{3.3}$$

It should be noted that if there are two random variables $\xi$ and $\eta$ with values on $X$, when the probability distributions obeyed by $\xi$ and $\eta$ are not equal, then $(X, \xi)$ and $(X, \eta)$ are two different information spaces; at this point, we must distinguish the two different information spaces with $X_1 = (X, \xi)$ and $X_2 = (X, \eta)$.

**Definition 3.2** $X$ and $Y$ are two source state sets, and the random variables $\xi$ and $\eta$ are taken on $X$ and $Y$, respectively; if $\xi$ and $\eta$ are compatible random variables, the probability distribution of joint event $xy (x \in X, y \in Y)$ is defined as

$$p(xy) = P\{\xi = x, \eta = y\}, \ \forall\, x \in X, \ y \in Y. \tag{3.4}$$

Then, we call the joint event set

$$XY = \{xy | x \in X, y \in Y\}$$

Together with the corresponding random variables $\xi$ and $\eta$, it is called the product space of information space $(X, \xi)$ and $(Y, \eta)$, denote as $(XY, \xi, \eta)$, when $\xi$ and $\eta$ are clear, they can be abbreviated as $XY = (XY, \xi, \eta)$. If $X = Y$ are two identical source state sets, $\xi$ and $\eta$ have the same probability distribution, then the product space $XY$ is denoted as $X^2$ and is called a power space.

Since the information space is a complete set of events, defined by the product information space, we have the following full probability formula and probability product formula:

$$\begin{cases} \sum\limits_{x \in X} p(yx) = p(y), \ \forall\, y \in Y \\ \sum\limits_{y \in Y} p(xy) = p(x), \ \forall\, x \in X. \end{cases} \tag{3.5}$$

And

$$p(x)p(y|x) = p(xy), \ \forall\, x \in X, y \in Y.$$

Where $p(y|x)$ is the conditional probability of $y$ under the condition of $x$.

**Definition 3.3** Let $X_1, X_2, \ldots, X_n (n \geq 2)$ be $n$ source state sets, $\xi_1, \xi_2, \ldots, \xi_n$ be $n$ compatible random variables with values, respectively, in $X_i$, the probability distribution of joint event $x_1 x_2 \cdots x_n$ is

$$p(x_1 x_2 \cdots x_n) = P\{\xi_1 = x_1, \xi_2 = x_2, \ldots, \xi_n = x_n\}. \tag{3.6}$$

Then called

$$X_1 X_2 \cdots X_n = \{x_1 x_2 \cdots x_n | x_i \in X_i, \ 1 \leq i \leq n\}$$

are the product of $n$ information spaces, especially when $X_1 = X_2 = \cdots = X_n = X$, and each $\xi_i$ has the same probability distribution on $X$, define $X^n = X_1 X_2 \cdots X_n$, called the $n$-th power space of information space $X$.

Let us give some classic examples of information space.

***Example 3.1*** (*Two point information space with parameter* $\lambda$) Let $X = \{0, 1\} = \mathbb{F}_2$ be a binary finite field, the random variable $\xi$ taken on $X$ is subject to the two-point distribution with parameter $\lambda$, that is

$$\begin{cases} p(0) = P\{\xi = 0\} = \lambda, \\ p(1) = P\{\xi = 1\} = 1 - \lambda. \end{cases}$$

where $0 < \lambda < 1$, then $(X, \xi)$ is called a two-point information space with parameter $\lambda$, still denote as $X$.

***Example 3.2*** (*Equal probability information space*) Let $X = \{x_1, x_2, \ldots, x_n\}$ be a source state sets, the random variable $\xi$ on $X$ obeys the equal probability distribution, that is

$$p(x) = P\{\xi = x\} = \frac{1}{|X|}, \ \forall\, x \in X.$$

Then $(X, \xi)$ is called equal probability information space, still denote as $X$.

***Example 3.3*** (*Bernoulli information space*) Let $X_0 = \{0, 1\} = \mathbb{F}_2$. Let the random variable $\xi_i$ be the $i$-th Bernoulli test; therefore, $\{\xi_i\}_{i=1}^n$ is a set of independent and identically distributed random variables. We let the product space

$$X = (X_0, \xi_1)(X_0, \xi_2) \cdots (X_0, \xi_n) = X_0^n \subset \mathbb{F}_2^n,$$

the power space $X$ is called Bernoulli information space, also alled memoryless binary information space. The probability function $p(x)$ in $X$ is

$$p(x) = p(x_1 x_2 \cdots x_n) = \prod_{i=1}^{n} p(x_i), \ x_i = 0 \text{ or } 1. \tag{3.7}$$

where $p(0) = \lambda, \ p(1) = 1 - \lambda$.

***Example 3.4*** (*Degenerate information space*) If $X = \{x\}$, it contains only one character. $X$ is called a degenerate information space, or trivial information space. The random variable $\xi$ takes the value $x$ of probability 1, that is $P\{\xi = x\} = 1$. At this time, $\xi$ is a random variable with degenerate distribution in probability.

**Definition 3.4** Let $X = \{x_1, x_2, \ldots, x_n\}$ be a source state sets, if $X$ is an information space, the information entropy $H(X)$ of $X$ is defined as

$$H(X) = -\sum_{x \in X} p(x) \log p(x) = -\sum_{i=1}^{n} p(x_i) \log p(x_i), \tag{3.8}$$

if $p(x_i) = 0$ in the above formula, we agreed that $p(x_i) \log p(x_i) = 0$, the base of logarithm can be selected arbitrarily; if the base of the logarithm is $D(D \geq 2)$, then $H(X)$ is called $D$-ary entropy, sometimes denote as $H_D(X)$.

**Theorem 3.1** *For any information space $X$, always have*

$$0 \leq H(X) \leq \log |X|. \tag{3.9}$$

*And $H(X) = 0$ if and only if $X$ is a degenerate information space, $H(X) = \log |X|$ if and only if $X$ is a equal probability information space.*

***Proof*** $H(X) \geq 0$ is trivial. We only prove the inequality on the right of Eq. (3.9). Because $f(x) = \log x$ is a strictly convex real value, from the Lemma 1.7 in Chap. 1, thake $g(x) = \frac{1}{p(x)}$ is a positive function, $p(x) > 0$, thus let $X = \{x_1, x_2, \ldots, x_m\}$,

$$H(X) = \sum_{i=1}^{m} p(x_i) \log \frac{1}{p(x_i)} \le \log \sum_{i=1}^{m} \frac{p(x_i)}{p(x_i)} = \log m.$$

The above equal sign holds if and only if $p(x_1) = p(x_2) = \cdots = p(x_m) = \frac{1}{m}$, that is, $X$ is equal probability information space. If $X = \{x\}$ is a degenerate information space, because $p(x) = 1$, so $H(X) = 0$. Conversely, if $H(X) = 0$, let $X = \{x_1, x_2, \ldots, x_m\}$, suppose $\exists\, x_i \in X$, such that $0 < p(x_i) < 1$, then

$$0 < p(x_i) \log \frac{1}{p(x_i)} \le H(X).$$

So there is $p(x_i) = 1$, but $p(x_j) = 0 (j \ne i)$; at this time, $X$ degenerates into $X = \{x_i\}$, which is a trivial information space, the Lemma holds.

An information space is a dynamic code (which changes with the change of the random variable on it). For "dynamic code", that is, the code rate of information space $X$, Shannon replaces $\frac{1}{n} H(X)$ with information entropy, so information entropy $H(X)$ becomes the first mathematical quantity to describe dynamic code. From Theorem 3.1, when the code is degenerate, the minimum rate of a dynamic code is 0, when the code is equal probability, the maximum rate is the rate of the usual static code.

Next, we discuss the information entropy of several typical information spaces.

***Example 3.5*** (i) Let $X$ be the two-point information space of parameter $\lambda$, then

$$H(X) = -\lambda \log \lambda - (1 - \lambda) \log(1 - \lambda) = H(\lambda).$$

$H(\lambda)$ we defined it in Chap. 1, it was called binary information entropy function at that time. Now we know why it is called entropy function

(ii) $X = \{x\}$ is degraded information space, then $H(X) = 0$.
(iii) When $X$ is equal overview information space, then $H(X) = \log |X|$.

***Remark*** Most authors directly regard a random variable as an information space. Mathematically, it is convenient to do so and call it the information measurement of random variables. However, from the perspective of information, using the concept of information space can better understand and simplify Shannon's theory; the core idea of this theory is the random measurement of information, not the information measurement of random variables.

## 3.2 Joint Entropy, Conditional Entropy, Mutual Information

**Definition 3.5** Let $X$, $Y$ be two information spaces, and $\xi$, $\eta$ be random variables with corresponding values, respectively. If $\xi$ and $\eta$ are independent random variables, that is

$$P\{\xi = x, \eta = y\} = P\{\xi = x\} \cdot P\{\eta = y\}, \ \forall\, x \in X, y \in Y.$$

$X$ and $Y$ are called independent information space, and the probability distribution of joint events is

$$p(xy) = p(x)p(y), \ \forall x \in X, y \in Y.$$

**Definition 3.6** Let $X, Y$ be two information spaces, the information entropy $H(XY)$ of the product space $XY$ is called the joint entropy of $X$ and $Y$, that is

$$H(XY) = -\sum_{x\in X}\sum_{y\in Y} p(xy) \log p(xy). \tag{3.10}$$

The conditional entropy $H(X|Y)$ of $X$ versus $Y$ is defined as

$$H(X|Y) = -\sum_{x\in X}\sum_{y\in Y} p(xy) \log p(x|y). \tag{3.11}$$

**Lemma 3.2** (Addition formula of entropy) *For any two information spaces $X$ and $Y$, then we have*

$$H(XY) = H(X) + H(Y|X) = H(Y) + H(X|Y).$$

*Generally, for $n$ information spaces $X_1, X_2, \ldots, X_n$, we have*

$$H(X_1X_2\cdots X_n) = \sum_{i=1}^{n} H(X_i|X_{i-1}X_{i-2}\cdots X_1). \tag{3.12}$$

***Proof*** By (3.10) and probability multiplication formula,

$$\begin{aligned} H(XY) &= -\sum_{x\in X}\sum_{y\in Y} p(xy)\log p(xy) \\ &= -\sum_{x\in X}\sum_{y\in Y} p(xy)(\log p(x) + \log p(y|x)) \\ &= -\sum_{x\in X} p(x)\log p(x) + H(Y|X) \\ &= H(X) + H(Y|X). \end{aligned}$$

The same can be proved

$$H(XY) = H(Y) + H(X|Y).$$

We prove (3.12) by induction, when $n = 2$,

$$H(X_1X_2) = H(X_1) + H(X_2|X_1).$$

The proposition is true, and for general $n$, we have

$$\begin{aligned} H(X_1X_2\cdots X_n) &= H(X_1X_2\cdots X_{n-1}) + H(X_n|X_1X_2\cdots X_{n-1}) \\ &= \sum_{i=1}^{n-1} H(X_i|X_{i-1}X_{i-2}\cdots X_1) + H(X_n|X_1X_2\cdots X_{n-1}) \\ &= \sum_{i=1}^{n} H(X_i|X_{i-1}X_{i-2}\cdots X_1). \end{aligned}$$

The Lemma 3.2 holds.

**Theorem 3.2** *We have*

$$H(XY) \leq H(X) + H(Y). \tag{3.13}$$

*If and only if $X$ and $Y$ are statistically independent information spaces,*

$$H(XY) = H(X) + H(Y). \tag{3.14}$$

*Generally, we have*

$$H(X_1X_2\cdots X_n) \leq H(X_1) + H(X_2) + \cdots + H(X_n). \tag{3.15}$$

*If and only if $X_1, X_2, \ldots, X_n$ is an independent random process,*

$$H(X_1X_2\cdots X_n) = H(X_1) + H(X_2) + \cdots + H(X_n). \tag{3.16}$$

***Proof*** By definition and Jensen inequality, we have

$$\begin{aligned} H(XY) - H(X) - H(Y) &= \sum_{x\in X}\sum_{y\in Y} p(xy)\log\frac{p(x)p(y)}{p(xy)} \\ &\leq \log\sum_{x\in X}\sum_{y\in Y} p(x)p(y) \\ &= 0. \end{aligned}$$

The above equal sign holds, if and only if for all $x \in X$, $y \in Y$, $\frac{p(x)p(y)}{p(xy)} = c($ where $c$ is a constant), thus $p(x)p(y) = cp(xy)$. Both sides sum at the same time, we have

$$1 = \sum_{x\in X} p(x)\sum_{y\in Y} p(y) = c\sum_{x\in X}\sum_{y\in Y} p(xy),$$

thus $c = 1$, $p(xy) = p(x)p(y)$. So if and only if $X$ and $Y$ are independent information spaces, (3.14) holds. By induction, we have (3.15) and (3.16). Theorem 3.2 holds.

By (3.15), we have the following direct corollary; for any information space $X$ and $n \geq 1$, we have

$$H(X^n) \leq nH(X). \tag{3.17}$$

**Definition 3.7** Let $X$ and $Y$ be two information spaces, and say that $X$ is completely determined by $Y$, if there is always a subset $N_x \subset Y$ of $Y$ for any given $x \in X$, satisfies

$$\begin{cases} p(x|y) = 1, & \text{if } y \in N_x; \\ p(x|y) = 0, & \text{if } y \notin N_x. \end{cases} \tag{3.18}$$

With regard to conditional information entropy $H(X|Y)$, we have the following two important special cases.

**Lemma 3.3** *(i)* $0 \leq H(X|Y) \leq H(X)$.
*(ii) If the information space $X$ is completely determined by $Y$, then*

$$H(X|Y) = 0. \tag{3.19}$$

*(iii) If $X$ and $Y$ are two separate information spaces,*

$$H(X|Y) = H(X). \tag{3.20}$$

***Proof*** (i) is trivial. Let us prove (3.19) first. By Definition 3.7 and (3.18), for given $x \in X$, we have

$$p(xy) = p(y)p(x|y) = 0, \ y \notin N_x.$$

Thus

$$\begin{aligned} H(X|Y) &= -\sum_{x\in X}\sum_{y\in Y} p(xy)\log p(x|y) \\ &= -\sum_{x\in X}\sum_{y\in N_x} p(xy)\log p(x|y) = 0. \end{aligned}$$

The proof of the formula (3.20) is obvious. Because $X$ and $Y$ are independent, the conditional probability

$$p(x|y) = p(x), \ \forall\, x \in X, y \in Y.$$

Thus

$$\begin{aligned} H(X|Y) &= -\sum_{x\in X}\sum_{y\in Y} p(x)p(y)\log p(x) \\ &= -\sum_{x\in X} p(x)\log p(x) = H(X). \end{aligned}$$

The Lemma 3.3 holds.

Next, we define the mutual information $I(X, Y)$ of two information spaces $X$ and $Y$.

**Definition 3.8** Let $X$ and $Y$ be two information spaces, and then their mutual information $I(X, Y)$ is defined as

$$I(X, Y) = \sum_{x\in X}\sum_{y\in Y} p(xy)\log\frac{p(x|y)}{p(x)}. \tag{3.21}$$

From the multiplication formula of probability, for all $x \in X$, $y \in Y$,

$$p(x)p(y|x) = p(y)p(x|y) = p(xy).$$

We have

$$\frac{p(x|y)}{p(x)} = \frac{p(y|x)}{p(y)}.$$

Therefore, there is a direct conclusion from the definition of mutual information $I(X, Y)$

$$I(X, Y) = I(Y, X).$$

**Lemma 3.4**

$$I(X, Y) = H(X) - H(X|Y) = H(Y) - H(Y|X).$$

***Proof*** By definition,

$$\begin{aligned}
I(X, Y) &= \sum_{x\in X}\sum_{y\in Y} p(xy)\log\frac{p(x|y)}{p(x)} \\
&= \sum_{x\in X}\sum_{y\in Y} p(xy)\log p(x|y) - \sum_{x\in X}\sum_{y\in Y} p(xy)\log p(x) \\
&= -H(X|Y) - \sum_{x\in X} p(x)\log p(x) \\
&= H(X) - H(X|Y).
\end{aligned}$$

The same can be proved

$$I(X, Y) = H(Y) - H(Y|X).$$

**Lemma 3.5** *Assuming that $X$ and $Y$ are two information spaces, $I(X, Y)$ is the amount of mutual information, then*

$$H(XY) = H(X) + H(Y) - I(X, Y). \tag{3.22}$$

*Further, we have $I(X, Y) \geq 0$, if and only if $X$ and $Y$ are independent, $I(X, Y) = 0$.*

***Proof*** By the addition formula of Lemma 3.2,

$$\begin{aligned} H(XY) &= H(X) + H(Y|X) \\ &= H(X) + H(Y) - (H(Y) - H(Y|X)) \\ &= H(X) + H(Y) - I(X, Y). \end{aligned}$$

The conclusion about $I(X, Y) \geq 0$ can be deduced directly from Theorem 3.2.

Let us prove an equation about entropy commonly used in the statistical analysis of cryptography.

**Theorem 3.3** *If $X, Y, Z$ are three information spaces, then*

$$\begin{aligned} H(XY|Z) &= H(X|Z) + H(Y|XZ) \\ &= H(Y|Z) + H(X|YZ). \end{aligned} \tag{3.23}$$

***Proof*** By the definition, we have

$$H(XY|Z) = -\sum_{x\in X}\sum_{y\in Y}\sum_{z\in Z} p(xyz)\log p(xy|z).$$

By probability product formula,

$$p(xyz) = p(z)p(xy|z) = p(xz)p(y|xz).$$

Thus

$$p(xy|z) = \frac{p(xz)p(y|xz)}{p(z)} = p(x|z)p(y|xz).$$

So we have

$$\begin{aligned} H(XY|Z) &= -\sum_{x\in X}\sum_{y\in Y}\sum_{z\in Z} p(xyz)\log p(x|z)p(y|xz) \\ &= -\sum_{x\in X}\sum_{y\in Y}\sum_{z\in Z} p(xyz)(\log p(x|z) + \log p(y|xz)) \\ &= -\sum_{x\in X}\sum_{z\in Z} p(xz)\log p(x|z) - \sum_{x\in X}\sum_{y\in Y}\sum_{z\in Z} p(xyz)\log p(y|xz) \\ &= H(X|Z) + H(Y|XZ). \end{aligned}$$

Similarly, the second formula can be proved.

Finally, we extend the formula (3.15) to conditional entropy.

**Lemma 3.6** *Let $X_1, X_2, \ldots, X_n, Y$ be information spaces, then we have*

$$H(X_1X_2\cdots X_n|Y) \leq H(X_1|Y) + \cdots + H(X_n|Y). \tag{3.24}$$

*Specially, when $X_1 = X_2 = \cdots = X_n = X$,*

$$H(X^n|Y) \leq nH(X|Y). \tag{3.25}$$

***Proof*** We make an induction of $n$. The proposition is trivial when $n = 1$. Let the proposition be true when $n$, i.e.,

$$H(X_1X_2\cdots X_n|Y) \leq H(X_1|Y) + \cdots + H(X_n|Y).$$

Then when $n + 1$, we let $X = X_1X_2\cdots X_n$, then

$$\begin{aligned}&H(X_1X_2\cdots X_{n+1}|Y) = H(XX_{n+1}|Y)\\ &= -\sum_{x\in X}\sum_{z\in X_{n+1}}\sum_{y\in Y} p(xzy)\log p(xz|y).\end{aligned}$$

From the full probability formula,

$$H(X|Y) + H(X_{n+1}|Y) = -\sum_{x\in X}\sum_{z\in X_{n+1}}\sum_{y\in Y} p(xzy)\log p(x|y)p(z|y).$$

So by Jensen inequality,

$$\begin{aligned}&H(XX_{n+1}|Y) - H(X|Y) - H(X_{n+1}|Y)\\ &= \sum_{x\in X}\sum_{z\in X_{n+1}}\sum_{y\in Y} p(xzy)\log\frac{p(x|y)p(z|y)}{p(xz|y)}\\ &\leq \log\sum_{x\in X}\sum_{z\in X_{n+1}}\sum_{y\in Y} p(y)p(x|y)p(z|y).\end{aligned}$$

By product formula

$$\begin{aligned}&\sum_{x\in X}\sum_{z\in X_{n+1}}\sum_{y\in Y} p(y)p(x|y)p(z|y)\\ &= \sum_{x\in X}\sum_{y\in Y} p(x|y)p(y)\\ &= \sum_{x\in X} p(x) = 1.\end{aligned}$$

So by the inductive hypothesis,

$$H(XX_{n+1}|Y) \leq H(X_{n+1}|Y) + H(X|Y)$$
$$\leq H(X_1|Y) + H(X_2|Y) + \cdots + H(X_{n+1}|Y).$$

The proposition holds for $n + 1$. So the Lemma holds.

## 3.3 Redundancy

Select a alphabet $\mathbb{F}_q$ or a remaining class ring $\mathbb{Z}_m$ of module $m$, each element in the alphabet is called character, and in the field of communication, alphabet is also called source state, and character is also called transmission signal. If the length of a $q$-ary code is increased, redundant transmission signals or characters will appear in each codeword. The digital measurement of "redundant characters" is called redundancy, which is a technical means to improve the accuracy of codeword transmission, and redundancy is an important mathematical quantity to describe this technical means. Therefore, we start by proving the following lemma.

**Lemma 3.7** *Let $X$, $Y$, $Z$ be three information spaces, then*

$$H(X|YZ) \leq H(X|Z). \tag{3.26}$$

***Proof*** By total probability formula,

$$H(X|Z) = -\sum_{x\in X}\sum_{z\in Z} p(xz)\log p(x|z)$$
$$= -\sum_{x\in X}\sum_{z\in Z}\sum_{y\in Y} p(xyz)\log p(x|z).$$

So

$$H(X|YZ) - H(X|Z)$$
$$= \sum_{x\in X}\sum_{y\in Y}\sum_{z\in Z} p(xyz)\log\frac{p(x|z)}{p(x|zy)}$$
$$\leq \log\sum_{x\in X}\sum_{y\in Y}\sum_{z\in Z}\frac{p(xyz)p(x|z)}{p(x|zy)}$$
$$= \log\sum_{x\in X}\sum_{y\in Y}\sum_{z\in Z} p(yz)p(x|z)$$
$$= \log\sum_{x\in X}\sum_{z\in Z} p(z)p(x|z)$$
$$= 0.$$

Thus $H(X|YZ) \leq H(X|Z)$. The Lemma holds.

Let $X$ be a source state set and randomly select codewords to enter the channel of information transmission is a discrete random process. This mathematical model can be constructed and studied on $X$ by taking the value of a group of random variables $\{\xi_i\}_{i\geq 1}$. Firstly, we assume that $\{\xi_i\}_{i\geq 1}$ obeys the same probability distribution when taking value on $X$, and we get a set of information spaces $\{X^i\}_{i\geq 1}$, let $H_0 = \log |X|$ be the entropy of $X$ as the equal probability information space, for $n \geq 1$, we let

$$H_n = H(X|X^{n-1}),\ H_1 = H(X).$$

By Lemma 3.7, then $\{H_n\}$ constitutes a number sequence with monotonic descent and lower bound, so that its limit exists, that is

$$\lim_{n\to\infty} H_n = a\ (a \geq 0). \tag{3.27}$$

We will extend the above observation to the general case: Let $\{\xi_i\}_{i\geq 1}$ be any set of random variables valued on $X$, for any $n \geq 1$, we let

$$X_n = (X, \xi_n),\ n \geq 1.$$

**Definition 3.9** A source state set $X$ has a set of random variables $\{\xi_i\}_{i\geq 1}$ valued on $X$, then $X$ is called a source.

(i) If $\{\xi_i\}_{i\geq 1}$ is a group of independent and identically distributed random variables, $X$ is called a memoryless source.
(ii) If for any integers $k, t_1, t_2, \ldots, t_k$ and $h$, random vector

$$(\xi_{t_1}, \xi_{t_2}, \ldots, \xi_{t_k})(\xi_{t_1+h}, \xi_{t_2+h}, \ldots, \xi_{t_k+h})$$

obey the same joint probability distribution, then $X$ is called a stationary source.
(iii) If $\{\xi_i\}_{i\geq 1}$ is a $k$-order Markov process, that is, for $\forall\ m > k \geq 1$,

$$\begin{aligned} &p(x_m|x_{m-1}x_{m-2}\cdots x_1)\\ &= p(x_m|x_{m-1}x_{m-2}\cdots x_{m-k}),\ \forall\ x_1, x_2, \ldots, x_m \in X, \end{aligned}$$

Then $X$ is called $k$-order Markov source, specially, $k = 1$, i.e.,

$$p(x_m|x_{m-1}x_{m-2}\cdots x_1) = p(x_m|x_{m-1}),\ \forall\ x_1, x_2, \ldots, x_m \in X,$$

call $X$ Markov source.

The concept from information space to source changes from a single random variable taking value on $X$ to an infinite dimensional random vector, so that the transmission process of code $X$ constitutes a discrete random process. By definition, we have

**Lemma 3.8** *Let $X$ be a source state set, and $\{\xi_i\}_{i\geq 1}$ be a set of random variables valued on $X$, we write*

$$X_i = (X, \xi_i),\ i \geq 1. \tag{3.28}$$

*(i) If $X$ is a memoryless source, the joint probability distribution on $X$ satisfies*

$$p(x_1 x_2 \cdots x_n) = \prod_{i=1}^{n} p(x_i),\ x_i \in X_i,\ n \geq 1. \tag{3.29}$$

*(ii) If $X$ is a stationary source, then for all integers $t_1, t_2, \ldots, t_k (k \geq 1)$ and $h$, there is the following joint probability distribution,*

$$p(x_{t_1} x_{t_2} \cdots x_{t_k}) = p(x_{t_1+h} x_{t_2+h} \cdots x_{t_k+h}), \tag{3.30}$$

*where $x_i \in X_i,\ i \geq 1$.*

*(iii) If $X$ is a stationary Markov source, then the conditional probability distribution on $X$ satisfies for any $m \geq 1$ and $x_1 x_2 \cdots x_m \in X_1 X_2 \cdots X_m$, we have*

$$\begin{aligned} &p(x_m | x_1 \cdots x_{m-1}) = p(x_m | x_{m-1}) \\ &= P\{\xi_{i+1} = x_m | \xi_i = x_{m-1}\},\ \forall\, 1 \leq i \leq m-1. \end{aligned} \tag{3.31}$$

***Proof*** (i) and (ii) can be derived directly from the definition. We only prove (iii). By (ii) of the definition 3.9, for $\forall\, i \geq 1$, we have

$$P\{\xi_i = x_{m-1}, \xi_{i+1} = x_m\} = P\{\xi_{m-1} = x_{m-1}, \xi_m = x_m\}$$

and

$$P\{\xi_i = x_{m-1}\} = P\{\xi_{m-1} = x_{m-1}\}.$$

Thus

$$\begin{aligned} &P\{\xi_i = x_{m-1}\} P\{\xi_{i+1} = x_m | \xi_i = x_{m-1}\} \\ &= P\{\xi_{m-1} = x_{m-1}\} P\{\xi_m = x_m | \xi_{m-1} = x_{m-1}\}. \end{aligned}$$

We have

$$P\{\xi_{i+1} = x_m | \xi_i = x_{m-1}\} = p(x_m | x_{m-1}).$$

The Lemma holds.

**Corollary 3.1** *A memoryless source $X$ must be a stationary source.*

***Proof*** Derived directly from Definition 3.9.

Next, we extend the limit formula in memoryless sources revealed by formula (3.27) to general stationary sources. For this purpose, we first prove two lemmas.

**Lemma 3.9** *Let $\{f(n)\}_{n\geqslant 1}$ be a sequence of real numbers, which satisfies the following semi countable additivity,*

$$f(n+m) \leqslant f(n) + f(m), \quad \forall\, n \geqslant 1,\ m \geqslant 1.$$

*Then $\lim\limits_{n\to\infty} \frac{1}{n} f(n)$ exists, and*

$$\lim_{n\to\infty} \frac{1}{n} f(n) = \inf\left\{\frac{1}{n} f(n) | n \geqslant 1\right\}. \tag{3.32}$$

***Proof*** Let

$$\delta = \inf\left\{\frac{1}{n} f(n) | n \geqslant 1\right\},\ \delta \neq -\infty.$$

For any $\varepsilon > 0$, select a sufficiently large positive integer $m$ so that

$$\frac{1}{m} f(m) < \delta + \frac{\varepsilon}{2}.$$

Let $n = am + b$, where $a$ is an integer, $0 \leqslant b < m$, by semi countable additivity, we have

$$f(n) \leqslant af(m) + (n - am)f(1).$$

Divide $n$ on both sides, we have

$$\frac{1}{n} f(n) \leqslant \frac{a}{am+b} f(m) + \frac{b}{am+b} f(1).$$

For given $b$, when $m$ is large enough, we have

$$\frac{bf(1)}{am+b} < \frac{1}{2}\varepsilon.$$

So there is

$$\frac{1}{n} f(n) < \frac{1}{m} f(m) + \frac{1}{2}\varepsilon < \varepsilon + \delta. \tag{3.33}$$

Thus we have

$$\delta \leqslant \varliminf_{n\to\infty} \frac{1}{n} f(n) \leqslant \varlimsup_{n\to\infty} \frac{1}{n} f(n) < \delta + \varepsilon.$$

So

$$\lim_{n\to\infty} \frac{1}{n} f(n) = \delta.$$

If $\delta = -\infty$, by (3.33),

$$\overline{\lim_{n\to\infty}}\frac{1}{n}f(n)=-\infty,$$

so we still have

$$\lim_{n\to\infty}\frac{1}{n}f(n)=\delta=-\infty.$$

The Lemma holds.

**Lemma 3.10** *Let $\{a_n\}_{n\geqslant 1}$ be a sequence of real numbers, and the limit $\lim\limits_{n\to\infty}a_n=a$, then*

$$\lim_{n\to\infty}\frac{1}{n}\sum_{i=1}^{n}a_i=a.$$

***Proof***

$$\begin{aligned}\left|\frac{1}{n}\sum_{i=1}^{n}a_i-a\right|=\left|\frac{1}{n}\sum_{i=1}^{n}(a_i-a)\right|&\leqslant\frac{1}{n}\sum_{i=1}^{n}|(a_i-a)|\\&=\frac{1}{n}\sum_{i=1}^{N}|a_i-a|+\frac{1}{n}\sum_{i=N+1}^{n}|a_i-a|\\&<\frac{1}{n}\sum_{i=1}^{N}|a_i-a|+\frac{n-N}{n}\varepsilon\\&<\frac{1}{n}\sum_{i=1}^{N}|a_i-a|+\varepsilon.\end{aligned}$$

When $\varepsilon>0$ is given, $N$is also given accordingly, the first item of the above formula tends to 0, when $n\to\infty$. So for any $\varepsilon>0$, when $n>N_0$,

$$\left|\frac{1}{n}\sum_{i=1}^{n}a_i-a\right|<2\varepsilon.$$

Thus there is

$$\lim_{n\to\infty}\frac{1}{n}\sum_{i=1}^{n}a_i=a.$$

The Lemma holds.

With the above preparations, we now give the main results of this section.

**Theorem 3.4** *Let $X$ be any source, $\{\xi_i\}_{i\geqslant 1}$ is a set of random variables valued on $X$. For any positive integer $n\geqslant 1$, let*

$$X_n=(X,\xi_n),\quad n\geqslant 1.$$

*Then when X is a stationary source, we have the following two limits that exist and are equal, that is*

$$\lim_{n\to\infty}\frac{1}{n}H(X_1X_2\ldots X_n)=\lim_{n\to\infty}H(X_n|X_1X_2\ldots X_{n-1}).$$

*We denote the above common limit as* $H_\infty(X)$.

***Proof*** Because $X$ is a stationary source, for any $n\geqslant 1$, $m\geqslant 1$, then the joint event probability distribution of random vector $\{\xi_{n+1},\xi_{n+2},\ldots,\xi_{n+m}\}$ on $X$ is equal to the joint probability distribution of random vector $(\xi_1,\xi_2,\ldots,\xi_m)$; therefore, we have

$$H(X_1X_2\cdots X_m)=H(X_{n+1}X_{n+2}\cdots X_{n+m}). \tag{3.34}$$

By Theorem 3.2, then

$$\begin{aligned}H(X_1X_2\cdots X_nX_{n+1}\cdots X_{n+m})&\leqslant H(X_1\cdots X_n)+H(X_{n+1}\cdots X_{n+m})\\&=H(X_1\cdots X_n)+H(X_1\cdots X_m).\end{aligned}$$

Let $f(n)=H(X_1\cdots X_n)$, then $f(n+m)\leqslant f(n)+f(m)$, so $\{f(n)\}_{n\geqslant 1}$ is a non-negative real number sequence with semi countable additive property, by Lemma 3.9, we have

$$\lim_{n\to\infty}\frac{1}{n}H(X_1X_2\cdots X_n)=\inf\left\{\frac{1}{n}H(X_1X_2\cdots X_n)|n\geqslant 1\right\}\geqslant 0.$$

Next, we prove that there is a second limit, that is

$$\lim_{n\to\infty}H(X_n|X_1X_2\cdots X_{n-1})\text{exist}.$$

Firstly, we prove that the sequence is monotonically decreasing, because $X$ is a stationary source, so

$$H(X_1X_2\cdots X_{n-1})=H(X_2X_3\cdots X_n)$$

and

$$H(X_2X_3\cdots X_nX_{n+1})=H(X_1X_2\cdots X_n).$$

So we have

$$H(X_{n+1}|X_2X_3\cdots X_n)=H(X_n|X_1X_2\cdots X_{n-1}). \tag{3.35}$$

By Lemma 3.7,

$$\begin{aligned}H(X_{n+1}|X_1X_2\cdots X_n)&\leqslant H(X_{n+1}|X_2X_3\cdots X_n)\\&=H(X_n|X_1X_2\cdots X_{n-1}).\end{aligned}$$

So $\{H(X_n|X_1X_2\cdots X_{n-1})\}_{n\geqslant 1}$ is a monotonically decreasing sequence and has a lower bound, so $\lim\limits_{n\to\infty} H(X_n|X_1X_2\cdots X_{n-1})$ exist. Further, by the addition formula of Lemma 3.2,

$$\frac{1}{n}H(X_1X_2\cdots X_n)=\frac{1}{n}\sum_{i=1}^{n}H(X_i|X_1X_2\cdots X_{i-1}).$$

By Lemma 3.10, finally we have

$$\lim_{n\to\infty}\frac{1}{n}H(X_1X_2\cdots X_n)=\lim_{n\to\infty}H(X_n|X_1X_2\cdots X_{n-1})=H_\infty(X).$$

We completed the proof of the Theorem.

We call $H_\infty(X)$ the entropy rate of source $X$. obviously, there is the following corollary.

**Corollary 3.2** *(i) For any stationary source X, we have*

$$H_\infty(X)\leqslant H(X_1)\leqslant \log|X|.$$

*(ii) If X is a memoryless source, then*

$$H_\infty(X)=H(X_1).$$

*(iii) If X is a stationary Markov source, then*

$$H_\infty(X)=H(X_2|X_1).$$

***Proof*** Since $\{H(X_n|X_1\cdots X_{n-1})\}_{n\geqslant 1}$ is a monotonically decreasing sequence, then

$$H_\infty(X)\leqslant H(X_1).$$

That is, (i) holds. If $X$ is a memoryless source, then

$$\begin{aligned}H(X_1\cdots X_n)&=-\sum_{x_1\in X_1}\cdots\sum_{x_n\in X_n}p(x_1x_2\cdots x_n)\log p(x_1x_2\cdots x_n)\\&=-\sum_{x_1\in X_1}\cdots\sum_{x_n\in X_n}p(x_1\ldots x_n)\{\log p(x_1)+\cdots+\log p(x_n)\}\\&=nH(X_1).\end{aligned}$$

So we have

$$H_\infty(X)=H(X_1).$$

Similarly, we can prove (iii).

**Definition 3.10** Let $X$ be a stationary source, we define

$$\delta = \log|X| - H_\infty(X), \ r = 1 - \frac{H_\infty X}{\log|X|}, \tag{3.36}$$

$\delta$ is the redundancy of information space $X$, and $r$ is the relative redundancy of $X$.

We write

$$H_0 = \log|X|, \ H_n = H(X_n|X_1X_2\cdots X_{n-1}), \ \forall\, n \geq 1.$$

By Theorem 3.4, we have $H_\infty(X) = H_0 \leq H_n$, so

$$H_n \geq (1-r)H_0, \ \forall\, n \geq 1. \tag{3.37}$$

In information theory, redundancy is used to describe the effectiveness of the information carried by the source output symbol. The smaller the redundancy, the higher the effectiveness of the information carried by the source output symbol, and vice versa.

## 3.4 Markov Chain

Let $X, Y, Z$ be three information spaces, if there is the following conditional probability formula

$$p(xy|z) = p(x|z)p(y|z). \tag{3.38}$$

Say that $X$ and $Y$ are statistically independent under the given condition of $Z$.

**Definition 3.11** If the information space $X$ and $Y$ are statistically independent under condition $Z$, $X, Y, Z$ is called a Markov chain, denote as $X \to Z \to Y$.

**Theorem 3.5** *$X \to Z \to Y$ is a Markov chain if and only if the probability of occurrence of the joint event $xzy$ is*

$$p(xzy) = p(x)p(z|x)p(y|z), \tag{3.39}$$

*if and only if*

$$p(xzy) = p(y)p(z|y)p(x|z). \tag{3.40}$$

***Proof*** If $X \to Z \to Y$ is a Markov chain, then $p(xy|z) = p(x|z)p(y|z)$, thus

$$\begin{aligned} p(xzy) &= p(z)p(xy|z) \\ &= p(z)p(x|z)p(y|z) \\ &= p(x)p(z|x)p(y|z). \end{aligned}$$

Similarly,

$$\begin{aligned} p(xzy) &= p(z)p(y|z)p(x|z) \\ &= p(y)p(z|y)p(x|z). \end{aligned}$$

That is (3.39) and (3.40) holds. Conversely, if (3.39) holds, then

$$\begin{aligned} p(xzy) &= p(x)p(z|x)p(y|z) \\ &= p(z)p(x|z)p(y|z). \end{aligned}$$

On the other hand, the product formula

$$p(xzy) = p(z)p(xy|z).$$

So we have

$$p(xy|z) = p(x|z)p(y|z).$$

That is $X \to Z \to Y$ is a Markov chain. Similarly, if (3.40) holds, then $X \to Z \to Y$ also is a Markov chain. The Theorem holds.

According to the above Theorem, or by Definition 3.11, obviously, if $X \to Z \to Y$ is a Markov chain, then $Y \to Z \to X$ is also a Markov chain.

**Definition 3.12** Let $U, X, Z, Y$ be four information spaces, and the probability of joint event $uxzy$ is

$$p(uxzy) = p(u)p(x|u)p(z|x)p(y|z), \tag{3.41}$$

Call $U, X, Z, Y$ a Markov chain, denote as $U \to X \to Z \to Y$.

**Theorem 3.6** *If $U \to X \to Z \to Y$ is a Markov chain, then $U \to X \to Z$ and $U \to Z \to Y$ are also Markov chains.*

***Proof*** Assuming that $U \to X \to Z \to Y$ is a Markov chain, then

$$p(uxzy) = p(u)p(x|u)p(z|x)p(y|z),$$

Both sides sum $y \in Y$ at the same time, and notice that $\sum_{y \in Y} p(y|z) = 1$, then

$$p(uxz) = p(u)p(x|u)p(z|x).$$

By Theorem 3.5, $U \to X \to Z$ is a Markov chain. The left side of the above formula can be expressed as

$$p(uxz) = p(ux)p(z|ux).$$

So we have

$$p(z|ux) = p(z|x).$$

Because $U \to X \to Z \to Y$ is a Markov chain, then

$$\begin{aligned} p(uxzy) &= p(u)p(x|u)p(z|x)p(y|z) \\ &= p(ux)p(z|ux)p(y|z) \\ &= p(uxz)p(y|z). \end{aligned}$$

Both sides sum $x \in X$ at the same time, then we have

$$\begin{aligned} p(uzy) &= p(uz)p(y|z) \\ &= p(u)p(z|u)p(y|z). \end{aligned}$$

Thus $U \to Z \to Y$ is also a Markov chain. The Theorem holds.

In the previous section, we defined the mutual information $I(X, Y)$ of two information spaces $X$ and $Y$ as

$$I(X, Y) = \sum_{x\in X}\sum_{y\in Y} p(xy) \log \frac{p(xy)}{p(x)p(y)}.$$

Now we define the mutual information $I(X, Y|Z)$ of $X$ and $Y$ under condition $Z$ as

$$I(X, Y|Z) = \sum_{x\in X}\sum_{y\in Y}\sum_{z\in Z} p(xyz) \log \frac{p(xy|z)}{p(x|z)p(y|z)}. \tag{3.42}$$

By definition, we have

$$I(X, Y|Z) = I(Y, X|Z). \tag{3.43}$$

$I(X, Y|Z)$ is called the conditional mutual information of $X$ and $Y$.

For conditional mutual information, we first prove the following formula.

**Theorem 3.7** *Let $X, Y, Z$ be three information spaces, then*

$$I(X, Y|Z) = H(X|Z) - H(X|YZ) \tag{3.44}$$

*and*

$$I(X, Y|Z) = H(Y|Z) - H(Y|XZ). \tag{3.45}$$

***Proof*** We only prove (3.44), the same is true for equation (3.45). Because

$$\begin{aligned} H(X|Z) - H(X|YZ) &= \sum_{x\in X}\sum_{y\in Y}\sum_{z\in Z} p(xyz) \log \frac{p(x|yz)}{p(x|z)} \\ &= \sum_{x\in X}\sum_{y\in Y}\sum_{z\in Z} p(xyz) \log \frac{p(xy|z)}{p(x|z)p(y|z)} \\ &= I(X, Y|Z). \end{aligned}$$

So (3.44) holds.

**Corollary 3.3** *We have $I(X, Y|Z) \geq 0$, if and only if $X \to Z \to Y$ is a Markov chain $I(X, Y|Z) = 0$.*

***Proof*** By Theorem 3.7,

$$I(X, Y|Z) = H(X|Z) - H(X|YZ) \geq 0.$$

If $X \to Z \to Y$ is a Markov chain, by (3.42),

$$\log \frac{p(xy|z)}{p(x|z)p(y|z)} = \log 1 = 0,$$

that is $I(X, Y|Z) = 0$. Vice versa.

Conditional mutual information can be used to establish the addition formula of mutual information.

**Corollary 3.4** (Addition formula of mutual information) *If $X_1, X_2, \ldots, X_n, Y$ are information spaces, then*

$$I(X_1 X_2 \cdots X_n, Y) = \sum_{i=1}^{n} I(X_i, Y|X_{i-1} \cdots X_1). \tag{3.46}$$

*Specially, when $n = 2$, we have*

$$I(X_1 X_2, Y) = I(X_1, Y) + I(X_2, Y|X_1). \tag{3.47}$$

***Proof*** By Lemma 3.4, we have

$$\begin{aligned} I(X_1 X_2 \cdots X_n, Y) &= H(X_1 X_2 \cdots X_n) - H(X_1 X_2 \cdots X_n|Y) \\ &= \sum_{i=1}^{n} H(X_i|X_{i-1} \cdots X_1) - \sum_{i=1}^{n} H(X_i|X_{i-1} \cdots X_1 Y). \end{aligned}$$

Again by the chain rule of conditional entropy to get

$$I(X_1 X_2 \cdots X_n, Y) = \sum_{i=1}^{n} I(X_i, Y|X_1 X_2 \cdots X_{i-1}).$$

Therefore, the corollary holds.

Finally, we use Markov chain to prove the inequality of mutual information.

**Theorem 3.8** *Suppose $X \to Z \to Y$ is a Markov chain, then we have*

$$I(X, Y) \leq I(X, Z) \tag{3.48}$$

*and*

$$I(X, Y) \le I(Y, Z). \tag{3.49}$$

***Proof*** We only prove (3.48), the same is true for equation (3.49). From equation (3.47) and corollary 3.3:

$$I(YZ, X) = I(Y, X) + I(X, Z|Y).$$

Thus we have

$$\begin{aligned} I(X, Y) &= I(X, YZ) - I(X, Z|Y) \\ &\le I(X, YZ) \\ &= I(X, Z) + I(X, Y|Z) \\ &= I(X, Z). \end{aligned}$$

In the last step, we use the Markov chain condition, thus $I(X, Y|Z) = 0$. The Theorem holds.

**Theorem 3.9** (Data processing inequality) *Suppose* $U \to X \to Y \to V$ *is a Markov chain, then we have*

$$I(U, V) \le I(X, Y).$$

***Proof*** According to the conditions, $U \to X \to Y$ and $U \to Y \to V$ is a Markov chain, respectively, by Theorem 3.8,

$$I(U, Y) \le I(X, Y)$$

and

$$I(U, V) \le I(U, Y).$$

Thus

$$I(U, V) \le I(X, Y).$$

The Theorem holds.

## 3.5 Source Coding Theorem

The information coding theory is usually divided into two parts: channel coding and source coding. The so-called channel coding is to ensure the success rate of decoding by increasing the length of codewords. Channel coding, also known as error correction code, is discussed in detail in Chap. 2. Source coding is to compress the data with redundant information to improve the success rate of decoding and recovery after information or data is stored. Another important result of Shannon's theory is that there are so-called good codes in source coding, which is characterized by fewer codewords as much as possible. To improve the storage space efficiency, and the error of decoding and restoration can be arbitrarily small. Source coding is also called typical code. Shannon first proved the asymptotic bisection property of 'block code' for memoryless source, and drew the statistical characteristics of typical code from now on. At Shannon's suggestion, McMillan (1953) and Breiman (1957) also proved a similar asymptotic bisection property for stationary ergodic sources. This is the very famous Shannon–McMillan–Breiman theorem in source coding, which constitutes the core content of modern typical code theory. The main purpose of this section is to strictly prove the asymptotic bisection of memoryless sources, so as to derive the source coding theorem for data compression (see Theorem 3.10). For the more general Shannon–McMillan–Breiman theorem, Chap. 2 of Ye Zhongxing's fundamentals of information theory (see Zhongxing, 2003 in reference 3) gives a proof under the condition of stationary ergodic Markov source, interested readers can refer to it or refer to more original documents (see McMillan, 1953; Moy, 1961; Shannon, 1959 in reference 3).

Firstly, let $X = (X, \xi)$ be an information space, and the entropy $H(X)$ of $X$ essentially depends only on the probability function $p(x)(x \in X)$ of random variable $\xi$. We can define the random variable taking value on $X$ according to $p(x)$.

$$\eta_1 = p(X), \ \eta_2 = \log p(X). \tag{3.50}$$

The probability function is

$$P\{\eta_1 \text{ value } x\} = P\{\eta_2 \text{ value } x\} = p(x). \tag{3.51}$$

It is easy to see the expected value of $\eta_2$

$$\begin{aligned} -E(\eta_2) &= -E(\log p(X)) \\ &= -\sum_{x \in X} p(x) \log p(x) = H(X). \end{aligned} \tag{3.52}$$

Therefore, we can regard the entropy $H(X)$ of $X$ as the mathematical expectation of random variable $\log \frac{1}{p(X)}$.

**Lemma 3.11** *Let $X$ be a memoryless source, $p(X^n)$ and $\log p(X^n)$ be two random variables whose values are on the power space $X^n$, then $-\frac{1}{n} \log p(X^n)$ converges to $H(X)$ according to probability, that is*

$$-\frac{1}{n} \log p(X^n) \xrightarrow{P} H(X).$$

***Proof*** Since $X$ is a memoryless source, $\{\xi_i\}_{i \geq 1}$ is a group of independent and identically distributed random variables, $X_i = (X, \xi_i)(i \geq 1)$, $X^n = X_1 X_2 \cdots X_n (n \geq 1)$ is a power space, then there is

$$\begin{cases} p(X^n) = p(X_1)p(X_2)\cdots p(X_n) \\ \log p(X^n) = \sum_{i=1}^{n} \log p(X_i). \end{cases}$$

Because $\{\xi_i\}_{i\geq 1}$ is independent and identically distributed, $\{p(X^n)\}$ and $\{\log p(X^n\}$ is also a group of independent and identically distributed random variables. According to Chebyshev's law of large numbers (see Theorem 1.3 of Chap. 1),

$$-\frac{1}{n}\log p(X^n) = \frac{1}{n}\sum_{i=1}^{n}\log\frac{1}{p(X_i)}$$

converges to the common expected value $H(X)$, that is

$$E\left(\log\frac{1}{p(X_i)}\right) = E\left(\log\frac{1}{p(X)}\right) = H(X).$$

For any $\varepsilon > 0$, for any codeword $x = x_1x_2\cdots x_n \in X^n$, there is

$$P\{|-\frac{1}{n}\log p(X^n) - H(X)| < \varepsilon\} > 1-\varepsilon. \tag{3.53}$$

The proof is completed.

**Definition 3.13** Let $X$ be a memoryless source, power space $X^n$, also known as block code,

$$X^n = \{x = x_1\cdots x_n | x_i \in X,\ 1\leq i\leq n\}, n\geq 1. \tag{3.54}$$

For any given $\varepsilon > 0$, $n \geq 1$, we define a typical code or a typical sequence $W_\varepsilon^{(n)}$ in the power space $X^n$ as

$$W_\varepsilon^{(n)} = \{x = x_1\cdots x_n \mid |-\frac{1}{n}\log p(x) - H(X)| < \varepsilon\}. \tag{3.55}$$

Because the definition, and $\varepsilon > 0$, $n \geq 1$, we have

$$W_\varepsilon^{(n)} \subset X^n,\ |X^n| = |X|^n. \tag{3.56}$$

**Lemma 3.12** (Progressive bisection) $|W_\varepsilon^{(n)}|$ *represents the number of codewords in typical code* $W_\varepsilon^{(n)}$, *then for any* $\varepsilon > 0$, *in binary channels, we have*

$$(1-\varepsilon)2^{n(H(X)-\varepsilon)} \leq |W_\varepsilon^{(n)}| \leq 2^{n(H(X)+\varepsilon)}. \tag{3.57}$$

***Proof*** By Lemma 3.11 and (3.53), then for any $x \in X^n$, we have

$$P\{|-\frac{1}{n}\log p(x) - H(X)| < \varepsilon\} > 1-\varepsilon.$$

In other words, for all codewords $x = x_1 x_2 \cdots x_n \in W_\varepsilon^{(n)}$, we have

$$H(X) - \varepsilon < -\frac{1}{n} \log p(x) < H(X) + \varepsilon.$$

Equivalent in binary channel,

$$2^{-n(H(X)+\varepsilon)} \leq p(x) \leq 2^{-n(H(X)-\varepsilon)}, \tag{3.58}$$

Denote the probability of occurrence of $W_\varepsilon^{(n)}$ as $P\{W_\varepsilon^{(n)}\}$, then

$$P\{W_\varepsilon^{(n)}\} = P\{x \in X^n : x \in W_\varepsilon^{(n)}\} > 1 - \varepsilon.$$

On the other hand,

$$P\{W_\varepsilon^{(n)}\} = \sum_{x \in W_\varepsilon^{(n)}} p(x),$$

by (3.58),

$$|W_\varepsilon^{(n)}| \cdot 2^{-n(H(X)+\varepsilon)} \leq P\{W_\varepsilon^{(n)}\} \leq 1.$$

So

$$|W_\varepsilon^{(n)}| \leq 2^{n(H(X)+\varepsilon)}.$$

Again by (3.58), there is

$$|W_\varepsilon^{(n)}| \cdot 2^{-n(H(X)-\varepsilon)} \geq P\{W_\varepsilon^{(n)}\} > 1 - \varepsilon.$$

So we have

$$|W_\varepsilon^{(n)}| > (1 - \varepsilon) 2^{n(H(X)-\varepsilon)}.$$

Combined with the above inequalities on both sides, we have

$$(1 - \varepsilon) 2^{n(H(X)-\varepsilon)} \leq |W_\varepsilon^{(n)}| \leq 2^{n(H(X)+\varepsilon)}.$$

We completed the proof.

By Lemma 3.12, for memoryless source $X$, the probability distribution $p(x)$ of its power space $X^n$ is approximate to

$$p(x) \sim 2^{-nH(X)}, \ \forall\, x \in X^n.$$

The number of codewords $|W_\varepsilon^{(n)}|$ in typical code $W_\varepsilon^{(n)}$ is approximately

$$|W_\varepsilon^{(n)}| \sim 2^{nH(X)}.$$

Further analysis shows that the proportion of typical code $W_\varepsilon^{(n)}$ in block code $X^n$ is very small, which can be summarized as the following Lemma.

**Lemma 3.13** *For a sufficiently small $\varepsilon > 0$ given, when X is not an equal probability information space, we have*

$$\lim_{n\to\infty} \frac{|W_\varepsilon^{(n)}|}{|X|^n} = 0.$$

***Proof*** By Lemma 3.12, we have

$$\frac{|W_\varepsilon^{(n)}|}{|X|^n} \le \frac{2^{n(H(X)+\varepsilon)}}{|X|^n}.$$

So

$$\frac{|W_\varepsilon^{(n)}|}{|X|^n} \le 2^{-n(\log|X|-H(X)-\varepsilon)}.$$

By Theorem 3.1, since $X$ is not an equal probability information space, when $\varepsilon$ is sufficient, we have

$$H(X) + \varepsilon < \log|X|.$$

Therefore, when $n$ is sufficiently large, the ratio of $\frac{|W_\varepsilon^{(n)}|}{|X|^n}$ can be arbitrarily small. The Lemma 3.13 holds.

Combining Lemmas 3.11, 3.12 and 3.13, we can describe that the typical codes in block codes have the following statistical characteristics.

**Corollary 3.5** *Assuming that X is a memoryless source and the typical sequence (or typical code) $W_\varepsilon^{(n)}$ in block code $X^n$ is defined by formula (3.55), then for any $\varepsilon > 0$, $n \ge 1$, we have*

*(i) (Progressive bisection)*

$$(1-\varepsilon)2^{n(H(X)-\varepsilon)} \le |W_\varepsilon^{(n)}| \le 2^{n(H(X)+\varepsilon)}.$$

*(ii) The occurrence probability $P\{W_\varepsilon^{(n)}\}$ of $W_\varepsilon^{(n)}$ is infinitely close to 1, that is*

$$P\{W_\varepsilon^{(n)}\} = P\{x \in X^n : x \in W_\varepsilon^{(n)}\} > 1-\varepsilon.$$

*(iii) When X is not equal to almost information space, the proportion of $W_\varepsilon^{(n)}$ in block code $X^n$ is any smaller, that is,*

$$\lim_{n\to\infty} \frac{|W_\varepsilon^{(n)}|}{|X|^n} = 0.$$

The above description of the statistical characteristics of typical codes is an important theoretical basis for source coding or data compression. Therefore, we find an effective way to compress the packet code information, so that the rearranged codewords are as few as possible, and the error probability of decoding and recovery is as small as possible. An effective method is to divide the codeword in block code $X^n$ into two parts; the codeword of typical code $W_\varepsilon^{(n)}$ is uniformly numbered from 1 to $M$. That is, the codeword in $W_\varepsilon^{(n)}$ forms one-to-one correspondence with the following positive integer set $I$,

$$I = \{1, 2, \ldots, M\}, \ M = |W_\varepsilon^{(n)}|.$$

For codewords that do not belong to $W_\varepsilon^{(n)}$, we uniformly number them as 1: Obviously, for $i, i \neq 1, 1 \leq i \leq n$, there is a unique codeword $x^{(i)} \in W_\varepsilon^{(n)}$ in $W_\varepsilon^{(n)}$, so we can accurately restore $i$ to $x^{(i)}$, that is $i \overset{\text{decode}}{\longrightarrow} x^{(i)}$ is the correct decoding. For $i = 1$, we will not be able to decode correctly, resulting in decoding recovery error. We denote the code rate of the typical code $W_\varepsilon^{(n)}$ as $\frac{1}{n}\log M$, by Lemma 3.12,

$$(1-\varepsilon)2^{n(H(X)-\varepsilon)} \leq M \leq 2^{n(H(X)+\varepsilon)}.$$

Equivalently,

$$\log(1-\varepsilon) + n(H(X)-\varepsilon) \leq \log M \leq n(H(X)+\varepsilon),$$

Therefore, the bit rate of typical code $W_\varepsilon^{(n)}$ is estimated as follows

$$\frac{1}{n}\log(1-\varepsilon) + H(X) - \varepsilon \leq \frac{1}{n}\log M \leq H(X) + \varepsilon, \tag{3.59}$$

when $0 < \varepsilon < 1$ given, we have

$$H(X) - \varepsilon \leq \lim_{n\to\infty} \frac{1}{n}\log M \leq H(X) + \varepsilon.$$

In other words, the code rate is typically close to $H(X)$. Let us look at the decoding error probability $P_e$ after this number, where

$$P_e = P\{x \in X^n : x \notin W_\varepsilon^{(n)}\}.$$

Because

$$P_e + P\{W_\varepsilon^{(n)}\} = 1,$$

According to the statistical characteristics $(ii)$ of the typical code $W_\varepsilon^{(n)}$,

$$P_e = 1 - P\{W_\varepsilon^{(n)}\} < \varepsilon. \tag{3.60}$$

From this, we derive the main result of this section, the so-called source coding theorem.

**Theorem 3.10** *(Shannon, 1948) Assuming that $X$ is a memoryless source, then*

*(i) When the code rate $R = \frac{1}{n}\log M_1 > H(X)$, there is an encoding with the code rate of $R$, so that when $n \to \infty$, the error probability of decoding recovery is $P_e \to 0$.*
*(ii) When the code rate $R = \frac{1}{n}\log M_1 < H(X) - \delta, \delta > 0$ and does not change with $n \to \infty$, then any coding with $R$ as the code rate has $\lim\limits_{n\to\infty} P_e = 1$.*

***Proof*** The above analysis has given the proof of $(i)$. In fact, if

$$R = \frac{1}{n}\log M_1 > H(X),$$

then when $\varepsilon$ is sufficiently small, by (3.59). Typical codes in block code $X_n$ are

$$R > \frac{1}{n}\log |W_\varepsilon^{(n)}|,\ M_1 > |W_\varepsilon^{(n)}|.$$

Therefore, we construct a code $C \subset X^n$, which satisfies

$$W_\varepsilon^{(n)} \subset C, |C| = M_1.$$

Thus, the code rate of $C$ is just equal to $R$, and the decoding error probability $P_e(C)$ after compression coding satisfies $P_e(C) < \varepsilon$. Because the probability of occurrence of $C$

$$P\{C\} + P_e(C) = 1.$$

But

$$P\{C\} \geq P\{W_\varepsilon^{(n)}\} > 1 - \varepsilon,$$

$(i)$ holds. To prove $(ii)$, we note that, $\forall\, x \in W_\varepsilon^{(n)}$, then

$$|-\frac{1}{n}\log p(x) - H(X)| < \varepsilon.$$

The above formula contains $\forall\, x \in W_\varepsilon^{(n)}$,

$$p(x) < 2^{-n(H(X)-\varepsilon)}.$$

Thus, the probability of occurrence of $W_\varepsilon^{(n)}$ satisfies

$$P\{W_\varepsilon^{(n)}\} = \sum_{x\in W_\varepsilon^{(n)}} p(x) \leq |W_\varepsilon^{(n)}| \cdot 2^{-n(H(X)-\varepsilon)}. \tag{3.61}$$

If we use $R$ as the bit rate, because

$$R = \frac{1}{n} \log M < H(X) - \delta,$$

then we have

$$|W_{\varepsilon}^{(n)}| < M = 2^{n(H(X)-\delta)}.$$

By (3.61),

$$P\{W_{\varepsilon}^{(n)}\} < 2^{-n(\delta-\varepsilon)}, \tag{3.62}$$

when $0 < \varepsilon < \delta$, we have

$$1 - P_e = P\{W_{\varepsilon}^{(n)}\} < \varepsilon.$$

Thus

$$\lim_{n\to\infty} P_e = 1,$$

Thus the theorem holds.

## 3.6 Optimal Code Theory

Let $X$ be a source state set, $x = x_1x_2 \cdots x_n \in X^n$ be a message sequence, and $x$ be output as a codeword $u = u_1u_2 \cdots u_k \in \mathbb{Z}_D^k$ of length $k$ after compression coding, where $D \geq 1$ is a positive integer, $\mathbb{Z}_D$ is the remaining class ring of mod $D$, $u = u_1u_2 \cdots u_k \in \mathbb{Z}_D^k$ is called a $D$- ary codeword of length $k$. $u$ is decoded and translated into message $x$, that is $u \to x$. The purpose of source coding is to find a good coding scheme to make the code rate as small as possible under the requirement of sufficiently small decoding error. Below, we give the strict mathematical definitions of equal length code and variable length code.

**Definition 3.14** Let $X$ be a source state set, $\mathbb{Z}_D$ is the remaining class ring of mod $D$, $n$, $k$ are positive integers. The mapping $f : X^n \to \mathbb{Z}_D^k$ is called equal length code coding function; $\mathbb{Z}_D^k \xrightarrow{\psi} X^n$ is called the corresponding decoding function. For $\forall\, x = x_1 \cdots x_n \in X^n$, $f(x) = u = u_1 \cdots u_k \in \mathbb{Z}_D^k$, $u = u_1 \cdots u_k$ is called a codeword of length $k$.

$$C = \{f(x) \in \mathbb{Z}_D^k | x \in X^n\}, \tag{3.63}$$

call

Call $C$ is the code coded by $f$, and $R = \frac{k}{n} log D$ is the coding rate of $f$, also known as the code rate of $C$. $C$ is called equal length code; it is sometimes called a block code with a packet length of $k$.

By Definition 3.14, the error probability of an equal length code coding scheme $(f, \psi)$ is

$$P_e = P\{\psi(f(x)) \neq x,\quad x \in X^n\}. \tag{3.64}$$

Let us first consider error free coding, that is $P_e = 0$. Obviously, $P_e = 0$ if and only if $f$ is a injection, $\psi = f^{-1}$ is the left inverse mapping of $f$. select a coding function $f : X^n \to \mathbb{Z}_D^k$ as a injection if and only if $|\mathbb{Z}_D^k| \geq |X^n|$, that is $D^k \geq N^n$, where $N = |X|$, take logarithms on both sides,

$$R = \frac{k}{n} logD \geq logN = \log |X|. \tag{3.65}$$

Therefore, the code rate of error free compression coding $f$ is at least $\log_2 |X|$ bits or $\ln |X|$ naits.

We consider progressive error free coding, that is, for any given $\varepsilon > 0$, required decoding error probability $P_e \leq \varepsilon$. By Theorem 3.10, only the code rate $R \geq H(X)$ is needed. In fact, take $X$ as an information space and encode the $n$-lengthen message column $x = x_1 x_2 \cdots x_n \in X^n$, if $x \in W_\varepsilon^{(n)}$ is a typical sequence (typical code), $x$ corresponds to a number in $M = |W_\varepsilon^{(n)}|$, if $x \notin W_\varepsilon^{(n)}$, uniformly code $x$ as 1. If the $M$ codewords in $W_\varepsilon^{(n)}$ are represented by $D$-ary digits, let $D^k = M$ (the insufficient part can be supplemented), and the code rate $R$ is

$$R = \frac{1}{n} \log M = \frac{k}{n} \log D.$$

Since $M$ is approximately $2^{nH(X)}$, $R$ is approximately $H(X)$, that is $R = \frac{1}{n} \log M \sim H(X)$. From the asymptotic bisection, the error probability of such coding is

$$P_e = P\{x = x_1 \cdots x_n \notin W_\varepsilon^{(n)}\} < \varepsilon, \text{ When } n \text{ is sufficiently large.}$$

However, in practical application, $n$ cannot increase infinitely, which requires us to find the best coding scheme when given a finite $n$, so that the code rate is as close as possible to the theoretical value $H(X)$. However, in application, we find that equal length code is not an efficient coding scheme, while variable length code is more practical. For example,

***Example 3.6*** Let $X = \{1, 2, 3, 4\}$ be an information space, and the probability distribution of random variable $\xi$ taking value on $X$ is

$$\xi \sim \begin{pmatrix} 1 & 2 & 3 & 4 \\ \frac{1}{2} & \frac{1}{4} & \frac{1}{8} & \frac{1}{8} \end{pmatrix}.$$

The entropy $H(X)$ of information space $X$ is

$$H(X) = -\frac{1}{2}\log_2 \frac{1}{2} - \frac{1}{4}\log_2 \frac{1}{4} - \frac{1}{8}\log_2 \frac{1}{8} - \frac{1}{8}\log_2 \frac{1}{8} = 1.75\text{bits}.$$

If equal length code is used for coding, the code length is 2, and the code is

| Source letter | Codeword |
|---|---|
| 1 | 00 |
| 2 | 01 |
| 3 | 10 |
| 4 | 11 |

Then the code rate $R(k=2, n=1)$ is

$$R = 2\log_2 2 = 2 > 1.75\text{bits}.$$

Obviously, the use efficiency of equal length codes is not high. If the above codes are replaced with unequal length codes, such as

| Source letter | Codeword |
|---|---|
| 1 | 0 |
| 2 | 10 |
| 3 | 110 |
| 4 | 111 |

We use $l(x)$ to represent the code length after the source letter $x$ is encoded, then the average code length $L$ required for $X$ encoding is

$$L = \sum_{i=1}^{4} p(x_i)l(x_i) = \frac{1}{2} \times 1 + \frac{1}{4} \times 2 + \frac{1}{8} \times 3 + \frac{1}{8} \times 3 = 1.75 \text{ bits} = H(X).$$

It can be seen that using unequal length code to compile $X$ has higher efficiency. This example also explains the following compression coding principle: for characters with high probability of occurrence, a shorter codeword is prepared, and for characters with low probability of occurrence, a longer codeword is prepared to ensure that the average coding length is as small as possible.

Next, we give the mathematical definition of variable length coding. For this purpose, let $X^*$ and $\mathbb{Z}_D^*$ be the set of finite length sequences, respectively. That is $X^* = \bigcup_{1\leq k<\infty} X^k$.

**Definition 3.15** (i) $X^n \xrightarrow{f} \mathbb{Z}_D^*$ is called a variable length code function, if any $x \in X^n$, $f(x) \in \mathbb{Z}_D^*$, When $x$ is different, the code length of $f(x)$ is not necessarily the same. We use $l(x)$ to table the length of $f(x)$, which is called the coding length of $x$. $C = \{f(x) \in \mathbb{Z}_D^* | x \in X^n\}$ is called variable length codeword set.

(ii) Let $f : X^* \longrightarrow \mathbb{Z}_D^*$ be a amapping, call $f$ is a coding mapping, $f(X^*)$ is called a code.

(iii) $f : X^* \longrightarrow \mathbb{Z}_D^*$ is called a block code mapping, if there is a mapping $g : X \longrightarrow \mathbb{Z}_D^*$, so that for any $x \in X^n (n \geq 1)$, write $x = x_1x_2\cdots x_n$, there is $f(x) = g(x_1)g(x_2)\cdots g(x_n)$.

(iv) $f : X^* \longrightarrow \mathbb{Z}_D^*$ is called a uniquely decodable map, if $f$ is a block code mapping and $f$ is a injection.

(v) $f : X^* \longrightarrow \mathbb{Z}_D^*$ is called a real-time code mapping. If $f$ is a block code mapping, and for any $x, y \in X^*$, $f(x)$ and $f(y)$ cannot be prefixes to each other.

***Remark 3.1*** $a = a_1 a_2 \cdots a_n \in \mathbb{Z}_D^n$, $b = b_1 b_2 \cdots b_m \in \mathbb{Z}_D^m$, call codeword $a$ the prefix of $b$, if $m \geq n$, and for any $1 \leq i \leq n$, there is $a_i = b_i$.

**Lemma 3.14** *Block code mapping $f : X^* \longrightarrow \mathbb{Z}_D^*$ is called a uniquely decodable mapping if and only if for $\forall\, n \geq 1$, $X^n \longrightarrow \mathbb{Z}_D^*$, $f$ is restricted to a injection on $X^n$.*

***Proof*** The necessity is obvious and the adequacy is proved. That is to prove for $\forall\, x = x_1 x_2 \cdots x_n \in X^n$, $y = y_1 y_2 \cdots y_m \in X^m$, $x \neq y$, there is $f(x) \neq f(y)$. Suppose there is $f(x) = f(y)$, because $f$ is a block code mapping, there is a mapping $g : X \longrightarrow \mathbb{Z}_D^*$, we have

$$f(x) = g(x_1)g(x_2)\cdots g(x_n) = g(y_1)g(y_2)\cdots g(y_m) = f(y).$$

Then

$$\begin{aligned} f(xy) &= g(x_1)g(x_2)\cdots g(x_n)g(y_1)g(y_2)\cdots g(y_m) \\ &= g(y_1)g(y_2)\cdots g(y_m)g(x_1)g(x_2)\cdots g(x_n) \\ &= f(yx). \end{aligned}$$

But $xy \neq yx$, this contradicts the fact that $f$ is restricted to a injection on $X^{n+m}$.

**Lemma 3.15** *A real-time code is uniquely decodable, and vice versa.*

***Proof*** Suppose $f : X^* \longrightarrow \mathbb{Z}_D^*$ as an instant code mapping, and for $x, y \in X^*$, $x \neq y$, there is $f(x) = a_1 a_2 \cdots a_n \in \mathbb{Z}_D^n$, $f(y) = b_1 b_2 \cdots b_m \in \mathbb{Z}_D^m (m \geq n)$. Because $f(x)$ is not a prefix of $f(y)$, it exists $i (1 \leq i \leq n)$, there is $a_i \neq b_i$, thus $f(x) \neq f(y)$, that is $f$ is an injection. In turn, let us take a counter example,

| Source letter | Codeword |
|---|---|
| 1 | 0 |
| 2 | 01 |
| 3 | 011 |
| 4 | 111 |

where $X = \{1, 2, 3, 4\}$ is the information space and $f : X \rightarrow \mathbb{Z}_2^*$ is a variable length code. $f(1)$ is the prefix of $f(2)$, that is, $f$ is not a real-time code map, but obviously $f$ is the only decodeable map. The Lemma holds.

What are the conditions for the code length of a real-time code? The following Kraft inequality gives a satisfactory answer.

**Lemma 3.16** *For the uniquely decodable code $C$ value in $\mathbb{Z}_D^*$, $|C| = m$, the code lengths are $l_1, l_2, \ldots, l_m$, then there is the following McMillan–Kraft inequality.*

$$\sum_{i=1}^{m} D^{-l_i} \leq 1. \tag{3.66}$$

*On the contrary, if $l_i$ satisfies the above conditions, there is a code length set of real-time code $C$ such that $\{l_1, l_2, \ldots, l_m\}$ is $C$.*

***Proof*** Consider

$$\left(\sum_{i=1}^{m} D^{-l_i})^n = (D^{-l_1} + D^{-l_2} + \cdots + D^{-l_m}\right)^n,$$

the form of each item is $D^{-l_{i_1}-l_{i_2}-\cdots-l_{i_n}} = D^{-k}$, where $l_{i_1} + l_{i_2} + \cdots + l_{i_n} = k$. Suppose $l = \max\{l_1, l_2, \ldots, l_m\}$, then the range of $k$ is from $n$ to $nl$. Define the number of items where $N_k$ is $D^{-k}$, then

$$\left(\sum_{i=1}^{m} D^{-l_i}\right)^n = \sum_{k=n}^{nl} N_k D^{-k}.$$

Note that $N_k$ can be regarded as the number of codeword sequences with a total length of $k$ just assembled by $n$ codewords in $C$, i.e.,

$$N_k = |\{(c_1, c_2, \ldots, c_n) \mid |c_1 c_2 \cdots c_n| = k, c_i \in C\}|.$$

The codeword is still in $\mathbb{Z}_D^*$, and because $f : X^* \longrightarrow \mathbb{Z}_D^*$ is an injection, so $N_k \leq D^k$. then we have

$$\left(\sum_{i=1}^{m} D^{-l_i}\right)^n = \sum_{k=n}^{nl} N_k D^{-k} \leq \sum_{k=n}^{nl} D^k D^{-k} = nl - n + 1 \leq nl.$$

If $x \geq 1$, and when $n$ Is Sufficiently Large, $x^n > nl$. But the above formula holds for all arbitrary $n$. That is $\sum_{i=1}^{m} D^{-l_i} \leq 1$.

On the contrary, assuming that Kraft inequality exists, that is, there is a given length $l_i (1 \leq i \leq m)$ satisfying formula (3.66), now we need to construct a real-time code with these lengths, and $l_i (1 \leq i \leq m)$ may not be completely different. Definition $n_j$ is the number of codewords with length j, if $l = \max\{l_1, l_2, \ldots, l_m\}$, then

$$\sum_{j=1}^{l} n_j = m.$$

(3.66) equivalent to

$$\sum_{j=1}^{l} n_j D^{-j} \leq 1.$$

Multiply both sides by $D^l$, then $\sum_{j=1}^{l} n_j D^{l-j} \leq D^l$. There is

$$n_l \leq D^l - n_1 D^{l-1} - n_2 D^{l-2} - \cdots - n_{l-1} D,$$

$$n_{l-1} \leq D^{l-1} - n_1 D^{l-2} - n_2 D^{l-3} - \cdots - n_{l-2} D,$$

$$\cdots$$

$$n_3 \leq D^3 - n_1 D^2 - n_2 D,$$

$$n_2 \leq D^2 - n_1 D,$$

$$n_1 \leq D.$$

Because $n_1 \leq D$, we can choose these $n_1$ codes arbitrarily, and the remaining $D - n_1$ codes with length 1 can be used as the prefix of other codewords. Therefore, there are $(D - n_1)D$ options for codewords with length of 2. That is $n_2 \leq D^2 - n_1 D$. Similarly, $(D - n_1)D - n_2$ codewords can be used as prefixes of subsequent codewords. Therefore, there are at most $((D - n_1)D - n_2)D$ options for codewords with length of 3. That is $n_3 \leq D^3 - n_1 D^2 - n_2 D$. ..., in this way, we can always construct a real-time code with length $\{l_1, l_2, \ldots, l_m\}$. The Lemma holds!

Let us give an example that is not the only one that can be decoded.

***Example 3.7*** Let $X = \{1, 2, 3, 4\}$, $\mathbb{Z}_D = \mathbb{F}_2$, the coding scheme is

| Source letter | Codeword |
|---|---|
| 1 | 0=f(1) |
| 2 | 1=f(2) |
| 3 | 00=f(3) |
| 4 | 11=f(4) |

Because the encoder inputs and the decoder receives continuous codeword symbols, if the character received by the decoder is 001101, there may be two decoding results, 112212 and 3412. This shows that $f^*$ is not an injection, that is, the code written by $f$ is not uniquely decodable.

By Lemma 3.16, real-time codes or, more generally, uniquely decodable codes must satisfy Kraft inequality. However, the variable length code compiled according to kraft inequality is not the optimal code, because from the perspective of random coding, an optimal code not only requires the accuracy of decoding, but also ensures the efficiency, that is, the average random code length requires the shortest. We summarize the strict mathematical definition of the optimal code as.

**Definition 3.16** Let $X = \{x_1, x_2, \ldots, x_m\}$ is an information space, a real-time code $C = \{f(x_1), f(x_2), \ldots, f(x_m)\}$ is called an optimal code if its average random code length

$$L = \sum_{i=1}^{m} p_i l_i \tag{3.67}$$

is the smallest, where $p_i = p(x_i)$ is the occurrence probability of $x_i$ and $l_i$ is the code length of $x_i$.

For a source state set $X$, when its statistical characteristics are determined, that is, after $X$ becomes an information space, the probability distribution $\{p(x)|x \in X\}$ is given. Therefore, to find the optimal compression coding scheme for an information space $X$ is to find the optimal solution $\{l_1, l_2, \ldots, l_m\}$ of (3.67) under the condition of kraft inequality. Usually, we use the Lagrange multiplier method to find the optimal solution. Let

$$J = \sum_{i=1}^{m} p_i l_i + \lambda \left( \sum_{i=1}^{m} D^{-l_i} \right),$$

Find the partial derivative of $l_i$

$$\frac{\partial J}{\partial l_i} = p_i - \lambda D^{-l_i} \log D.$$

Thus

$$D^{-l_i} = \frac{p_i}{\lambda \log D}.$$

By Kraft inequality, that is

$$\sum_{i=1}^{m} D^{-l_i} \leq 1.$$

We get

$$1 \geq \sum_{i=1}^{m} D^{-l_i} = \frac{1}{\lambda \log D} \sum_{i=1}^{m} p_i \Rightarrow \lambda \geq \frac{1}{\log D}.$$

Thus, the optimal code length $l_i$ is

$$l_i \geq -\log_D p_i, \; p_i \geq D^{-l_i}. \tag{3.68}$$

The corresponding optimal average code length $L$ is

$$L = \sum_{i=1}^{m} p_i l_i \geq -\sum_{i=1}^{m} p_i \log_D p_i = H_D(X). \tag{3.69}$$

That is, $L$ is the $D$-ary information entropy $H_D(X)$ of $X$. from this, we get the main results of this section.

**Theorem 3.11** *The average length L of any D-ary real-time code in an information space X shall satisfies*

$$L \geq H_D(X).$$

*The equal sign holds if and only if* $p_i = D^{-l_i}$.

Next, we will give another proof of Theorem 3.11. Therefore, we consider that there are two random variables $\xi$ and $\eta$ on a source state set $X$, and their probability distributions are

$$p(x) = P\{\xi = x\},\ q(x) = P\{\eta = x\},\ \forall\, x \in X.$$

The relative entropy of random variables is defined as

$$D(p||q) = \sum_{x \in X} p(x) \log \frac{p(x)}{q(x)}. \tag{3.70}$$

**Lemma 3.17** *The relative entropy* $D(p||q)$ *of two random variables on* $X$ *satisfies*

$$D(p||q) \geq 0,\ \textit{and}\ D(p||q) = 0 \iff p(x) = q(x), \forall\, x \in X.$$

***Proof*** If the real number $x > 0$ is expanded by the power series of $e^x$, it can be obtained

$$e^{x-1} = 1 + (x-1) + \frac{1}{2}(x-1)^2 + \cdots.$$

Thus $e^{x-1} \geq x$, there is $\log x \leq x - 1$, by (3.70), then

$$\begin{aligned} -D(p||q) &= \sum_{x \in X} p(x) \log \frac{q(x)}{p(x)} \\ &\leq \sum_{x \in X} p(x)(\frac{q(x)}{p(x)} - 1) = 0. \end{aligned}$$

Thus, there is $D(p||q) \geq 0$, $D(p||q) = 0$'s conclusion is obvious.

***Proof*** (Another proof of theorem 3.11) Investigate $L - H_D(X)$,

$$\begin{aligned} L - H_D(X) &= \sum_{i=1}^{m} p_i l_i - \sum_{i=1}^{m} p_i \log_D \frac{1}{p_i} \\ &= -\sum_{i=1}^{m} p_i \log_D D^{-l_i} + \sum_{i=1}^{m} p_i \log_D p_i. \end{aligned} \tag{3.71}$$

Define

$$r_i = \frac{D^{-l_i}}{c},\ c = \sum_{j=1}^{m} D^{-l_i}.$$

By Kraft inequality, we have

$$c \leq 1, \text{ and } \sum_{i=1}^{m} r_i = 1.$$

Therefore, $\{r_i, 1 \leq i \leq m\}$ is a probability distribution on $X$, by (3.71),

$$L - H_D(X) = -\sum_{i=1}^{m} p_i \log_D cr_i + \sum_{i=1}^{m} p_i \log_D p_i = \sum_{i=1}^{m} p_i \left( \log_D \frac{p_i}{r_i} + \log_D \frac{1}{c} \right).$$

By Lemma 3.17 and $c \leq 1$, we have

$$L - H_D(X) \geq 0, \text{ and } L = H_D(X) \text{ if and only if } c = 1 \text{ and } r_i = p_i,$$

that is

$$p_i = D^{-l_i}, \text{ or } l_i = \log_D \frac{1}{p_i}.$$

We complete the proof of theorem 3.11.

By Theorem 3.11, coding according to probability, then the code length of $D$-ary optimal code is

$$l_i = \log_D \frac{1}{p_i}, \quad 1 \leq i \leq m.$$

But in general, $\log_D \frac{1}{p_i}$ is not an integer, we use $\lceil a \rceil$ to represent the smallest integer not less than the real number $a$. Take

$$l_i = \left\lceil \log_D \frac{1}{p_i} \right\rceil, \quad 1 \leq i \leq m. \tag{3.72}$$

Then

$$\sum_{i=1}^{m} D^{-l_i} \leq \sum_{i=1}^{m} D^{-\log_D \frac{1}{p_i}} = \sum_{i=1}^{m} p_i = 1.$$

Then the code length defined by formula (3.72) is $\{l_1, l_2, \ldots, l_m\}$ and satisfies Kraft inequality. From Lemma 3.16, we can define the corresponding real-time code.

**Definition 3.17** Let $X = \{x_1, x_2, \ldots, x_m\}$ be an information space, $p_i = p(x_i)$,

$$l(f(x_i)) = l_i = \left\lceil \log_D \frac{1}{p_i} \right\rceil, \quad 1 \leq i \leq m.$$

Then the real-time code corresponding to $\{l_1, l_2, \ldots, l_m\}$ is called Shannon code.

**Corollary 3.6** *The code length $l(f(x_i))$ of a Shannon code $C = \{f(x_i) | 1 \leq i \leq m\}$ satisfies*

$$l_i = \left\lceil \log_D \frac{1}{p(x_i)} \right\rceil, \log_D \frac{1}{p(x_i)} \le l_i < \log_D \frac{1}{p(x_i)} + 1 \tag{3.73}$$

*and*

$$H_D(X) \le L < H_D(X) + 1.$$

*Where L is the average code length of C.*

***Proof*** According to the definition of $\lceil a \rceil$, $a \le \lceil a \rceil < a + 1$, thus

$$\log_D \frac{1}{p(x_i)} \le l_i < \log_D \frac{1}{p(x_i)} + 1.$$

So both sides multiply by $p(x_i)$ and sum $1 \le i \le m$, then there is

$$\sum_{i=1}^{m} p(x_i) \log_D \frac{1}{p(x_i)} \le \sum_{i=1}^{m} p(x_i) l_i < \sum_{i=1}^{m} p(x_i) \left( \log_D \frac{1}{p(x_i)} + 1 \right).$$

That is

$$H_D(X) \le L < H_D(X) + 1.$$

The Corollary holds.

## 3.7 Several Examples of Compression Coding

### *3.7.1 Morse Codes*

In variable length codes, in order to make the average code length as close to the source entropy as possible, the code length should match the occurrence probability of the corresponding coded characters as much as possible. The principle of probabilistic coding is that the characters with high occurrence probability are configured with short codewords, and the characters with low occurrence probability are configured with long codewords, So as to make the average code length as close to the source entropy as possible. This idea has existed long before Shannon theory. For example, Morse code invented in 1838 uses three symbols of dot, dash and space to encode 26 letters in English. It is expressed in binary, one dot is 10, a total of 2 bits, one dash is 1110, a total of 4 bits and the space is 000. There are three bits in total. For example, the commonly used English letter E is represented by a dot, while the infrequently used letter Q is represented by two dashes, one dot and one dash, which can make the average length of the codeword of the English text shorter. However, Morse code does not completely match the occurrence probability, so it is not the optimal code, and it is basically not used now. The following table is the coding table of Morse code (Fig. 3.1)

**Fig. 3.1** The coding table of Morse code

| | |
|---|---|
| A | • — |
| B | — • • • |
| C | — • — • |
| D | — • • |
| E | • |
| F | • • — • |
| G | — — • |
| H | • • • • |
| I | • • |
| J | • — — — |
| K | — • — |
| L | • — • • |
| M | — — |
| N | — • |
| O | — — — |
| P | • — — • |
| Q | — — • — |
| R | • — • |
| S | • • • |
| T | — |
| U | • • — |
| V | • • • — |
| W | • — — |
| X | — • • — |
| Y | — • — — |
| Z | — — • • |

It is worth noting that Morse code appeared as a kind of password in the early stage, which is widely used in the transmission and storage of sensitive politics (such as military intelligence). The early cryptosystem compilers were also manufactured based on the principle of Morse code, which quickly mechanized the compilation and translation of passwords. In this sense, Morse code has played an important role in promoting the development of cryptography.

### 3.7.2 Huffman Codes

Shannon, Fano and Huffman have all studied the coding methods of variable length codes, among which Huffman codes have the highest coding efficiency. We focus on the coding methods of Huffman binary and ternary codes.

Let $X = \{x_1, x_2, \ldots, x_m\}$ be the source letter set of $m$ symbols, arrange the $m$ symbols in the order of occurrence probability, take the two letters with the lowest probability to prepare the numbers "0" and "1," respectively, then add their probabilities as a new letter and rearrange them in the order of probability with the source letters without binary numbers. Then take the two letters with the lowest probability to prepare the numbers "0" and "1," respectively, add the probabilities of the two letters as the probability of a new letter, and re queue; continue the above process until the probability of the remaining letters is added to 1. At this time, all source letters correspond to a string of "0" and "1," and we get a variable length code, which is called Huffman code. Taking $X = \{1, 2, 3, 4, 5\}$ as the information space as an example, the corresponding probability distribution is

$$\xi \sim \begin{pmatrix} 1 & 2 & 3 & 4 & 5 \\ 0.25 & 0.25 & 0.2 & 0.15 & 0.15 \end{pmatrix}.$$

Binary information entropy $H_2(X)$ and ternary information entropy $H_3(X)$ are

$$\begin{aligned} H_2(X) &= -0.25 \log_2 0.25 - 0.25 \log_2 0.25 - 0.2 \log_2 0.2 \\ &\quad - 0.15 \log_2 0.15 - 0.15 \log_2 0.15 \\ &= 2.28 \text{ bits}, \\ H_3(X) &= -0.25 \log_3 0.25 - 0.25 \log_3 0.25 - 0.2 \log_3 0.2 \\ &\quad - 0.15 \log_3 0.15 - 0.15 \log_3 0.15 \\ &= 1.44 \text{ bits}, \end{aligned}$$

respectively. The binary Huffman coding diagram of $X$ is (Fig. 3.2).

The ternary Huffman coding diagram of $X$ is (Fig. 3.3).

In summary, Huffman code has the following characteristics. Assuming that the occurrence probability of the $i$-th source letter is $p_i$ and the corresponding code length is $l_i$, then

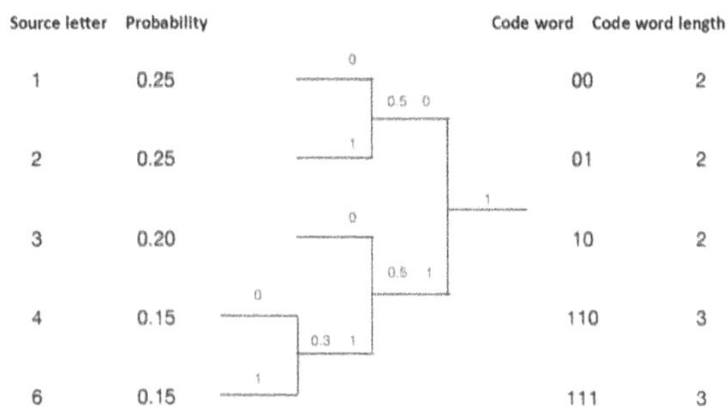


**Fig. 3.2** The binary Huffman coding

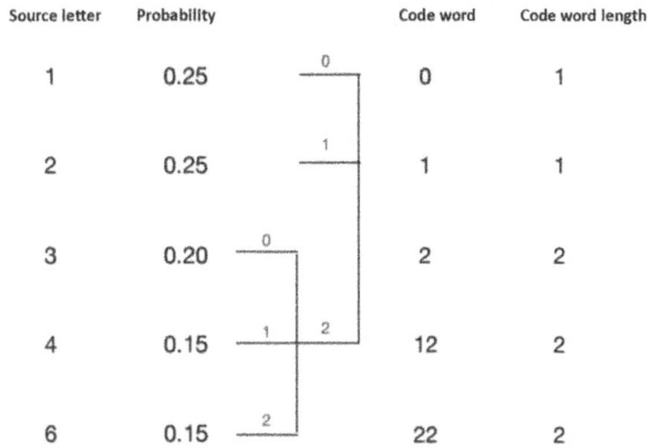


**Fig. 3.3** The ternary Huffman coding

(1) If $p_i > p_j$, then $l_i \le l_j$, that is, the source letter with low probability has a longer codeword;
(2) The longest two codewords have the same code length;
(3) The codeword letters of the two longest codewords are only different from the last letter, and the front ones are the same;
(4) In real-time codes, the average code length of Huffman code is the smallest. In this sense, Huffman code is the optimal code.

Huffman code has been applied in practice, which is mainly used in the compression standard of fax image. However, in the actual data compression, the statistical characteristics of some sources change before and after. In order to make the statistical characteristics based on the coding adapt to the changes of the actual statistical characteristics of the source, an adaptive coding technology has been developed. In each step of coding, the coding of a new message is based on the statistical characteristics of previous messages. For example, R. G. Gallager first proposed the step-by-step updating technology of Huffman code in 1978, and D.E. Knuth made this technology a practical algorithm in 1985. Adaptive Huffman coding technology requires complex data structure and continuous updating of codeword set according to the statistical characteristics of source, We would not go into details here.

### *3.7.3 Shannon–Fano Codes*

Shannon–Fano code is an arithmetic code. Let $X$ be an information space. It can be inferred from Corollary 3.6 in the previous section that the code length of Shannon code on $X$ is

$$l(x) = \left\lceil \log \frac{1}{p(x)} \right\rceil, \forall\, x \in X.$$

Here, we introduce a constructive coding method using cumulative distribution function to allocate codewords, commonly known as Shannon–Fano coding method. Without losing generality, let each letter $x$ in $X$, there is $p(x) > 0$, and define the cumulative distribution function $F(x)$ and the modified distribution function $\bar{F}(x)$ as

$$F(x) = \sum_{a \le x} p(a),\ \ \bar{F}(x) = \sum_{a < x} p(a) + \frac{1}{2}p(x), \tag{3.74}$$

where $X = \{1, 2, \ldots, m\}$ is a given information space. Without losing generality, let $p(1) \le p(2) \le \cdots \le p(m)$.

As can be seen from the definition, if $x \in X$, then $p(x) = F(x) - F(x-1)$, specially, if $x, y \in X$, then we have

$$\bar{F}(x) \ne \bar{F}(y).$$

So when we know $\bar{F}(x)$, we can find the corresponding $x$. The basic idea of Shannon–Fano arithmetic code is to use $\bar{F}(x)$ to encode $x$. Because $\bar{F}(x)$ is a real number, its binary decimal represents the first $l(x)$ bits, denote as $\{\bar{F}(x)\}_{l(x)}$, there is

$$\bar{F}(x) - \{\bar{F}(x)\}_{l(x)} < 2^{-l(x)}. \tag{3.75}$$

Take $l(x) = \left\lceil \log \frac{1}{p(x)} \right\rceil + 1$, then we have

$$\frac{1}{2^{l(x)}} = \frac{1}{2 \cdot 2^{\left\lceil \log \frac{1}{p(x)} \right\rceil}} < \frac{p(x)}{2} = \bar{F}(x) - F(x-1), \tag{3.76}$$

Now let the binary decimal of $\bar{F}(x)$ be expressed as

$$\bar{F}(x) = 0.a_1 a_2 \cdots a_{l(x)} a_{l(x)+1} \cdots,\ \ \forall\, a_i \in \mathbb{F}_2.$$

Then Shannon–Fano code is

$$f(x) = a_1 a_2 \cdots a_{l(x)},\ \text{that is } x \overset{\textbf{encode}}{\longrightarrow} a_1 a_2 \cdots a_{l(x)} \in \mathbb{F}_2^{l(x)}. \tag{3.77}$$

**Lemma 3.18** *The binary Shannon Fano code is a real-time code, and its average length L is at most two bits different from the theoretical optimal value $H(X)$.*

***Proof*** By (3.76),

$$2^{-l(x)} < \frac{1}{2}p(x) = \bar{F}(x) - F(x-1).$$

Let the binary decimal of $\bar{F}(x)$ be expressed as

$$\bar{F}(x) = 0.a_1 a_2 \cdots a_{l(x)} \cdots,\ \ \forall\, a_i \in \mathbb{F}_2.$$

We use $[A, B]$ to represent a closed interval on the real axis, so

$$\bar{F}(x) \in [0.a_1a_2\cdots a_{l(x)}, 0.a_1a_2\cdots a_{l(x)} + \frac{1}{2^{l(x)}}].$$

If $y \in X, x \neq y$, and $f(x)$ is the prefix of $f(y)$, then we have

$$\bar{F}(y) \in [0.a_1a_2\cdots a_{l(x)}, 0.a_1a_2\cdots a_{l(x)} + \frac{1}{2^{l(x)}}].$$

But

$$\bar{F}(y) - \bar{F}(x) \geq \frac{1}{2}p(y) \geq \frac{1}{2}p(x) > \frac{1}{2^{l(x)}},$$

This is contrary to the fact that $\bar{F}(x)$ and $\bar{F}(y)$ are in the same interval. Therefore, we have $f$ as real-time code, that is, Shannon–Fano code is real-time code. Considering its average code length $L$,

$$L = \sum_{x\in X} p(x)l(x) = \sum_{x\in X} p(x)\left(\left\lceil \log\frac{1}{p(x)}\right\rceil + 1\right) < \sum_{x\in X} p(x)\left(\log\frac{1}{p(x)} + 2\right) = H(X) + 2.$$

We complete the proof of the Lemma.

Let $n \geq 1$, $X^n$ is the power space of the information space, $x = x_1\cdots x_n \in X^n$ is called a message column of length $n$. In order to improve the coding efficiency, it is often necessary to compress the power space $X^n$, which is called arithmetic coding. Shannon–Fano code can also be used as arithmetic coding. Its basic method is to find a fast algorithm for calculating joint probability distribution $p(x_1x_2\cdots x_n)$ and cumulative distribution function $F(x)$, and then use Shannon–Fano method to encode $x = x_1\cdots x_n$. We will not introduce the specific details here.

## 3.8 Channel Coding Theorem

Let $X$ be the input alphabet and $Y$ the output alphabet, and let $\xi$ and $\eta$ be two random variables with values on $X$ and $Y$. The probability functions $p(x)$ and $p(y)$ of $X$ and $Y$ and the conditional probability function $p(y|x)$ are

$$p(x) = P\{\xi = x\},\ p(y) = P\{\eta = y\},\ p(y|x) = P\{\eta = y|\xi = x\}\text{respectively.}$$

From the full probability formula,

$$\begin{cases} p(y|x) \geq 0, & \forall\, x \in X, y \in Y. \\ \sum\limits_{y\in Y} p(y|x) = 1, & \forall\, x \in X. \end{cases} \tag{3.78}$$

If $X$ and $Y$ are finite sets, the conditional probability matrix $T = (p(y|x))_{|X|\times|Y|}$ is called the transition probability matrix from $X$ to $Y$, i.e.,

$$T = \begin{pmatrix} p(y_1|x_1) & p(y_2|x_1) & \dots & p(y_N|x_1) \\ p(y_1|x_2) & p(y_2|x_2) & \dots & p(y_N|x_2) \\ \dots & & & \\ p(y_1|x_M) & p(y_2|x_M) & \dots & p(y_N|x_M) \end{pmatrix}, \tag{3.79}$$

where $|X| = M$, $|Y| = N$. By (3.78), each row of the transition probability matrix $T$ is added to 1.

**Definition 3.18** (i) A discrete channel is composed of a finite information space $X$ as the input alphabet, a finite information space $Y$ as the output alphabet, and a transition probability matrix $T$ from $X$ to $Y$, denote that this discrete channel is $\{X, T, Y\}$. If $X = Y = \mathbb{F}_q$ is $q$ -element finite field, then $\{X, T, Y\}$ is a discrete $q$-ary channel. In particular, if $q = 2$, then $\{X, T, Y\}$ is called discrete binary channel.

(ii) If $\{X, T, Y\}$ is a discrete $q$-ary channel and $T = I_q$ is the $q$-order identity matrix, $\{X, I_q, Y\}$ is called a noise free channel.

(iii) If $\{X, T, Y\}$ is a discrete $q$-ary channel and $T = T'$ is a $q$-order symmetric matrix, $\{X, T, Y\}$ is called a symmetric channel.

In discrete channel $\{X, T, Y\}$, codeword spaces $X^n$ and $Y^n$ with length $n$ are defined as

$$X^n = \{x = x_1 \cdots x_n | x_i \in X\}, Y^n = \{y = y_1 \cdots y_n | y_i \in Y\}, n \geq 1.$$

The probabilities of joint events $x = x_1 \cdots x_n$ and $y = y_1 \cdots y_n$ are defined as

$$p(x) = p(x_1 \cdots x_n) = \prod_{i=1}^{n} p(x_i), \quad p(y) = p(y_1 \cdots y_n) = \prod_{i=1}^{n} p(y_i), \tag{3.80}$$

then $X$ and $Y$ become a memoryless source, $X^n$ and $Y^n$ are power spaces, respectively.

**Definition 3.19** Discrete channel $\{X, T, Y\}$ is called a memoryless channel if for any positive integer $n \geq 1$, $x = x_1 \cdots x_n \in X^n$, $y = y_1 \cdots y_n \in Y^n$, we have

$$\begin{cases} p(y|x) = \prod_{i=1}^{n} p(y_i|x_i), \\ p(x_i y_i) = p(x_1 y_1), \forall\, i \geq 1. \end{cases} \tag{3.81}$$

From the joint event probability $p(x_i y_i) = p(x_1 y_1)$ in equation (3.81), then there is

$$p(y_i|x_i) = \frac{p(x_1)}{p(x_i)} p(y_1|x_1). \tag{3.82}$$

The above formula shows that in a memoryless channel, the conditional probability $p(y_i|x_i)$ does not depend on $y_i$.

Definition 3.19 is the statistical characteristic of a memoryless channel. The following lemma gives a mathematical characterization of a memoryless channel.

**Lemma 3.19** *A discrete channel $\{X, T, Y\}$ is a memoryless channel if and only if the product information space $XY$ is a memoryless source, and a power space $(XY)^n = X^nY^n$.*

***Proof*** If $XY$ is a memoryless source (see Definition 3.9), thn for any $n \geq 1$, and $x = x_1 \cdots x_n \in X^n$, $y = y_1 \cdots y_n \in Y^n$, $xy \in X^nY^n$, there is

$$p(xy) = p(x_1 \cdots x_n y_1 \cdots y_n) = p(x_1y_1 \cdots x_ny_n) = \prod_{i=1}^{n} p(x_iy_i).$$

Thus

$$p(x)p(y|x) = p(x)\prod_{i=1}^{n} p(y_i|x_i),$$

so we have

$$p(y|x) = \prod_{i=1}^{n} p(y_i|x_i).$$

$p(x_iy_i) = p(x_1y_1)$ is given by the definition of memoryless source, so $\{X, T, Y\}$ is a memoryless channel. Conversely, if $\{X, T, Y\}$ is a memoryless channel, by (3.81), there are

$$p(xy) = \prod_{i=1}^{n} p(x_iy_i)$$

and $p(x_iy_i) = p(x_1y_1)$, then for any $a = a_1a_2 \cdots a_n \in (XY)^n$, where $a_i = x_iy_i$, we have

$$p(a) = p(x_1 \cdots x_n y_1 \cdots y_n) = p(xy) = \prod_{i=1}^{n} p(x_iy_i) = \prod_{i=1}^{n} p(a_i)$$

and $p(a_i) = p(a_1)$, therefore, $XY$ is a memoryless source, that is, a group of independent and identically distributed random vectors $\xi = (\xi_1, \xi_2, \ldots, \xi_n, \ldots)$ take value on $XY$, and $(XY)^n = X^nY^n$ is called power space. The Lemma holds.

The following lemma further characterizes the statistical characteristics of a memoryless channel.

**Lemma 3.20** *If $\{X, T, Y\}$ is a discrete memoryless channel, the conditional entropy $H(Y^n|X^n)$ and information $I(X^n, Y^n)$ of information space $X^n$ and $Y^n$ satisfy $\forall\, n \geq 1$,*

$$\begin{cases} H(Y^n|X^n) = nH(Y|X). \\ I(X^n, Y^n) = nI(X, Y). \end{cases} \tag{3.83}$$

***Proof*** Because $XY$ is a memoryless source, we have

$$H(X^nY^n) = H((XY)^n) = nH(XY) = nH(X) + nH(Y|X).$$

On the other hand, by the addition formula of entropy, there is

$$H(X^nY^n) = H(X^n) + H(Y^n|X^n) = nH(X) + H(Y^n|X^n).$$

The combination of the above two formulas has

$$H(Y^n|X^n) = nH(Y|X).$$

According to the definition of mutual information,

$$\begin{aligned} I(X^n, Y^n) &= H(Y^n) - H(Y^n|X^n) \\ &= nH(Y) - nH(Y|X) \\ &= n(H(Y) - H(Y|X)) = nI(X, Y). \end{aligned}$$

The Lemma holds.

Let us define the channel capacity of a discrete channel, this concept plays an important role in channel coding. First, we note that the joint probability distribution $p(xy)$ in the product space $XY$ is uniquely determined by the probability distribution $p(x)$ on $X$ and the probability transformation matrix $T$, that is $p(xy) = p(x)p(y|x)$; therefore, the mutual information $I(X, Y)$ of $X$ and $Y$ is also uniquely determined by $p(x)$ and $T$. In fact,

$$\begin{aligned} I(X, Y) &= \sum_{x\in X}\sum_{y\in Y} p(xy)\log\frac{p(xy)}{p(x)p(y)} \\ &= \sum_{x\in X} p(x)\sum_{y\in Y} p(y|x)\log\frac{p(y|x)}{\sum\limits_{x\in X} p(x)p(y|x)}. \end{aligned}$$

**Definition 3.20** The channel capacity $B$ of a discrete memoryless channel $\{X, T, Y\}$ is defined as

$$B = \max_{p(x)} I(X, Y), \tag{3.84}$$

where formula (3.84) is the maximum of all probability distributions $p(x)$ on $X$.

**Lemma 3.21** *The channel capacity $B$ of a discrete memoryless channel $\{X, T, Y\}$ is estimated as follows:*

$$0 \leq B \leq \min\{\log|X|,\ \log|Y|\}.$$

***Proof*** The amount of mutual information between the two information spaces is $I(X, Y) \geq 0$ (see Lemma 3.5), so there is $B \geq 0$. By Lemma 3.4,

$$I(X, Y) = H(X) - H(X|Y) \leq H(X) \leq \log|X|$$

and

$$I(X, Y) = H(Y) - H(Y|X) \leq H(Y) \leq \log|Y|,$$

so we have

$$0 \leq B \leq \min\{\log|X|,\ \log|Y|\}.$$

The calculation of information capacity is a problem of solving the conditional extremum of constrained convex function. We will not discuss it in detail here but calculate its channel capacity for two simple channels.

***Example 3.8*** The channel capacity of noiseless channel $\{X, T, Y\}$ is $B = \log|X|$.

***Proof*** Let $\{X, T, Y\}$ be a noise free channel, then $|X| = |Y|$, and the probability transfer matrix $T$ is the identity matrix, so

$$\begin{aligned} I(X, Y) &= \sum_{x \in X} \sum_{y \in Y} p(xy) \log \frac{p(y|x)}{p(y)} \\ &= \sum_{x \in X} p(x) \sum_{y \in Y} p(y|x) \log \frac{p(y|x)}{p(y)}. \end{aligned}$$

Because $p(y|x) = 0$, if $y \neq x$; $p(y|x) = 1$, if $x = y$. So there is

$$I(X, Y) = \sum_{x \in X} p(x) \log \frac{1}{p(x)} = H(X) \leq \log|X|.$$

Thus

$$B = \max_{p(x)} I(X, Y) = \log|X|.$$

***Example 3.9*** The channel capacity $B$ of binary symmetric channel $\{X, T, Y\}$ is

$$B = 1 - p\log p - (1 - p)\log(1 - p) = 1 - H(p),$$

where $p < \frac{1}{2}$, $H(p)$ is the binary entropy function.

***Proof*** In binary symmetric channel $\{X, T, Y\}$, $X = Y = \mathbb{F}_2 = \{0, 1\}$, $T$ is a second-order symmetric matrix

$$T = \begin{pmatrix} 1-p & p \\ p & 1-p \end{pmatrix}, \quad p < 1.$$

Let $a$ be the random variable in the input space $\mathbb{F}_2$ and $b$ be the random variable in the output space $\mathbb{F}_2$, all of which obey the two-point distribution, and then the transfer matrix $T$ of the symmetric binary channel can be represented by the following clearer schematic diagram:

$$\begin{cases} P\{b=1|a=0\} = P\{b=0|a=1\} = p \\ P\{b=0|a=0\} = P\{b=1|a=1\} = 1-p. \end{cases}$$

Calculate mutual information $I(X,Y)$, there is

$$I(X,Y) = H(X) - H(X|Y),$$

however,

$$\begin{aligned} H(X|Y) &= \sum_{x\in\mathbb{F}_2}\sum_{y\in\mathbb{F}_2} p(xy)\log p(x|y) \\ &= -p\log p - (1-p)\log(1-p) = H(p). \end{aligned}$$

Thus

$$B = \max\{I(X,Y)\} = \max\{H(X) - H(p)\} = 1 - H(p).$$

In order to state and prove the channel coding theorem, we introduce the concept of joint typical sequence. By the Definition 3.13 of Sect. 5 this chapter, if $X$ is a memoryless source, for any small $\varepsilon > 0$ and positive integer $n \geq 1$, in the power space $X^n$, we define the typical sequence $W_\varepsilon^{(n)}$ as

$$W_\varepsilon^{(n)} = \{x = x_1\cdots x_n \in X^n || -\frac{1}{n}\log p(x) - H(X)| < \varepsilon\}.$$

If $\{X,T,Y\}$ is a memoryless channel, by Lemma 3.19, $XY$ is a memoryless source, in the power space $(XY)^n = X^nY^n$, we define the joint canonical sequence $W_\varepsilon^{(n)}$ as (Fig. 3.4)

$$\begin{aligned} W_\varepsilon^{(n)} = \Bigg\{xy \in X^nY^n \Bigg| |-\frac{1}{n}\log p(x) - H(X)| < \varepsilon,\ |-\frac{1}{n}\log p(y) - H(Y)| < \varepsilon, \\ |-\frac{1}{n}\log p(xy) - H(XY)| < \varepsilon\Bigg\}. \end{aligned} \tag{3.85}$$

**Lemma 3.22** (Progressive bisection) *In memoryless channel $\{X,T,Y\}$, the joint typical sequence $W_\varepsilon^{(n)}$ satisfies the following asymptotic bisection properties:*

*(i)* $\lim\limits_{n\to\infty} P\{xy \in W_\varepsilon^{(n)}\} = 1;$

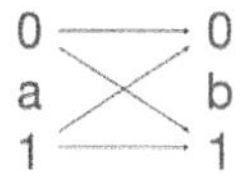


**Fig. 3.4** The transfer matrix

*(ii)* $(1-\varepsilon)\,2^{n(H(XY)-\varepsilon)} \leq |W_\varepsilon^{(n)}| \leq 2^{n(H(XY)+\varepsilon)}$;
*(iii)* *If* $x \in X^n$, $y \in Y^n$,*and* $p(xy) = p(x)p(y)$, *then*

$$(1-\varepsilon)\,2^{-n(I(X,Y)+3\varepsilon)} \leq P\{xy \in W_\varepsilon^{(n)}\} \leq 2^{-n(I(X,Y)-3\varepsilon)}.$$

***Proof*** By Lemma 3.13, we have

$$-\frac{1}{n}\log p(X^n) \to H(X), \text{ Convergence according to probability when } n \to \infty;$$

$$-\frac{1}{n}\log p(Y^n) \to H(Y), \text{ Convergence according to probability when } n \to \infty;$$

$$-\frac{1}{n}\log p(X^nY^n) \to H(XY), \text{ Convergence according to probability when } n \to \infty.$$

So when $\varepsilon$ is given, as long as $n$ is sufficiently large, there is

$$P_1 = P\left\{|-\frac{1}{n}\log p(x) - H(X)| > \varepsilon\right\} < \frac{1}{3}\varepsilon,$$

$$P_2 = P\left\{|-\frac{1}{n}\log p(y) - H(Y)| > \varepsilon\right\} < \frac{1}{3}\varepsilon,$$

$$P_3 = P\left\{|-\frac{1}{n}\log p(xy) - H(XY)| > \varepsilon\right\} < \frac{1}{3}\varepsilon,$$

where $x \in X^n$, $y \in Y^n$. Thus, it can be obtained

$$P\left\{xy \notin W_\varepsilon^{(n)}\right\} \leq P_1 + P_2 + P_3 < \varepsilon.$$

Thus

$$P\left\{xy \in W_\varepsilon^{(n)}\right\} > 1 - \varepsilon,$$

in other words,

$$\lim_{n\to\infty} P\{xy \in W_\varepsilon^{(n)}\} = 1.$$

Property $(i)$ holds. To prove $(ii)$, let $x \in X^n$, $y \in Y^n$, and $xy \in W_\varepsilon^{(n)}$, then

$$H(XY) - \varepsilon < -\frac{1}{n} \log p(xy) < H(XY) + \varepsilon.$$

Equivalently,

$$2^{-n(H(XY)+\varepsilon)} < p(xy) < 2^{-n(H(XY)-\varepsilon)}.$$

By total probability formula,

$$1 = \sum_{xy \in X^n Y^n} p(xy) \geq \sum_{xy \in W_\varepsilon^{(n)}} p(xy) \geq |W_\varepsilon^{(n)}| \, 2^{-n(H(XY)+\varepsilon)}.$$

So there is

$$|W_\varepsilon^{(n)}| \leq 2^{n(H(XY)+\varepsilon)}.$$

On the other hand, when $n$ is sufficiently large,

$$\begin{aligned} 1 - \varepsilon < P\{xy \in W_\varepsilon^{(n)}\} &= \sum_{xy \in W_\varepsilon^{(n)}} p(xy) \\ &\leq |W_\varepsilon^{(n)}| \, 2^{-n(H(XY)-\varepsilon)}. \end{aligned}$$

So there is

$$(1 - \varepsilon) \, 2^{n(H(XY)-\varepsilon)} \leq |W_\varepsilon^{(n)}| \leq 2^{n(H(XY)+\varepsilon)},$$

property $(ii)$ holds. Now let's prove property $(iii)$. If $p(xy) = p(x)p(y)$, then

$$\begin{aligned} P\{xy \in W_\varepsilon^{(n)}\} &= \sum_{xy \in W_\varepsilon^{(n)}} p(x)p(y) \\ &\leq |W_\varepsilon^{(n)}| 2^{-n(H(X)-\varepsilon)} 2^{-n(H(Y)-\varepsilon)} \\ &\leq 2^{n(H(XY)+\varepsilon-H(X)-H(Y)+2\varepsilon)} \\ &= 2^{-n(I(X,Y)-3\varepsilon)}. \end{aligned}$$

Similarity can prove its lower bound, so we have

$$(1 - \varepsilon) \, 2^{-n(I(X,Y)+3\varepsilon)} \leq P\{xy \in W_\varepsilon^{(n)}\} \leq 2^{-n(I(X,Y)-3\varepsilon)}.$$

We have completed the proof of Lemma.

The following lemma has important applications in proving the channel coding theorem. In fact, the conclusion of lemma is valid in general probability space.

**Lemma 3.23** *In memoryless channel* $\{X, T, Y\}$, *if codeword* $y \in Y^n$ *is uniquely determined by* $x \in X^n$, $x' \in X^n$, $x'$ *and* $x$ *are independent,* $y$ *and* $x'$ *are also independent.*

***Proof*** If $y$ is uniquely determined by $x$, then $p(x) = p(y) = p(xy)$, or $p(y|x) = 1$. Therefore, the probability of joint event $yxx'$ is

$$p(yxx') = p(xx') = p(x)p(x') = p(y)p(x').$$

on the other hand,

$$p(yxx') = p(yx').$$

Thus

$$p(yx') = p(y)p(x').$$

The Lemma holds.

In order to define the error probability of channel transmission, we first introduce the workflow of channel coding. After source compression coding, a source message input set is generated,

$$W = \{1, 2, \ldots, M\}, \ \ M \geq 1 \text{ is positive integers.}$$

Injection $f : W \to X^n$ is called coding function, $f$ encodes each input message $w \in W$ as $f(w) \in X^n$. Codeword $x = f(w) \in X^n$ receives codeword $y \in Y^n$ after transmission through channel $\{X, T, Y\}$, we write $x \xrightarrow{T} y$, or $y = T(x)$. Mapping $g : Y^n \to W$ is called decoding function. Therefore, the so-called channel coding is a pair of mapping $(f, g)$. Obviously,

$$C = f(W) = \{f(w)|w \in W\} \subset X^n$$

is a code with length $n$ in codeword space $X^n$, number of codewords is $|C| = |W| = M$. $C$ is the code of $f$. The code rate $R_C$ is

$$R_C = \frac{1}{n} \log |C| = \frac{1}{n} \log M.$$

For each input message $w \in W$, if $g(T(f(w))) \neq w$, it is said that the channel transmission is wrong, the transmission error probability $\lambda_w$ is

$$\lambda_w = P\{g(T(f(w))) \neq w\}, \quad w \in W. \tag{3.86}$$

The transmission error probability of codeword $x = f(w) \in C$ is recorded as $P_e(x)$, obviously, $P_e(x) = \lambda_w$, that is, $P_e(x)$ is the conditional probability

$$\begin{aligned} P_e(x) &= P\{g(T(x)) \neq w|x = f(w)\} \\ &= P\{g(T(f(w))) \neq w\} = \lambda_w. \end{aligned} \tag{3.87}$$

We define the transmission error probability of code $C = f(W) \subset X^n$ as $P_e(C)$,

$$P_e(C) = \frac{1}{M} \sum_{x \in C} P_e(x) = \frac{1}{M} \sum_{w=1}^{M} \lambda_w. \tag{3.88}$$

As before, a code $C$ with length $n$ and number of codewords $M$ is recorded as $C = (n, M)$.

**Theorem 3.12** (Shannon's channel coding theorem, 1948) *Let $\{X, T, Y\}$ be a memoryless channel and $B$ be the channel capacity, then*

*(i) When $R < B$, there is a column of codes $C_n = (n, 2^{[nR]})$, its transmission error probability $P_e(C_n)$ satisfies*

$$\lim_{n \to \infty} P_e(C_n) = 0; \tag{3.89}$$

*(ii) Conversely, if the transmission error probability of code $C_n = (n, 2^{[nR]})$ satisfies Eq. (3.89), there is an absolute normal number $N_0$, and we have the code rate $R_{C_n}$ of $C_n$ satisfies*

$$R_{C_n} \leq B, \text{ when } n \geq N_0.$$

If $C_n = (n, 2^{[nR]})$, by Lemma 2.27 of Chap. 2,

$$R - \frac{1}{n} < R_{C_n} \leq R. \tag{3.90}$$

so $(i)$ of Theorem 3.12 indicates that the code rate is sufficiently close to the channel capacity $B$, the "good code" with sufficiently small transmission error probability exists. $(ii)$ indicates that the bit rate of the so-called good code with sufficiently small transmission error probability does not exceed the channel capacity. Shannon's proof Theorem 3.12 uses random code technology; this idea of using random method to prove deterministic results is widely used in information theory. At present, it has more and more applications in other fields.

***Proof*** (Proof of theorem 3.12) Firstly, the probability function $p(x_i)$ is arbitrarily selected on the input alphabet $X$, and the joint probability in power space $X^n$ is defined as

$$p(x) = \prod_{i=1}^{n} p(x_i), \quad x = x_1 \cdots x_n \in X^n, \tag{3.91}$$

In this way, we get a memoryless source $X$ and power space $X^n$, which constitute the codeword space of channel coding. Then $M = 2^{[nR]}$ codewords are randomly selected in $X^n$ to obtain a random code $C_n = (n, 2^{[nR]})$. In order to illustrate the randomness of codeword selection, we borrow the source message set $W = \{1, 2, \ldots, M\}$, where $M = 2^{[nR]}$. For every message $w$, $1 \leq w \leq M$, the randomly generated codeword is marked as $X^{(n)}(w)$. So we get a random code

$$C_n = \{X^{(n)}(1), X^{(n)}(2), \ldots, X^{(n)}(M)\} \subset X^n.$$

The generation probability $P\{C_n\}$ of $C_n$ is

$$P\{C_n\} = \prod_{w=1}^{M} P\{X^{(n)}(w)\} = \prod_{w=1}^{M}\prod_{i=1}^{n} p(x_i(w)),$$

where $X^{(n)}(w) = x_1(w)x_2(w)\cdots x_n(w) \in X^n$.

We take $A_n = \{C_n\}$ as the set of all random codes $C_n$, which is called the random code set. The average transmission error probability on random code set $A_n$ is defined as

$$\bar{P}_e(A_n) = \sum_{C_n \in A_n} P\{C_n\}P_e(C_n). \tag{3.92}$$

If you want to prove that for any $\varepsilon > 0$, When $n$ is sufficiently large, $\bar{P}_e(A_n) < \varepsilon$, then there is at least one code $C_n \in A_n$ such that $P_e(C_n) < \varepsilon$, which proves the $(i)$. Therefore, we prove it in two steps.

(1) Principles of constructing random codes and encoding and decoding

We select each message in the source message set $W = \{1, 2, \ldots, M\}$ with equal probability, that is $w \in W$, the selection probability of $w$ is

$$p(w) = \frac{1}{M} = 2^{-[nR]}, \quad w = 1, 2, \ldots, M.$$

In this way, $W$ becomes an equal probability information space. For each input message $w$, it is randomly coded as $X^{(n)}(w) \in X^n$, where

$$X^{(n)}(w) = x_1(w)x_2(w)\cdots x_n(w) \in X^n.$$

Codeword $X^{(n)}(w)$ is transmitted through memoryless channel $\{X, T, Y\}$ with conditional probability

$$p(y|X^{(n)}(w)) = \prod_{i=1}^{n} p(y_i|x_i(w))$$

received codeword $y = y_1y_2\cdots y_n \in Y^n$. The decoding principle of $y$ is: If $X^{(n)}(w)$ is the only input codeword so that $X^{(n)}(w)y$ is joint typical, that is $X^{(n)}(w)y \in W_\varepsilon^{(n)}$, then decode $g(y) = w$; if there is no such codeword $X^{(n)}(w)$, or there are two or more codewords $X^{(n)}(w)$ and $y$ are joint typical, $y$ cannot be decoded correctly.

(2) Estimating the average error probability of random code set $A_n$

By (3.92) and (3.88),

$$\begin{aligned}\bar{P}_e(A_n) &= \sum_{C_n \in A_n} P\{C_n\} P_e(C_n) \\ &= \sum_{C_n \in A_n} P\{C_n\} \frac{1}{M} \sum_{x \in C_n} P_e(x) \\ &= \frac{1}{M} \sum_{w=1}^{M} \lambda_w \sum_{C_n \in A_n} P\{C_n\} \\ &= \frac{1}{M} \sum_{w=1}^{M} \lambda_w,\end{aligned} \tag{3.93}$$

where $\lambda_w$ is given by Eq. (3.86). Because $w$ is input with equal probability, in other words, $w$ is encoded with equal probability. Therefore, the transmission error probability $\lambda_w$ of $w$ does not depend on $w$, that is

$$\lambda_1 = \lambda_2 = \cdots = \lambda_M.$$

By (3.93), we have $\bar{P}_e(A_n) = \lambda_1$. To estimate $\lambda_1$, we define

$$E_i = \{y \in Y^n | X^n(i)y \in W_\varepsilon^{(n)}\},\ i = 1, 2, \ldots, M, \tag{3.94}$$

If $E_1^c = Y^n \backslash E_1$ is the remainder of $E_1$, because of the decoding principle,

$$\lambda_1 = P\{E_1^c \cup E_2 \cup \cdots \cup E_M\} \leq P\{E_1^c\} + \sum_{i=2}^{M} P\{E_i\}. \tag{3.95}$$

By property $(i)$ of Lemma 3.22,

$$\lim_{n \to \infty} P\{xy \notin W_\varepsilon^{(n)}\} = 0.$$

So there is

$$\lim_{n \to \infty} P\{X^{(n)}(1)y \notin W_\varepsilon^{(n)}\} = 0.$$

Therefore, when $n$ is sufficiently large,

$$P\{E_1^c\} < \varepsilon.$$

Obviously, codeword $X^{(n)}(1)$ and other codewords $X^{(n)}(i)$, $(i = 2, \ldots, M)$ are independent of each other (see 3.91). By Lemma 3.23, $y = T(X^{(n)}(1))$ and $X^{(n)}(i)$ $(i \neq 1)$ also are independent of each other. Then by the property $(iii)$ of Lemma 3.22,

$$P\{E_i\} = P\{X^{(n)}(i)y \in W_\varepsilon^{(n)}\} \leq 2^{-n(I(X,Y)-3\varepsilon)}\ (i \neq 1).$$

To sum up,

$$\begin{aligned}\bar{P}_e(A_n) =&\lambda_1 \leq \varepsilon + \sum_{i=2}^{M} 2^{-n(I(X,Y)-3\varepsilon)} \\ &\leq \varepsilon + 2^{[nR]} 2^{-n(I(X,Y)-3\varepsilon)} \\ &\leq \varepsilon + 2^{-n(I(X,Y)-R-3\varepsilon)}.\end{aligned}$$

If $R < I(X, Y)$, then $I(X, Y) - R - 3\varepsilon > 0$(when $\varepsilon$ is sufficiently small), so when $n$ is large enough, we have $\bar{P}_e(A_n) < 2\varepsilon$. Due to the channel capacity $B = \max\{I(X, Y)\}$, we can choose $p(x)$ to make $B = I(X, Y)$. So when $R < B$, we have $\bar{P}_e(A_n) < 2\varepsilon$, this completes the proof of $(i)$.

To prove $(ii)$, let's look at a special case first. If the error probability of $C = (n, 2^{[nR]})$ is $P_e(C) = 0$, then the bit rate of $C$ is $R_C < B + \frac{1}{n}$, so when $n$ is sufficiently large, there is $R_C \leq B$.

In fact, because $P_e(C) = 0$, decoding function $g : Y^n \to W$ only determines $W$, there is $H(W|Y^n) = 0$. Because $W$ is equal probability information space, so

$$H(W) = \log |W| = [nR].$$

Using the decomposition of mutual information, there are

$$I(W, Y^n) = H(W) - H(W|Y^n) = H(W) = [nR]. \tag{3.96}$$

on the other hand, $W \to X^n \to Y^n$ forms a Markov chain, by data inequality (see Theorem 3.8)

$$I(W, Y^n) \leq I(X^n, Y^n).$$

By Lemma 3.20,

$$I(W, Y^n) \leq I(X^n, Y^n) = nI(X, Y) \leq nB.$$

By (3.96), there is $[nR] \leq nB$. Because $nR - 1 < [nR] \leq nR$, so $nR < nB + 1$, that is $R < B + \frac{1}{n}$, by (3.90), we have

$$R_C \leq R < B + \frac{1}{n},$$

thus

$$R_C \leq B, \text{ when } n \text{ is sufficiently large.}$$

The above formula shows that when the transmission error probability is 0, as long as $n$ is sufficiently large, there is $R_C \leq B$. Secondly, if the transmission error is allowed, that is, the error probability of $C_n$ is $P_e(C_n) < \varepsilon$, where $C_n = (n, 2^{[nR]})$. Then when $n$ is sufficiently large, we still have $R_{C_n} \leq B$.

In order to prove the above conclusion, we note the error probability of random code $C_n$ is

$$P_e(C_n) = \lambda_w, \tag{3.97}$$

where $w \in W$ is any given message. When $w$ is given, we define a random variable $\xi_w$ with a value on $\{0, 1\}$ as

$$\xi_w = \begin{cases} 1, & \text{if } g(T(f(w))) \neq w; \\ 0, & \text{if } g(T(f(w))) = w. \end{cases}$$

Let $E = (\mathbb{F}_2, \xi_w)$ be a binary information space, by (3.97), then we have

$$P_e(C_n) = P\{\xi_w = 1\}.$$

By Theorem 3.3,

$$\begin{aligned} H(EW|Y^n) &= H(W|Y^n) + H(E|WY^n) \\ &= H(E|Y^n) + H(W|EY^n). \end{aligned} \tag{3.98}$$

Note that $E$ is uniquely determined by $Y^n$ and $W$, so $H(E|WY^n) = 0$, at the same time, $E$ is a binary information space, $H(E) \leq \log 2 = 1$, there is

$$H(E|Y^n) \leq H(E) \leq 1.$$

On the other hand, the random variable $\xi_w$ is only related to $w \in W$, so

$$H(W|EY^n) = P_e(C_n) \log(|W| - 1) \leq nRP_e(C_n).$$

By (3.98), we have

$$H(W|Y^n) \leq 1 + nRP_e(C_n).$$

Because $f(W) = X^n(W)$ is a function of $W$, we have the following Fano inequality

$$H(f(W)|Y^n) \leq H(W|Y^n) \leq 1 + nRP_e(C_n).$$

Finally,

$$\begin{aligned} = H(W) &= H(W|Y^n) + I(W, Y^n) \\ &\leq H(W|Y^n) + I(f(W), Y^n) \\ &\leq 1 + nRP_e(C_n) + I(X^n, Y^n) \\ &\leq 1 + nRP_e(C_n) + nB, \end{aligned}$$

because of $nR - 1 < [nR]$, then we have

$$nR < 2 + nRP_e(C_n) + nB.$$

Thus

$$R_{C_n} \leq R < B + \frac{2}{n} + \varepsilon,$$

When $n$ is sufficiently large, we obtain $R_{C_n} \leq B$, which completes the proof of the theorem.

It can be seen from Example 3.9 that the channel capacity $B = 1 - H(p)$ of a binary symmetric channel. Therefore, Theorem 3.12 extends Theorem 2.10 in the previous chapter to a more general memoryless channel; at the same time, it is also proved that the code rate of a good code does not exceed the capacity of the channel.

## Exercise 3

1. The joint probability functions of the two information spaces $X$ and $Y$ are as follows:

| $Y$ | $X$ | |
|---|---|---|
| | 0 | 1 |
| 0 | $\frac{1}{4}$ | $\frac{1}{4}$ |
| 1 | $\frac{1}{12}$ | $\frac{5}{12}$ |

   Solve $H(X)$, $H(Y)$, $H(XY)$, $H(X|Y)$, $H(Y|X)$, and $I(X, Y)$.
2. Let $X_1, X_2, X_3$ be three information spaces on $\mathbb{F}_2$, Known $I(X_1, X_2) = 0$, $I(X_1, X_2, X_3) = 1$, prove:

$$H(X_3) = 1, \text{ and } H(X_1X_2X_3) = 2.$$

3. Give an example to illustrate $I(X, Y|Z) \geq I(X, Y)$.
4. Can $I(X, Y|Z) = 0$ be derived from $I(X, Y) = 0$? In turn, can $I(X, Y|Z) = 0$ deduce $I(X, Y) = 0$? Please prove or give examples.
5. Let $X$, $Y$, $Z$ be three information spaces, prove:

   (i) $H(XY|Z) \geq H(X|Z)$;
   (ii) $I(XY, Z) \geq I(X, Z)$;
   (iii) $H(XYZ) - H(XY) \leq H(XZ) - H(X)$;
   (iv) $I(X, Z|Y) = I(Z, Y|Z) - I(Z, Y) + I(X, Z)$.

   It also explains under what conditions the equality sign holds.
6. Can $I(X, Y) = 0$ deduce $I(X, Z) = I(X, Z|Y)$?
7. Let the information space be $X = \{0, 1, 2, \ldots\}$ and the value probability $p(n)$ of random variable $\xi$ be

$$p(n) = P\{\xi = n\}, n = 0, 1, \ldots.$$

   Given the mathematical expectation $E\xi = A > 0$ of $\xi$, find the maximum probability distribution $\{p(n)|n = 0, 1, \ldots\}$ of $H(X)$ and the corresponding maximum information entropy.

8. Let the information space be $X = \{0, 1, 2, \ldots\}$, and take an example of the random variable $\xi$ taken from $X$, so that $H(X) = \infty$.
9. Let $X_1 = (X, \xi)$, $X_2 = (X, \eta)$ be two information spaces and $\xi$ be a function of $\eta$, prove $H(X_1) \leq H(X_2)$, and explain this result.
10. Let $X_1 = (X, \xi)$, $X_2 = (X, \eta)$ be two information spaces and $\eta = f(\xi)$, prove

    (i) $H(X_1) \geq H(X_2)$, give the conditions under which the equal sign holds.
    (ii) $H(X_1|X_2) \geq H(X_2|X_1)$, give the conditions under which the equal sign holds.

## References

Bassoli, R., Marques, H., & Rodriguez, J. (2013). Network coding theory, a survey. *IEEE Commun. Surveys Tutor., 15*(4), 1950–1978.
Berger, T. (1971). *Rate distortion theory: a mathematical basis for data compression*. Prentice-Hall.
Blahut, R. E. P. (1965). *Ergodic theory and informtion*. Wiley.
Chung, K. L. (1961). A note on the ergodic theorem of information theory. *Addison. Math Statist., 32*, 612–614.
Cover, T. M., & Thomas, J. A. (1991). *Elements of information theory*. Wiley.
Csiszár, I., & Körner, J. (1981). *Informaton theory: Coding theorems for discrete memoryless systems*. Academic Press.
El Gamal, A., & Kim, Y. H. (2011). *Network information theory*. Cambrige University Press
Fragouli, C., Le Boudec, J. Y., & Widmer, J. (2006). Network coding: An instant primer. *ACMSIG-COMM Computer Communication Review, 36*, 63–68.
Gallager, R. G. (1968). *Information theory and reliable communication*. Wiley.
Gray, R. M. (1990). *Entropy and information theory*. Springer.
Guiasu, S. (1977). *Inormation theory with applications*. McGraw-Hill.
Ho, T., & Lun, D. Network coding: An introduction. *Computer Journal*.
Hu, X. H., & Ye, Z. X. (2006). Generalized quantum entropy. *Journal of Mathematical Physics, 47*(2), 1–7.
Ihara, S. (1993). *Information theory for continuous systems*. World Scientific.
Kakihara, Y. (1999). *Abstract methods in information theory*. World Scientific.
McMillan, B. (1953). The basic theorems of information theory. *Annals of Mathematical Statistics, 24*(2), 196–219.
Moy, S. C. (1961). A note on generalizations of Shannon-McMilllan theorem. *Pacific Journal of Mathematics, 11*, 705–714.
Nielsen, M. A., & Chuang, I. L. (2000). *Quantum computation and quantum information*. Cambridge University Press.
Shannon, C. E. (1948). A mathematical theory of communication. *Bell Labs Technical Journal, 27*(4), 379–423, 623–656.
Shannon, C. E. (1959). Coding theorem for a discrete source with a fidelity criterion. *IRE National Convention Record, 4*, 142–163.
Shannon, C. E. (1958). Channels with side information at the transmitter. *IBM Journal of Research and Development, 2*(4), 189–193.
Shannon, C. E. (1961). Two-way communication channels. *Proceedings of the Fourth Berkeley Symposium on Mathematical Statistics and Probability, 1*, 611–644.
Thomasian, A. J. (1960). An elementary proof of the AEP of information theory. *Annals of Mathematical Statistics, 31*(2), 452–456.
Wolfowitz, J. (1978). *Coding theorems of information theory* (3rd ed.). Springer-Verlag.

Ye, Z. X., & Berger, T. (1998). *Information measures for discrete random fields*. Science Press.
Yeung, R. W. (2002). *A first course in information theory*. Kluwer Academic.
Qiu, P. (2003). *Information theory and coding*. Higher Education Press. (in Chinese).
Qiu, P., Zhang, C., Yang, S., et al. (2012). *Multi user information theory*. Science Press. (in Chinese).
Ye, Z. (2003). *Fundamentals of information theory*. Higher Education Press. (in Chinese).
Zhang, Z., & Lin, X. (1993). *Information theory and optimal coding*. Shanghai Science and Technology Press. (in Chinese).

# Chapter 4
# Cryptosystem and Authentication System

In 1949, Shannon published a famous paper entitled "communication theory of secure systems" in the technical bulletin of Bell laboratory. Based on the mathematical theory of information established by him in 1948 (see Chap. 3), this paper makes a comprehensive discussion on the problem of secure communication and establishes the mathematical theory of secure communication system. It has a great impact on the later development of cryptography. It is generally believed that Shannon transformed cryptography from art (creative ways and methods) to science, so he is also known as the father of modern cryptography. The main purpose of this chapter is to introduce Shannon's important ideas and results in cryptography theory, which is the cornerstone of the whole modern cryptography.

## 4.1 Definition and Statistical Characteristics of Cryptosystem

Let $X = \{a_1, a_2, \ldots, a_q\}$ be the plaintext alphabet and a source. $\{\xi_i\}_{i=1}^{\infty}$ is a set of random variables valued on $X$, for any given positive integer $n \geq 1$, we define the plaintext space $P$ as the product information space $X_1 X_2 \cdots X_n$, that is

$$P = X_1 X_2 \cdots X_n, \text{ where } X_i = (X, \xi_i), \ 1 \leq i \leq n.$$

If $m = m_1 m_2 \cdots m_n \in P (m_i \in X_i)$, $m$ is called a plaintext information column of alphabet length $n$, or a plaintext string of length $n$, the joint probability $p(m)$ is defined as

$$p(m) = p(m_1 m_2 \cdots m_n) = P\{\xi_1 = m_1, \xi_2 = m_2, \ldots, \xi_n = m_n\}. \tag{4.1}$$



Z. Zheng, *Modern Cryptography Volume 1*, Financial Mathematics and Fintech,
https://doi.org/10.1007/978-981-19-0920-7_4

Let $Z = \{b_1, b_2, \ldots, b_s\}$ be the key alphabet, which is also a memoryless source (see Definition 3.9 of Chap. 3), let $\{\eta_i\}_{i=1}^{\infty}$ be a group of random variables with independent values on $Z$ and equal probability distribution, then for any $b \in Z$,

$$p(b) = P\{\eta_i = b\} = \frac{1}{|Z|} = \frac{1}{s}, \ \forall\, i \geq 1. \tag{4.2}$$

We define the power space $Z^r$ as a key space, denoted by $K$, that is

$$K = Z^r = \{k = k_1 k_2 \cdots k_r | k_i \in Z, 1 \leq i \leq r\}.$$

Each $k = k_1 k_2 \cdots k_r \in K$ is called a key of length $r$, and the joint probability $p(k)$ is

$$p(k) = p(k_1 k_2 \cdots k_r) = \prod_{i=1}^{r} p(k_i) = \frac{1}{|Z|^r} = \frac{1}{|K|}. \tag{4.3}$$

This shows that the $r$-dimensional random vector $\eta = (\eta_1, \eta_2, \ldots, \eta_r)$ taking value on the key space $K$ is also equally almost distributed on $K$. Unless otherwise specified, we generally stipulate that the plaintext space $P$ and the key space $K$ are independent information spaces, that is

$$p(mk) = p(m)p(k), \forall\, m \in P, k \in K. \tag{4.4}$$

For every $k \in K$, $k$ defines or controls an encryption transform $E_k$, denote by

$$E = \{E_k | k \in K\}.$$

$E$ is called encryption algorithm. When $k \in K$ is given, the encryption transformation $E_k$ acts on the plaintext $m \in P$ to produce a cryptosystemtext $E_k(m)$, each encryption transformation $E_k$ is an injection, and its left inverse mapping is recorded as $D_k$, which is called decryption transformation. Taking $1_P$ as the identity transformation of plaintext space, that is $1_P(m) = m, \forall\, m \in P$, then we have

$$D_k E_k = 1_P, \text{ or } D_k(E_k(m)) = m, \forall\, m \in P. \tag{4.5}$$

Define cryptosystemtext space $C$ as

$$C = \{E_k(m) | m \in P, k \in K\} \subset X_1 X_2 \cdots X_n. \tag{4.6}$$

That is, cryptosystemtext space $C$ and plaintext space $P$ have the same alphabet and the same letter length.

For each cryptosystemtext $c \in C$, $c = E_k(m)$, then $c$ is uniquely determined by plaintext $m$ and key $k$, so we can define the occurrence probability $p(c)$ of cryptosystemtext $c$ as

$$p(c) = p(E_k(m)) = p(km) = p(k)p(m). \tag{4.7}$$

Obviously,

$$\sum_{c \in C} p(c) = \sum_{k \in K} \sum_{m \in P} p(km) = \sum_{k \in K} p(k) \sum_{m \in P} p(m) = 1,$$

Therefore, the cryptosystemtext space $C$ defined by formula (4.7) is also an information space.

When $k \in K$ is given, we let

$$A_k = \{E_k(m) | m \in P\} \subset C. \tag{4.8}$$

Then the encryption transformation $E_k$ is the full mapping of $P \to A_k$, so $E_k$ is a 1–1 correspondence of $P \to A_k$, and its inverse mapping is the decryption transformation $D_k$, that is

$$D_k E_k = 1_P, \; E_k D_k = 1_{A_k}, k \in K.$$

We denote $D$ as all decryption transformations, that is

$$D = \{D_k | k \in K\}. \tag{4.9}$$

$D$ is called decryption algorithm.

**Definition 4.1** Under the above provisions, $\mathfrak{R} = \{P, C, K, E, D\}$ is called a cryptosystem, where $P, C, K$ is the information space, $K$ and $P$ are statistically independent, $E$ is the encryption algorithm and $D$ is the decryption algorithm.

The statistical characteristics of a cryptosystem are attributed to the following theorem.

**Theorem 4.1** *If* $\mathfrak{R} = \{P, C, K, E, D\}$, *then*

*(1)* $\forall\, c \in C$, *we have*

$$p(c) = \sum_{\substack{k \in K \\ c \in A_k}} p(k)p(D_k(c)). \tag{4.10}$$

*(2)* $c \in C, m \in P$, *then*

$$p(c|m) = \sum_{\substack{k \in K \\ E_k(m)=c}} p(k). \tag{4.11}$$

*(3)* $c \in C, m \in P$, *then*

$$p(m|c) = \frac{p(m) \sum_{\substack{k \in K \\ D_k(c)=m}} p(k)}{\sum_{\substack{k \in K \\ c \in A_k}} p(k)p(D_k(c))}, \tag{4.12}$$

*where $A_k$ is given by equation (4.8).*

***Proof*** By (4.7), if $c \in C$, then

$$\begin{aligned} p(c) &= \sum_{\substack{k\in K, m\in P \\ E_k(m)=c}} p(km) = \sum_{\substack{k\in K, m\in P \\ E_k(m)=c}} p(k)p(m) \\ &= \sum_{k\in K} p(k) \sum_{\substack{m\in P \\ E_k(m)=c}} p(m) = \sum_{\substack{k\in K \\ c\in A_k}} p(k)p(D_k(c)). \end{aligned}$$

(1) holds. (2) is trivial. Because when $m \in P$ is given, the occurrence probability $p(c|m)$ of cryptosystemtext $c$ has

$$p(c|m) = \sum_{\substack{k\in K, \\ E_k(m)=c}} p(k).$$

To prove (3), by (1.24),

$$p(m|c) = \frac{p(m)p(c|m)}{\sum_{m'\in P} p(m')p(c|m')},$$

the items in the denominator are

$$\begin{aligned} \sum_{m'\in P} p(m')p(c|m') &= \sum_{m'\in P} p(m') \sum_{\substack{k\in K \\ E_k(m')=c}} p(k) \\ &= \sum_{k\in K} p(k) \sum_{\substack{m'\in P \\ E_k(m')=c}} p(m') \\ &= \sum_{\substack{k\in K \\ c\in A_k}} p(k)p(D_k(c)). \end{aligned}$$

So in the end

$$p(m|c) = \frac{p(m)\sum_{\substack{k\in K \\ E_k(m)=c}} p(k)}{\sum_{\substack{k\in K \\ c\in A_k}} p(k)p(D_k(c))},$$

Theorem 4.1 holds!

By Theorem 4.1, the statistical characteristics of a cryptosystem can be summarized as follows: The probability distribution of cryptosystemtext space and the conditional probability distribution of plaintext about cryptosystemtext are completely determined by the probability distribution of plaintext space and key space. That is, anyone who knows the probability distribution of plaintext space and key space will know the probability distribution of cryptosystemtext and the conditional probability distribution of plaintext about cryptosystemtext.

It is assumed that the plaintext space and the key space are statistically independent, by (3.14) of Theorem 3.2 of Chap. 3, we have

$$H(PK) = H(P) + H(K). \tag{4.13}$$

It has been previously specified that the key source alphabet $Z$ is an equal probability information space without memory, and the probability $p(k)$ of the key space $K = \{k = k_1 k_2 \cdots k_r | k_i \in Z\}$ is

$$p(k) = \prod_{i=1}^{r} p(k_i) = \frac{1}{|Z|^r} = \frac{1}{|K|}. \tag{4.14}$$

Therefore, the key space is also an equal probability information space.

From the definition of cryptosystem, when the plaintext space and key space are given, the cryptosystemtext space is completely determined. On the contrary, when the cryptosystemtext space and key space are known, the plaintext space is also known, combined with Lemma 3.3 in the previous chapter, we have

**Theorem 4.2** *In a cryptosystem* $\mathfrak{R} = \{P, C, K, E, D\}$, *we have*

$$H(P|KC) = 0, \ \ H(C|KP) = 0, \ \ H(K|PC) = 0.$$

***Proof*** We only prove $H(P|KC) = 0$, similarly to $H(C|KP) = 0$ and $H(K|PC) = 0$. For given $m \in P$, let

$$N_m = \{kc | c \in C, k \in K \text{and } E_k(m) = c\}.$$

Thus $N_m \subset KC$, and

$$\begin{cases} p(m|kc) = 1, & \text{if } kc \in N_m; \\ p(m|kc) = 0, & \text{if } kc \notin N_m. \end{cases}$$

Because assuming $kc \in N_m$ is selected, then $E_k(m) = c$, thus $m = D_k(c)$, $m$ will be determined. Conversely, if $kc \notin N_m$, when the $kc$-joint event occurs and $m$ cannot occur, thus $p(m|kc) = 0$. By Lemma 3.3 from the previous chapter, $H(P|KC) = 0$, we complete the proof of Theorem 4.2.

**Corollary 4.1** *In a cryptosystem* $\mathfrak{R} = \{P, C, K, E, D\}$, *we always have*

$$H(P) \leq H(C).$$

***Proof*** It is stipulated that $P$ and $K$ are statistically independent, so there are

$$\begin{aligned} H(P) = H(P|K) &= H(P|K) + H(C|PK) \\ &= H(PC|K) \\ &= H(C|K) + H(P|KC) \\ &= H(C|K) \\ &\leq H(C). \end{aligned}$$

The Corollary holds.

The Corollary shows that the uncertainty of plaintext is less than that of cryptosystemtext in cryptosystem.

## 4.2 Fully Confidential System

Generally speaking, the mutual information $I(P, C)$ between plaintext space and cryptosystemtext space (see Definition 3.8 in the previous chapter) reflects the information of plaintext space contained in cryptosystemtext space, so $I(P, C)$ minimization is an important design goal of cryptosystem. If the cryptosystemtext does not provide any information about the plaintext, or the analyst cannot obtain any information about the plaintext by observing the cryptosystemtext, such a cryptosystem is called completely confidential.

**Definition 4.2** A cryptosystem $\mathfrak{R} = \{P, C, K, E, D\}$, if $H(P|C) = H(P)$, or $I(P, C) = 0$, $\mathfrak{R}$ is called complete secrecy system, or unconditional secrecy system.

**Theorem 4.3** *For any cryptosystem* $\mathfrak{R} = \{P, C, K, E, D\}$, *we have*

$$I(P, C) \geq H(P) - H(K). \tag{4.15}$$

***Proof*** By Theorem 4.2, we have $H(P|KC) = 0$, and

$$\begin{aligned} H(P|C) &= H(P|C) + H(K|PC) \\ &= H(PK|C) \\ &= H(K|C) + H(P|KC). \end{aligned}$$

So we have

$$H(P|C) = H(K|C) \leq H(K).$$

By definition,

$$I(P, C) = H(P) - H(P|C) \geq H(P) - H(K).$$

So the Theorem holds.

From the previous chapter, we know the amount of mutual information $I(X, Y) \geq 0$, there is

**Corollary 4.2** *In a completely confidential cryptosystem* $\mathfrak{R} = \{P, C, K, E, D\}$, *there is always*

$$H(P) \leq H(K) = \log_2 |K|. \tag{4.16}$$

***Proof*** Defined by $\mathfrak{R} = \{P, C, K, E, D\}$ as a completely confidential system, so $I(P, C) = 0$. From the above theorem, there are

$$H(P) - H(K) \leq I(P, C) = 0,$$

Thus there is $H(P) \leq H(K)$. By (4.14), $K$ is equipotential distribution, so there is

$$H(P) \leq \log_2 |K|.$$

It can be seen from the above that the larger the scale $|K|$ of the key space, the better the confidentiality of the system!

**Definition 4.3** A cryptosystem $\mathfrak{R} = \{P, C, K, E, D\}$ is called a "one secret at a time" system, if there is a unique key $k \in K$ for a given $m \in P$ and $c \in C$, such that $c = E_k(m)$.

As can be seen from the above definition, for given $m \in P$, if $k \neq k'$, then $E_k(m) \neq E_{k'}(m)$. In other words, we only use a unique key $k$ to encrypt the same set of plaintext and cryptosystemtext. This is also the origin of the concept of "one secret at a time". Thus, for any given plaintext $m \in P$ and cryptosystemtext $c \in C$, there happens to be a unique key $k \in K$ such that $E_k(m) = c$. Therefore, when $k$ traverses the key space $K$, $m$ traverses the plaintext space $P$, and each $m$ appears only once. Thus, for $c \in C$, we have

$$\begin{aligned} p(c) &= \sum_{\substack{k \in K \\ E_k(m)=c}} \sum_{m \in P} p(k)p(m) \\ &= \sum_{\substack{k \in K \\ E_k(m)=c}} p(k) \sum_{m \in P} p(m) \\ &= \frac{1}{|K|} \sum_{m \in P} p(m) = \frac{1}{|K|}. \end{aligned} \tag{4.17}$$

That is to say, in a one-time cryptosystem, the cryptosystemtext space $C$ is also an equal probability information space.

**Theorem 4.4** *The one-time password system is a completely confidential system.*

***Proof*** When $c, m$ given, by (4.11),

$$p(c|m) = \sum_{\substack{k\in K \\ E_k(m)=c}} p(k) = \frac{1}{|K|}.$$

By (4.12) and (4.17),

$$\begin{aligned} p(m|c) &= \frac{p(m)}{\sum_{\substack{k\in K \\ E_{m'}(k)=c}} \sum_{m'\in P} p(m')} \\ &= \frac{p(m)}{\sum_{m'\in P} p(m')} \\ &= p(m). \end{aligned}$$

Thus

$$\begin{aligned} H(P|C) &= -\sum_{m\in P}\sum_{c\in C} p(mc)\log_2 p(m|c) \\ &= -\sum_{m\in P}\sum_{c\in C} p(mc)\log_2 p(m) \\ &= -\sum_{m\in P} p(m)\log_2 p(m) \\ &= H(P). \end{aligned}$$

Therefore, $\mathfrak{R} = \{P, C, K, E, D\}$ is a completely confidential system.

## 4.3 Ideal Security System

In order to introduce Shannon's concepts of unique solution distance and ideal cryptosystem, we first consider the scenario of secret only attack. In the scenario of secret only attack, when the cryptanalyzer intercepts cryptosystemtext $c$, he may decrypt $c$ with all decryption keys $D_k$ to obtain

$$m' = D_k(c),\ k \in K.$$

Therefore, he records the keys corresponding to all meaningful messages $m'$, only one of the set of these keys is a correct key, while other incorrect keys are called pseudo keys. A large number of cryptosystemtexts are required as samples in secret only attacks. Therefore, we will consider the product space $P^n$ of plaintext and cryptosystemtext and the joint events in $C^n$, $P^n$ and $C^n$ as plaintext string and cryptosystemtext string.

**Definition 4.4** For cryptosystemtext string $y \in C^n$ with given length $n$, let

$$K(y) = \{k \in K | \exists\, x \in P^n \text{ such that } E_k(x) = y\}. \tag{4.18}$$

Then the number of pseudo keys is $|K(y)| - 1$. The mathematical expectation $S_n$ of the pseudo key is defined as

$$S_n = \sum_{y \in C^n} p(y)(|K(y)| - 1). \tag{4.19}$$

Therefore, the mathematical expectation of pseudo key is the weighted average of the number of pseudo keys of each cryptosystemtext string. We first prove the following two theorems.

**Theorem 4.5** *If* $\mathfrak{R} = \{P, C, K, E, D\}$ *is a cryptosystem, there are*

$$H(K|C) = H(K) + H(P) - H(C). \tag{4.20}$$

***Proof*** From the addition formula of information entropy (see Theorem 3.2 in the previous chapter),

$$H(KPC) = H(KP) + H(C|KP) = H(KC) + H(P|KC).$$

By Theorem 4.2, we have

$$H(C|KP) = H(P|KC) = 0.$$

thus,

$$H(KP) = H(KC).$$

Again, from the addition formula and note that $K$ and $P$ are statistically independent, so

$$H(KP) = H(P) + H(K|P) = H(P) + H(K)$$

and

$$H(P) + H(K) = H(KC) = H(C) + H(K|C).$$

So we have

$$H(K|C) = H(P) + H(K) - H(C).$$

The Theorem holds.

**Theorem 4.6** *Let* $\mathfrak{R} = \{P, C, K, E, D\}$ *be a cryptosystem, and* $|C| = |P|$, *let* $r$ *be the redundancy of* $P$, *then the pseudo key mathematical expectation* $S_n$ *of a cryptosystemtext string with a given length of* $n$ *satisfies*

$$S_n \geq \frac{2^{H(K)}}{|P|^{nr}} - 1. \tag{4.21}$$

***Proof*** From the definition and properties of product space,

$$\mathfrak{R}_n = \{P^n, C^n, K, E_n, D_n\}$$

also constitutes a cryptosystem. By Theorem 4.5, then

$$H(K|C^n) = H(K) + H(P^n) - H(C^n).$$

By (3.9), we have

$$H(C^n) \leq n \log_2 |C|, \ |C| = |P|.$$

Replace information space $X$ with $P$, then we have

$$\begin{aligned} H(P^n) &= H(P^{n-1}) + H(P|P^{n-1}) \\ &= H(P^{n-1}) + H_n \\ &\geq H(P^{n-1}) + H_\infty. \end{aligned}$$

So we have

$$H(P^n) \geq nH_\infty = n(1-r)H_0 = n(1-r)\log_2 |P|. \tag{4.22}$$

Combined with the above formula, we have an estimate

$$H(K|C^n) \geq H(K) + n(1-r)\log_2 |P| - n \log_2 |P|. \tag{4.23}$$

Because of the definition,

$$\begin{aligned} H(K|C^n) &= -\sum_{y\in C^n}\sum_{k\in K} p(ky)\log_2 p(k|y) \\ &= -\sum_{y\in C^n} p(y)\sum_{k\in K} p(k|y)\log_2 p(k|y) \\ &= -\sum_{y\in C^n} p(y)\sum_{k\in K(y)} p(k|y)\log_2 p(k|y). \end{aligned}$$

We get

$$\sum_{k\in K(y)} p(k|y) = \sum_{k\in K} p(k) = 1.$$

Then by Jensen inequality,

$$
\begin{aligned}
H(K|C^n) &\le \sum_{y \in C^n} p(y) \log_2 |k(y)| \\
&\le \log_2 \sum_{y \in C^n} p(y)|k(y)| \\
&= \log_2(S_n + 1).
\end{aligned}
$$

Finally, (4.21) can be obtained from form (4.23) to complete the proof!

When the mathematical expectation of the number of pseudo keys is greater than 0, the secret only attack cannot break the password in theory, so we define the unique solution distance of a cryptosystem as the value of $n$ of $S_n = 0$.

**Definition 4.5** A cryptosystem whose unique solution distance is infinite is called an ideal security system.

From Theorem 4.6, we can obtain an approximate value of the distance of the unique solution.

$$
n_0 \approx \frac{H(k)}{r \log_2 |P|}.
$$

The unique solution distance indicates the minimum amount of cryptosystemtext that may be decrypted successfully when an exhaustive attack is carried out. Generally speaking, the greater the unique solution distance, the better the confidentiality of the system. However, Shannon only gives the existence of the unique solution distance, but does not give a specific calculation program. In practice, the amount of cryptosystemtext required to decryptosystem a cryptosystemtext is far greater than the theoretical value of the unique solution distance.

## 4.4 Message Authentication

Authentication system, also known as authentication code, is an important tool to ensure the authenticity and integrity of messages. In 1984, Simmons systematically put forward the information theory of authentication system for the first time. He used mathematics to study the theoretical and practical security of authentication system. This paper puts forward the performance limit of authentication system and the mathematical principles that should be followed in the design of authentication code. Although Simmons' theory is not mature and perfect, its position in authentication system is as important as Shannon's theory in cryptosystem, which lays a theoretical foundation for the research of mathematical theory of authentication system.

In cryptography, authentication system includes entity authentication and message authentication. We mainly discuss message authentication system. At present, there are two main models of authentication system. One is the arbiter-free authentication system model. In this model, the participants of the system are mainly message sender, message receiver and attacker, in which the message sender and receiver trust each other. They share the same key information; another model is the authentication system model with arbiter. In this model, the participants of the system have arbiters in addition to the information sender, receiver and attacker. At this time, the sender and receiver of the message do not trust each other, but they all trust the arbiter. The arbiter shares the key information with the sender and receiver.

An authentication system without privacy and confidentiality function and without arbiter is composed of four parts: a finite set $S$ of source states, called the source set, a finite set $A$ of authentication tags, called the tag set, a key space composed of all solvable keys, and an authentication rule set $E = \{e_k(s)|k \in K, s \in S\}$, where for any $k \in K$, $s \in S$, $e_k(s)$ is the authentication rule. It is a mapping of $S \to A$.

**Definition 4.6** An authentication system is $T = \{S, A, K, E\}$, where $S, A, K$ is the information space, $S$ is the source space or source set, $A$ is the label space or label set, and $K$ is the key space, where $S$ and $K$ are statistically independent,

$$E = \{e_k(s)|k \in K, s \in S\}.$$

Each $e_k(s)$ is an injection of $S \to A$, which is called an authentication rule.

**Definition 4.7** The product space $SA$ is called the message space, and $M$ represents $SA$.

Authentication protocol: The sender and receiver of the message use the following protocol to transmit information. First, they secretly select and share the random key $k \in K$; if the sender wants to transmit an information source state $s \in S$ to the receiver, the sender calculates $a = e_k(s)$ and sends the message $sa \in M$ to the receiver. When the receiver receives message $sa$, he calculates $a' = e_k(s)$ again, if $a' = a$, he confirms that the message is reliable and receives the message, otherwise he refuses to receive the message $sa$.

**Definition 4.8** Matrix $[e_k(s)]_{|K|\times|S|}$ is called authentication matrix. Its rows are marked by key $k \in K$ and columns by source state $s \in S$. It is a $|K| \times |S|$-order matrix, the element intersecting row $k$ and column $s$ is $e_k(s)$.

Authentication matrix is an important tool in authentication theory research. Our detailed list is as follows:
Let $K = \{k_1, k_2, \ldots, k_n\}$, $S = \{s_1, s_2, \ldots, s_m\}$. Then the authentication matrix is an $n \times m$-order matrix, which is listed as follows:

$$\begin{bmatrix} e_{k_1}(s_1) & e_{k_1}(s_2) & \cdots & e_{k_1}(s_m) \\ e_{k_2}(s_1) & e_{k_2}(s_2) & \cdots & e_{k_2}(s_m) \\ \vdots & & & \\ e_{k_n}(s_1) & e_{k_n}(s_2) & \cdots & e_{k_n}(s_m) \end{bmatrix}_{n\times m}.$$

## 4.5 Forgery Attack

In the process of message authentication, the attacker is an intermediate intruder. We usually consider two types of attacks, one is forgery attack and the other is substitution attack, which correspond to secret only attack and plaintext attack in cryptosystem. In forgery attack, the attacker sends message $sa \in M$ in the channel and wants the receiver to confirm that it is true and receive it; in the substitution attack, the attacker first observes a message $sa \in M$ in the channel, so he analyzes the coding rules currently used, then he tampers the message $sa$ with $s'a' \in M$, where $s' \neq s$, and wants the receiver to receive it as a real message.

We assume that the attacker adopts the optimal deception strategy. $p_{d_0}$ represents the probability that the forgery attacker is most likely to succeed in deception, and $p_{d_1}$ represents the probability that the attacker is most likely to succeed in deception. The probability $p_d$ that the attacker is successful in deception is defined as

$$p_d = \max\{p_{d_0}, p_{d_1}\}. \tag{4.24}$$

Simmons' theory is mainly to estimate the lower bound of $p_d$, so as to provide a theoretical basis for constructing authentication codes with attack success probability $p_d$ as small as possible.

First, let's look at the definition and estimation of the maximum probability $p_{d_0}$ of successful deception by forgery attackers.

$A = \{a_1, a_2, \ldots, a_r\}$ represents the authentication tag space. The attacker first selects a source state $s \in S$ and an authentication tag $a \in A$. Let $k_0 \in K$ represent the shared key selected by the sender and receiver, if $a = e_{k_0}(s)$, the forgery attacker can successfully deceive the receiver. We use pay off $(s, a)$ to represent the probability that the message receiver receives $sa$ as a true message, that is

$$\text{pay off}\,(s, a) = p(a = e_{k_0}(s)) = \sum_{\substack{k \in K \\ e_k(s)=a}} p(k). \tag{4.25}$$

If the attacker adopts the optimal strategy, then

$$p_{d_0} = \max\{\text{pay off}\,(s, a) | s \in S, a \in A\}. \tag{4.26}$$

**Theorem 4.7** *If the scale of authentication tag space A is set to $|A| = r$, for any fixed source state $s \in S$, there will always be an authentication tag $a \in A$ such that*

$$\textit{pay off}\,(s, a) \geq \frac{1}{r}, \textit{ thus } p_{d_0} \geq \frac{1}{r}.$$

***Proof*** By the definition of pay off $(s, a)$,

$$\sum_{a\in A} \text{pay off}\,(s,a) = \sum_{a\in A} \sum_{\substack{k\in K\\ e_k(s)=a}} p(k).$$

When $a$ runs through the $s$ column of the authentication matrix, $k$ traverses the whole key space, so

$$\sum_{a\in A} \text{pay off}\,(s,a) = \sum_{k\in K} p(k) = 1.$$

Therefore, there is at least one $a \in A$ such that

$$\text{pay off}\,(s,a) \geq \frac{1}{|A|} = \frac{1}{r}.$$

**Theorem 4.8** *Let $T = \{S, A, K, E\}$ be a message authentication system,*

$$p_{d_0} = \max\{\textit{pay off}\,(s,a) | s \in S, a \in A\}$$

*is the maximum probability of successful forgery attack, then*

$$\log_2 p_{d_0} \geq H(K|SA) - H(K)$$

*and*

$$p_{d_0} \geq \frac{1}{2^{H(K)-H(K|SA)}}.$$

***Proof*** By definition, we know that $p_{d_0}$ is not less than the mathematical expectation of pay off $(s,a)$, that is

$$p_{d_0} \geq \sum_{s\in S, a\in A} p(sa)\text{pay off}\,(s,a).$$

Then by Jensen inequality, we have

$$\begin{aligned}\log_2 p_{d_0} &\geq \log_2 \sum_{s\in S, a\in A} p(sa)\text{pay off}\,(s,a)\\ &\geq \sum_{s\in S, a\in A} p(sa) \log_2 \text{pay off}\,(s,a).\end{aligned}$$

Obviously,

$$\text{pay off}\,(s,a) = p(a|s).$$

Thus

$$p(sa) = p(s)p(a|s) = p(s)\text{pay off}\,(s,a). \tag{4.27}$$

So

$$\begin{aligned}\log_2 p_{d_0} &\geq \sum_{s\in S}\sum_{a\in A} p(sa)\log_2 \text{pay off }(s,a)\\ &= \sum_{s\in S}\sum_{a\in A} p(sa)\log_2 p(a|s)\\ &= -H(A|S).\end{aligned}$$

Because the source space $S$ and the key space $K$ are statistically independent, so

$$H(SK) = H(K) + H(S).$$

Also, the tag space $A$ is completely determined by the source space $S$ and the key space $K$, so

$$H(A|KS) = 0.$$

By the addition formula of information space,

$$\begin{aligned}H(KAS) &= H(AS) + H(K|AS)\\ &= H(S) + H(A|S) + H(K|AS).\end{aligned}$$

On the other hand,

$$\begin{aligned}H(KAS) &= H(KS) + H(A|KS)\\ &= H(KS) = H(K) + H(S).\end{aligned}$$

On the whole, we have

$$-H(A|S) = H(K|AS) - H(K).$$

Thus

$$\log_2 p_{d_0} \geq -H(A|S) = H(K|AS) - H(K).$$

We completed the proof of the theorem.

$M = SA$ is called message space, it can be seen from theorem 4.8 that the maximum success probability $p_{d_0}$ of forgery attack satisfies

$$p_{d_0} \geq \frac{1}{2^{I(K,M)}},$$

where $I(K, M)$ is the average amount of mutual information between the key space and the information space. If the amount of mutual information $I(K, M)$ is larger, the probability of the most successful forgery attack is lower. On the contrary, if the amount of mutual information is smaller, the success rate of forgery attack is higher.

## 4.6 Substitute Attack

The so-called substitution attack is that the attacker first observes a message $(s, a)$ on the message, and then replaces $(s, a)$ with message $(s', a')$, hoping that the receiver will receive $(s', a')$ as a real message. Considering the maximum success probability $p_{d_1}$ of substitution attack, it is more difficult than forgery attack, the main reason is that $p_{d_1}$ depends on both the probability distribution of source state space $S$ and the probability distribution of key space $K$.

Let $(s', a')$ and $(s, a)$ be two messages, where $s \neq s'$. We use pay off $(s', a', s, a)$ to express the probability that using $(s', a')$ instead of $(s, a)$ can cheat success, then

$$\text{pay off } (s', a', s, a) = p(a' = e_{k_0}(s')|a = e_{k_0}(s)), \ \ k_0 \in K.$$

The above formula represents the conditional probability of $a' = e_{k_0}(s')$ under the condition of $a = e_{k_0}(s)$ under the same key $k_0$, so

$$\begin{aligned} \text{pay off } (s', a', s, a) &= \frac{p(a' = e_{k_0}(s'), a = e_{k_0}(s))}{p(a = e_{k_0}(s))} \\ &= \frac{\sum\limits_{\substack{k\in K, \\ e_{k_0}(s')=a', e_{k_0}(s)=a}} p(k)}{\text{pay off } (s, a)}. \end{aligned} \tag{4.28}$$

When the message $(s, a) \in M$ is given, the attacker uses the optimal strategy to maximize the success probability of the deceiver, so let

$$p_{s,a} = \max\{\text{pay off } (s', a', s, a)|s' \in S, s' \neq s, a' \in A\}, \tag{4.29}$$

Taking $p_{s,a}$ as a random variable, its mathematical expectation on message set $M = SA$ is

$$p_{d_1} \geq \sum_{s\in S, a\in A} p(sa)p_{s,a}. \tag{4.30}$$

The above formula is the formal definition of $p_{d_1}$, which is the weighted average of the maximum success probability of pay off $(s', a', s, a)$ in message space $M$.

Like Theorem 4.7, we have

**Theorem 4.9** *Let $T = \{S, A, K, E\}$ be an authentication code, $|A| = r$, then for any given $s' \in S$, $s \in S$, $s \neq s'$ and $a \in A$, there is a label $a' \in A$ such that*

$$\textit{pay off } (s', a', s, a) \geq \frac{1}{|A|} = \frac{1}{r}.$$

*So we have*

$$p_{d_1} \geq \frac{1}{r}.$$

***Proof*** By (4.28),

$$\sum_{a' \in A} \text{pay off}\,(s', a', s, a) = \frac{1}{\text{pay off}\,(s, a)} \sum_{a' \in A} \sum_{\substack{k \in K \\ e_k(s)=a, e_k(s')=a'}} p(k)$$

$$= \frac{1}{\text{pay off}\,(s, a)} \sum_{\substack{k \in K \\ e_k(s)=a}} p(k) = 1.$$

So at least one $a' \in A$ such that

$$\text{pay off}\,(s', a', s, a) \geq \frac{1}{|A|} = \frac{1}{r}.$$

By the definition of $p_{s,a}$, for $\forall\, s \in S$ and $a \in A$, we have

$$p_{s,a} \geq \frac{1}{|A|} = \frac{1}{r}.$$

Thus

$$p_{d_1} \geq \sum_{s \in S, a \in A} p(sa) p_{s,a} \geq \frac{1}{r} \sum_{a \in A} p(a) = \frac{1}{r}.$$

**Theorem 4.10** *Let $T = \{S, A, K, E\}$ be an authentication code, for any $(s, a) \in M$, when using $(s', a')$ instead of attack, let $p_{d_1}$ be the mathematical expectation of $p_{s,a}$ in space $M$, then*

$$\log_2 p_{d_1} \geq H(K|M^2) - H(K|M)$$

*and*

$$p_{d_1} \geq \frac{1}{2^{H(K|M)-H(K|M^2)}}. \tag{4.31}$$

***Proof*** By (4.29), $p_{s,a}$ will not be less than the mathematical expectation of pay off $(s', a', s, a)$ on $s' \in S$, $a' \in A$, that is

$$p_{s,a} \geq \sum_{s' \in S, a' \in A} p(s'a'|sa)\text{pay off}\,(s', a', s, a).$$

By (4.30) and Jensen inequality, we have

$$\begin{aligned}\log_2 p_{d_1} &\geq \sum_{s\in S,a\in A} p(sa)\log_2 p_{s,a}\\ &\geq \sum_{s\in S,a\in A} p(sa) \sum_{s'\in S,a'\in A} p(s'a'|sa)\log_2 \text{pay off }(s',a',s,a)\\ &= \sum_{s\in S,a\in A}\sum_{s'\in S,a'\in A} p(sas'a')\log_2 \text{pay off }(s',a',s,a)\\ &= \sum_{s\in S,a\in A}\sum_{s'\in S,a'\in A} p(sas'a')\log_2 p(a's'|as)\\ &= -H(M|M).\end{aligned}$$

In addition,

$$\begin{aligned}H(KM^2) &= H(M|M) + H(K|M^2)\\ &= H(K|M) + H(M|KM).\end{aligned}$$

So there are

$$-H(M|M) = H(K|M^2) - H(K|M) - H(M|KM).$$

It can be proved that

$$H(M|KM) = 0.$$

So there are

$$-H(M|M) = H(K|M^2) - H(K|M).$$

Thus

$$\log_2 p_{d_1} \geq H(K|M^2) - H(K|M).$$

That is

$$p_{d_1} \geq \frac{1}{2^{H(K|M)-H(K|M^2)}}.$$

The Theorem holds!

**Definition 4.9** An authentication code $\{S, A, K, E\}$ is called perfect if

$$p_d = 2^{H(K|M)-H(K)}.$$

**Theorem 4.11** *Perfect certification system exists.*

***Proof*** The theorem is proved directly by the construction method. First, let the source state space be $S = \{0, 1\}$. Let $N$ be a positive even number, and define the label space $A$ and the key space $K$ as follows:

$$A = \mathbb{Z}_2^{\frac{N}{2}} = \{a_1a_2\cdots a_{\frac{N}{2}} | a_i \in \mathbb{Z}_2, 1 \leq i \leq \frac{N}{2}\}$$

and

$$K = \mathbb{Z}_2^N = \{k_1k_2\cdots k_N | k_i \in \mathbb{Z}_2, 1 \le i \le N\}.$$

The authentication rule $e_k(s)$ determined by $k = k_1k_2\cdots k_{\frac{N}{2}}k_{\frac{N}{2}+1}\cdots k_N$ is defined as

$$e_k(0) = k_1k_2\cdots k_{\frac{N}{2}}$$

and

$$e_k(1) = k_{\frac{N}{2}+1}\cdots k_N.$$

Assuming that all $2^N$ keys $k$ are equitably selected, so for $s \in S$ and $a \in A$, we have

$$\text{pay off }(s, a) = p(a = e_k(s)) = 2^{-\frac{N}{2}}.$$

So there is $p_{d_0} = 2^{-\frac{N}{2}}$, similarly to $p_{d_1} = 2^{-\frac{N}{2}}$, so

$$p_d = 2^{-\frac{N}{2}}.$$

Easy to calculate

$$H(K|M) - H(K) = \frac{N}{2} - N = -H(K|M).$$

So

$$p_d = 2^{H(K|M)-H(K)}.$$

Therefore, $\{S, A, K, E\}$ is a perfect authentication system.

## 4.7 Basic Algorithm

### 4.7.1 Affine Transformation

Encryption with matrix comes from the classical *Vigenère* password. Let $X = \{a_1, a_2, \ldots, a_N\}$ be a plaintext alphabet of $N$ characters, we replace the characters in $\mathbb{Z}_N$ and $X$ with numerical values, where $\mathbb{Z}_N$ is the remaining class ring of mod $N$. Let $P = \mathbb{Z}_N^k$ be the plaintext space, $x = x_1x_2\cdots x_k \in P$ is called a plaintext unit or a plaintext message of length $k$. Let $M_k(\mathbb{Z}_N)$ be a $k$-order full matrix ring over $\mathbb{Z}_N$, $A \in M_k(\mathbb{Z}_N)$ is a invertible matrix of order $k$, $b = b_1b_2\cdots b_k \in \mathbb{Z}_N^k$ is a given directional quantity, each plaintext unit $x = x_1x_2\cdots x_k$ in $P$ is encrypted by affine transformation $(A, b)$:

$$\begin{pmatrix} x_1' \\ x_2' \\ \vdots \\ x_k' \end{pmatrix} = A \begin{pmatrix} x_1 \\ x_2 \\ \vdots \\ x_k \end{pmatrix} + \begin{pmatrix} b_1 \\ b_2 \\ \vdots \\ b_k \end{pmatrix}. \tag{4.32}$$

where $x = x_1 x_2 \cdots x_k$ is clear text, $x' = x_1' x_2' \cdots x_k'$ is cryptosystemtext. The decryption algorithm is:

$$\begin{pmatrix} x_1 \\ x_2 \\ \vdots \\ x_k \end{pmatrix} = A^{-1} \begin{pmatrix} x_1' \\ x_2' \\ \vdots \\ x_k' \end{pmatrix} - A^{-1} \begin{pmatrix} b_1 \\ b_2 \\ \vdots \\ b_k \end{pmatrix}. \tag{4.33}$$

Because affine transformation $(A, b)$ is a 1–1 correspondence on $\mathbb{Z}_N^k \longrightarrow \mathbb{Z}_N^k$, its inverse transformation is $(A^{-1}, -A^{-1}b)$; therefore, using affine transformation $(A, b)$, we obtain the so-called high-order affine cryptosystem. This cryptosystem was first proposed by mathematician Lester hill in American Mathematics monthly in 1929, so it is also called Hill cryptosystem.

Hill cryptosystem divides the plaintext into a group of $k$ characters and then encrypts each plaintext unit in turn by using $k$-order affine transformation $(A, b)$ on $\mathbb{Z}_N$. The advantage of this password is that it hides the statistical characteristics of a single character (such as 26 letters in English), which can better resist the statistical analysis of the occurrence frequency of characters, and has strong ability to resist cryptosystemtext only attacks. However, on the basis of mastering a large amount of plaintext, it is not difficult to find the key $(A, b)$, so the hill password is not strong against the attack of known plaintext.

The mathematical principles used by Hill cryptosystem are the following two conclusions.

**Lemma 4.1** *The set of all k-order affine transformations on $\mathbb{Z}_N$ is written as $G_k$, that is*

$$G_k = \{(A, b) | A \textit{ is a } k\textit{-order reversible square matrix}, b \in \mathbb{Z}_N^k\}.$$

*Then $G_k$ forms a group under the multiplication of transformation, which is called the k-order affine transformation group of ring $\mathbb{Z}_N$.*

***Proof*** Take $A$ as the $k$-order identity matrix $E$ and $b = 0$ as the $k$-dimensional zero vector, then $(E, 0)$ is the identity transformation of $\mathbb{Z}_N^k \longrightarrow \mathbb{Z}_N^k$ and the unit element of $G_k$. Secondly, we look at the product of two affine transformations $(A_1, b_1)$ and $(A_2, b_2)$,

$$(A_1, b_1)(A_2, b_2) = (A_1 A_2, A_1 b_2 + b_1) \in G_k.$$

Obviously, the inverse transformation of $(A, b)$ is

$$(A, b)^{-1} = (A^{-1}, -A^{-1}b) \in G_k.$$

Therefore, $G_k$ is a group. The Lemma holds.

From the above lemma, any group element $(A, b) \in G_k$ of affine transformation group will form a Hill cryptosystem. If we select $n$ group elements $(A_1, b_1)$, $(A_2, b_2)$, ..., $(A_n, b_n)$ in $G_k$ and let

$$(A, b) = \prod_{i=1}^{n}(A_i, b_i).$$

Using $(A, b)$ to encrypt, we get a more complex Hill cryptosystem.

**Lemma 4.2** *$A \in M_k(\mathbb{Z}_N)$, $|A| = D$ is the determinant of $A$, then $A$ is reversible if and only if $D$ and $N$ are coprime, that is $(D, N) = 1$.*

***Proof*** If $(D, N) = 1$, then there is $D_1$ such that $D_1 D \equiv 1 (\bmod N)$, let

$$A^* = D_1 \begin{bmatrix} A_{11} & A_{12} & \cdots & A_{1k} \\ A_{21} & A_{22} & \cdots & A_{2k} \\ \cdots & \cdots & \cdots & \cdots \\ A_{k1} & A_{k2} & \cdots & A_{kk} \end{bmatrix}, \tag{4.34}$$

where $A = (a_{ij})_{k \times k}$, $A_{ij}$ is the algebraic cofactor of $a_{ij}$. obviously, we have

$$A^*A = AA^* = \begin{bmatrix} 1 & 0 & \cdots & 0 \\ 0 & 1 & \cdots & 0 \\ \vdots & \vdots & & \vdots \\ 0 & 0 & \cdots & 1 \end{bmatrix},$$

So $A$ is reversible, $A^{-1} = A^*$. Let's take $k = 2$ as an example, if $|A| = D$, $(D, N) = 1$,

$$A = \begin{pmatrix} a & b \\ c & d \end{pmatrix} \Longrightarrow A^{-1} = \begin{pmatrix} D_1 d & -D_1 b \\ -D_1 c & D_1 a \end{pmatrix}.$$

Conversely, if $A$ is reversible and $A^{-1}$ is the inverse matrix, because $A^{-1}A = AA^{-1} = E$, we get

$$|AA^{-1}| = |A||A^{-1}| \equiv 1 (\bmod N).$$

So we have $(D, N) = 1$. The Lemma holds.

If $k = 1$, first-order affine cryptosystem $x' \equiv ax + b (\bmod N)$, where $(a, N) = 1$, contains many famous classical passwords, especially when $a = 1$, $b = 3$, $N = 26$, $x' = x + 3 (\bmod 26)$ is the famous Caesar code in history.

Next, we analyze the computational complexity of affine cryptography. We have the following Lemma.

**Lemma 4.3** *If $A = (a_{ij})_{k\times k}$ is a k-order reversible square matrix on $\mathbb{Z}_N$, the bit operation times of $A^{-1}$ are estimated as follows*

$$Time(A^{-1}) = O(k^4 k! \log^3 N).$$

*Therefore, when k is a fixed constant, the algorithm for finding $A^{-1}$ is polynomial. When N is a fixed constant, the algorithm for finding $A^{-1}$ is exponential. In other words, the greater the order of the matrix, the higher the computational complexity.*

***Proof*** Because $A = (a_{ij})_{k\times k}$ is reversible, then determinant

$$D = |A| = \sum_{j_1 j_2 \cdots j_k} (-1)^{\tau}(j_1 j_2 \cdots j_k) a_{1j_1} a_{2j_2} \cdots a_{kj_k},$$

where $j_1 j_2 \cdots j_k$ is an arrangement of $1, 2, \ldots, k$ and $\tau(j_1 j_2 \cdots j_k)$ is the reverse order number of the arrangement. The number of bit operations of each summation is $O(k^3 \log^2 N)$, and there are $k!$ summation terms in total, thus

$$\text{Time}(D) = O(k^3 k! \log^2 N).$$

By Lemma 1.5 of Chapter 1, find the multiplicative inverse of $D$ under mod $N$, $D^{-1} \bmod N = D_1$ is

$$\text{Time}(D_1) = O(\log^3 N).$$

The bit operation times of each algebraic cofactor $A_{ij}$ of the adjoint matrix $A^*$ of formula (4.34) is $O((k-1)^3(k-1)! \log^2 N)$, and there are $k^2$ algebraic cofactors, thus

$$\text{Time}(A^*) = O(k^4 k! \log^2 N).$$

So

$$\text{Time}(A^{-1}) = O(k^4 k! \log^3 N).$$

When $k$ is constant, the algorithm for finding $A^{-1}$ is polynomial. When $N$ is constant and $k \longrightarrow \infty$, it is obvious that the algorithm for finding $A^{-1}$ is exponential. The Lemma holds.

### *4.7.2 RSA*

In 1976, two mathematicians from Stanford University, Diffie and Hellman, put forward a new idea of cryptosystem design. In short, the encryption algorithm and decryption algorithm are designed based on the principle of asymmetry. We can use the following schematic diagram to illustrate

$$P \xrightarrow{f} C \xrightarrow{f^{-1}} P, \tag{4.35}$$

where $P$ is the plaintext space, $C$ is the cryptosystemtext space, $f$ encryption algorithm and $f^{-1}$ decryption algorithm. If $f$ and $f^{-1}$ are the same algorithm, such as the involution operation in binary system, or the encryption algorithm $f$ can easily deduce the decryption algorithm $f^{-1}$. For example, the matrix encryption algorithm mentioned in the previous section (the matrix order is very small), which is called symmetric cryptosystem. The essence of symmetric cryptosystem is that the encryption key and decryption key have the same confidentiality importance. Diffie and Hellman proposed that if $f \neq f^{-1}$ and $f$ are encryption algorithms that are easy to implement, while $f^{-1}$ is a decryption algorithm that is very difficult to calculate, the key can be divided into encryption key and decryption key. Even if the encryption key is made public to the public, the security of decryption key will not be affected. This encryption algorithm $f$ is called asymmetric or trapdoor one-way function. The password using asymmetric $f$ is called asymmetric password or public key cryptosystem. Due to the bold innovation of Diffie and Hellman, cryptography has ushered in a new era—the era of public key cryptography. Its basic feature is that passwords change from few users to many users, which greatly improves the efficiency and social value of passwords.

How to design asymmetric encryption algorithm? Rivest, Shamir and Adleman jointly put forward the first secure and practical one-way encryption algorithm, which is called RSA algorithm in academic circles. This public key cryptosystem has been widely used in cryptographic design and has become an international standard algorithm. In addition to its simplicity and practicality, its security completely depends on the difficulty of large prime factorization of huge integers.

Let $p, q$ be two large and relatively safe prime numbers, assume

$$10^{300} < p, q, \quad \text{its binary digits } k > 1024\text{bits}. \tag{4.36}$$

Let $n = pq$, $\varphi(n)$ be an Euler function, then

$$\varphi(n) = (p-1)(q-1) = n + 1 - p - q.$$

Randomly select a positive integer $e$ to satisfy

$$1 < e < \varphi(n), (e, \varphi(n)) = 1. \tag{4.37}$$

The large prime numbers $p$ and $q$ and $e$ satisfying formula (4.37) are randomly generated. The so-called random generation is to randomly select the $p, q$ and $e$ with the help of the computer random number generator (or pseudo-random number generator), and its computational complexity is

**Lemma 4.4** *Randomly generated large prime number $p$ and $q$, $n = pq$, $\varphi(n)$ is Euler function, $1 < e < \varphi(n)$, $(e, \varphi(n)) = 1$, then*

$$\begin{cases} \textit{Time (select out n)} = O(\log^4 n), \\ \textit{Time (find e)} = O(\log^2 n). \end{cases}$$

***Proof*** Use the random number generator to generate a huge integer $m$, such as $m > 10^{300}$, and then detect whether $m, m+1, m+2, \ldots,$ is a prime number. From the prime number theorem, we know that the frequency of prime numbers adjacent to $m$ is about $O(\frac{1}{\log m})$, so we only need about $O(\log m)$ tests to find the required prime number $p$, by Lemma 1.5 of Chapter 1,

$$\text{Time (find prime } p) = O(\log^2 m) = O(\log^2 n).$$

Similarly,

$$\text{Time (find prime q)} = O(\log^2 n).$$

Because $n = pq$, so

$$\text{Time (select out n)} = O(\log^4 n).$$

$n$ after confirmation, $\varphi(n) = (p-1)(q-1)$. A positive integer $a$, $1 < a < \varphi(n)$, is randomly generated by the random number generator, and then whether $a, a+1, a+2, \ldots$ and $\varphi(n)$ are mutually prime is detected in turn. Again, according to the prime number theorem, the frequency of the prime factor of $\varphi(n)$ appearing in the vicinity of $a$ is $O(\frac{1}{\log a})$, so we only need $O(\log a)$ tests to get the required $e$. Thus

$$\text{Time (select out e)} = O(\log^2 a) = O(\log^2 n).$$

The Lemma holds.

After randomly selecting $p, q, n = pq$, and $e$, because $(e, \varphi(n)) = 1$, then exist $d = e^{-1} \bmod \varphi(n)$, that is

$$de \equiv 1(\bmod\, \varphi(n)), 1 < d < \varphi(n). \tag{4.38}$$

**Definition 4.10** After randomly determining $n = pq$, let $P_e = (n, e)$ be called public key, $P_d = (n, d)$ be called private key, or $e$ be public key and $d$ be private key.

By Lemma 1.5 of chapter 1, calculate the number $\text{Time}(d) = O(\log^3 \varphi(n)) = O(\log^3 n)$ of bit operations required for $d = e^{-1} \bmod \varphi(n)$. By Lemma 4.4, we have

**Corollary 4.3** *The computational complexity of randomly generated public key $P_e = (n, e)$ and private key $P_d = (n, d)$ is polynomial.*

The key mathematical principle used in RSA cryptographic design is the generalized Euler congruence theorem. $n \geq 1$ is a positive integer, $(m, n) = 1$, from Euler theorem, it can be seen that

$$m^{\varphi(n)} \equiv 1(\bmod\, n), \Longrightarrow m^{\varphi(n)+1} \equiv m(\bmod\, n). \tag{4.39}$$

We will prove that under the condition that $n$ is a positive integer without square factor, there is formula (4.39) for all positive integers $m$, whether $(m, n) = 1$ or $(m, n) > 1$.

**Lemma 4.5** *If $n = pq$ is the product of two different prime numbers, then for all positive integers $m$, $k$, there are*

$$m^{k\varphi(n)+1} \equiv m(\bmod n). \tag{4.40}$$

***Proof*** If $(m, n) = 1$, then by Euler Theorem,

$$m^{k\varphi(n)} \equiv 1(\bmod n), \Longrightarrow m^{k\varphi(n)+1} \equiv m(\bmod n).$$

We only consider the case of $(m, n) > 1$, because $n = pq$, so $(m, n) = p$, $(m, n) = q$, or $(m, n) = n$. If $(m, n) = n$, then(4.40) holds. Might as well let $(m, n) = p$, then $m = pt$, where $1 \leq t < q$. By Euler theorem, because $(m, q) = 1$, so

$$m^{\varphi(q)} \equiv 1(\bmod q), \Longrightarrow m^{k\varphi(q)\varphi(p)} \equiv 1(\bmod q).$$

For $\forall\, k \geq 1$, there is

$$m^{k\varphi(n)} \equiv 1(\bmod q).$$

We write

$$m^{k\varphi(n)} = rq + 1.$$

Both sides are multiplied by $m$,

$$m^{k\varphi(n)+1} = rtn + m.$$

The above formula contains

$$m^{k\varphi(n)+1} \equiv m(\bmod n).$$

We have completed the proof of lemma.

With the above preparations, the workflow of RSA password can be divided into the following three steps:

(1) Suppose $A$ is a user of RSA, and $A$ randomly generates two huge prime numbers $p = p(A)$, $q = q(A)$, $n = n(A)$, where $n = pq$, $\varphi(n) = (p - 1)(q - 1)$. Then randomly generate positive integers $e = e(A)$, satisfies $1 < e < \varphi(n)$, $(e, \varphi(n)) = 1$, calculated $d \equiv e^{-1}(\bmod \varphi(n))$, and $1 < d < \varphi(n)$. User $A$ destroys two prime numbers $p$ and $q$, and only keeps three numbers $n, e, d$, after publishing $P_e = (n, e)$ as public key, he has private key $P_d = (n, d)$ and keeps it strictly confidential.

(2) User $B$ of another RSA sends encrypted information to user $A$ using the known public key $(n, e)$ of user $A$. $B$ selects $P = \mathbb{Z}_n$ as the plaintext space and encrypts each $m \in \mathbb{Z}_n$. The encryption algorithm $c = f(m)$ is defined as

$$c = f(m) \equiv m^e (\bmod n), \quad 1 \leq c \leq n. \tag{4.41}$$

where $c$ is cryptosystemtext.

(3) After receiving the cryptosystemtext $c$ sent by user $B$, user $A$ decrypts it with its own private key $(n, d)$. Decryption algorithm $f^{-1}$ is defined as:

$$m = f^{-1}(c) \equiv c^d (\bmod n), \quad 1 \leq m \leq n. \tag{4.42}$$

User $A$ gets the plaintext $m$ sent by user $B$. so far, RSA cryptosystem completes encryption and decryption.

The correctness and uniqueness of RSA password are guaranteed by the following Lemma.

**Lemma 4.6** *The encryption algorithm $f$ defined by equation (4.41) is a 1–1 correspondence of $\mathbb{Z}_n \longrightarrow \mathbb{Z}_n$, and $f^{-1}$ defined by equation (4.42) is the inverse mapping of $f$.*

***Proof*** By Lemma 4.5, for all $m \in \mathbb{Z}_n$, $k$ is a positive integer, then there is

$$m^{k\varphi(n)+1} \equiv m(\bmod n).$$

Because of $ed \equiv 1(\bmod \varphi(n))$, we can write

$$ed = k\varphi(n) + 1.$$

By (4.41), then there is

$$c^d \equiv m^{ed} \equiv m^{k\varphi(n)+1} \equiv m(\bmod n).$$

That is to say, for all $m \in \mathbb{Z}_n$,

$$f^{-1}(f(m)) = m.$$

In the same way, we have

$$m^e \equiv c^{ed} \equiv c^{k\varphi(n)+1} \equiv c(\bmod n).$$

In other words,

$$f(f^{-1}(c)) = c.$$

By Lemma 1.1 of Chap. 1, $f$ is a 1–1 correspondence of $\mathbb{Z}_n \longrightarrow \mathbb{Z}_n$, and $ff^{-1} = 1$, $f^{-1}f = 1$. Th Lemma holds.

Another very important application of RSA is for digital signature. From the workflow of RSA password, it can be seen that the encryption algorithm defined in formula (4.41) is based on the public key $(n_A, e_A)$ of user $A$, and we denote $f$ as $f_A$ and the decryption algorithm defined in formula (4.42) as $f_A^{-1}$. The workflow of RSA digital signature is: User $A$ sends his digital signature to user $B$, that is, $A$ sends an encrypted message to $B$. Let $P_e(A) = (n_A, e_A)$ be the public key of $A$ and $P_d(A) = (n_A, d_A)$ the private key of $A$. Similarly, $P_e(B) = (n_B, e_B)$ is the public key of $B$ and $P_d(B) = (n_B, d_B)$ is the private key of $B$. Then the digital signature sent by user $A$ to user $B$ is

$$\begin{cases} f_B f_A^{-1}(m), & \text{if } n_A < n_B \\ f_A^{-1} f_B(m), & \text{if } n_A > n_B. \end{cases} \tag{4.43}$$

where $m \in \mathbb{Z}_{n_A}$ is the digital signature published by user $A$. After receiving the above digital signature of user $A$, user $B$ adopts the following two different digital verification according to the two cases of $n_A < n_B$ and $n_A > n_B$, formula (4.43) is the real signature of user $A$.

(i) If $n_A < n_B$, user $B$ first decrypts with his private key $f_B^{-1} = (n_B, d_B)$ and then decrypts with user $A$'s public key $f_A = (n_A, e_A)$, the verification is as follows

$$f_A f_B^{-1}(f_B f_A^{-1}(m)) = f_A f_A^{-1}(m) = m.$$

(ii) If $n_A > n_B$, user $B$ uses user $A$'s public key $f_A = (n_A, e_A)$ first, then decrypt and verify with your own private key $f_B^{-1} = (n_B, d_B)$

$$f_B^{-1} f_A(f_A^{-1} f_B(m)) = f_B^{-1} f_B(m) = m.$$

The security of RSA is the difficulty of large prime factorization based on $n$. When all users select the large prime numbers $p$ and $q$, let $n = pq$, then destroy $p$ and $q$, only $(n, e)$ and its own secret $(n, d)$ key information are retained, even if $(n, e)$ is published to the public, outsiders only know $n$ and do not know $\varphi(n)$, so they cannot obtain the information of private key $(n, d)$. Because the calculation of $\varphi(n)$ must rely on the prime factorization of $n$, from the product formula of Euler, it is not difficult to see

$$\varphi(n) = n \prod_{p|n} (1 - \frac{1}{p}).$$

Because we have very little knowledge of prime numbers, we have not found a general term formula to give an infinite number of prime numbers, so it is undoubtedly a difficult problem to judge whether a huge integer $n$ is prime, not to mention the prime factorization of $n$.

### 4.7.3 Discrete Logarithm

Let $G$ be a finite group and $b,\ y \in G$ be two group elements of $G$, let $t$ be the minimum positive integer satisfying $b^t = 1$, $t$ is called the order of $b$, denote as $t = o(b)$. If there is one $x$, $1 \le x \le o(b)$ such that $y = b^x$, $x$ is called the discrete logarithm of $y$ under base $b$. Known $b \in G$, $0 \le x \le o(b)$, it's easy to calculate $y = b^x$. Conversely, for any group element $y$, it is very difficult to find the discrete logarithm of $y$ under base $b$. Therefore, using discrete logarithm to encrypt has become the most mainstream encryption algorithm in public key cryptosystem, including the famous ElGamal cryptosystem and elliptic curve cryptosystem. ElGamal cryptosystem uses the discrete logarithm on the multiplication group formed by all $\mathbb{F}_q^*$ of nonzero elements in finite field $\mathbb{F}_q$. Elliptic curve cryptography uses the discrete logarithm algorithm of Mordell group on elliptic curve. Here we mainly discuss ElGamal cryptography, and elliptic curve cryptography is discussed in Chap. 6. We first prove several basic conclusions in finite field.

**Lemma 4.7** *Let $\mathbb{F}_q$ be a finite field of $q$ elements and $q = p^n$ be the power of prime $p$. $\mathbb{F}_q^* = \mathbb{F}_q \backslash \{0\}$ is all the nonzero elements in $\mathbb{F}_q$, then $\mathbb{F}_q^*$ is a cyclic group of order $(q-1)$ under multiplication, and the generating element $g$ of $\mathbb{F}_q^*$ is called the generator of finite field $\mathbb{F}_q$.*

***Proof*** According to Lagrange theorem, the number of zeros of polynomials in any field is not greater than the degree of polynomials. The finite field $\mathbb{F}_q^*$ is a finite group of order $(q-1)$ under multiplication. To prove that $\mathbb{F}_q^*$ is a cyclic group, it is only proved that for any factor $d$ of $q-1$, $d|q-1$, the number of solutions of equation $x^d = 1$ in $\mathbb{F}_q^*$ is not greater than $d$. This point can be deduced from Lagrange's theorem, because the number of zeros of polynomial $x^d - 1$ in the whole field $\mathbb{F}_q$ is not greater than $d$, so the number of zeros in $\mathbb{F}_q^*$ is not greater than $d$. So $\mathbb{F}_q^*$ is a finite cyclic group. The Lemma holds.

**Lemma 4.8** *Let $\mathbb{F}_q$ be a $q$-element finite field, $q = p^n$, $\mathbb{F}_p \subset \mathbb{F}_q$ is a subfield, $\mathbb{F}_p^* < \mathbb{F}_q^*$ is a subgroup of $\mathbb{F}_q^*$, if $g$ is the generator of $\mathbb{F}_q^*$, then $g' = g^{\frac{q-1}{p-1}}$ is the generator of $\mathbb{F}_p^*$.*

***Proof*** $g$ is the generator of $\mathbb{F}_q^*$, then $o(g) = q-1$. Let $g' = g^{\frac{q-1}{p-1}}$, then

$$o(g') = \frac{o(g)}{(q-1, \frac{q-1}{p-1})} = p - 1.$$

Thus $(g')^{p-1} = 1$, that is $(g')^p = g'$, so $g' \in \mathbb{F}_p$. Because $\mathbb{F}_p^*$ is a cyclic group of order $p-1$, and $o(g') = p-1$, so $\mathbb{F}_p^* = < g' >$, $g'$ is the generator of $\mathbb{F}_p^*$. The Lemma holds.

**Lemma 4.9** *Let $\mathbb{F}_q$ be a $q$-element finite field, $q = p^n$, for any $d|n$, let*

$$A_d = \{p(x) \in \mathbb{F}_p[x] | \deg p(x) = d,\ p(x) \text{ is an irreducible monic polynomial}\}$$

*and*

$$f_d(x) = \prod_{p(x) \in A_d} p(x).$$

*Then we have*

$$x^q - x = x^{p^n} - x = \prod_{d|n} f_d(x). \tag{4.44}$$

***Proof*** We know

$$x^{p^d} - x | x^{p^n} - x \Longleftrightarrow d|n.$$

Let $p(x) \in A_d$, that is $p(x) \in \mathbb{F}_p[x]$, $\deg p(x) = d$, $p(x)$ is an irreducible monic polynomial. Let $\alpha$ be a root of $p(x)$, then add a finite extension field of $\alpha$ on $\mathbb{F}_p$ and $\mathbb{F}_p(\alpha)$ is a $d$-th finite extension on $\mathbb{F}_p$. If $d|n$, then

$$\mathbb{F}_p(\alpha) = \mathbb{F}_{p^d} \subset \mathbb{F}_q,$$

so there is $\alpha \in \mathbb{F}_q$. Because the zeros of $p(x)$ are all in $\mathbb{F}_q$, so there is $p(x)|x^q - x$. Any $p(x)$ in $A_d$ has $p(x)|x^q - x$, so

$$f_d(x) = \prod_{p(x) \in A_d} p(x),\ f_d(x)|x^q - x.$$

Conversely, $p(x)$ is the first irreducible polynomial, and $\deg p(x) = d$. If $p(x)|x^q - x$, then the zeros of $p(x)$ are all in $\mathbb{F}_q$. Let $\alpha$ be a zero point of $p(x)$, then there is $\mathbb{F}_p(\alpha) \subset \mathbb{F}_q$, that is $\mathbb{F}_{p^d} \subset \mathbb{F}_q = \mathbb{F}_{p^n}$, so $d|n$. Finally,

$$x^q - x = \prod_{d|n} f_d(x).$$

The Lemma holds.

**Lemma 4.10** *$N_p(d)$ represents the number of the first irreducible polynomial with degree $d$ in $\mathbb{F}_p[x]$, then*

$$N_p(n) = \frac{1}{n} \sum_{d|n} \mu(d) p^{\frac{n}{d}}, \tag{4.45}$$

*where $\mu$ is Möbius function.*

***Proof*** By Lemma 4.9 and (4.44),

$$x^q - x = x^{p^n} - x = \prod_{d|n} f_d(x).$$

Comparing the degree of polynomials on both sides, there is

$$p^n = \sum_{d|n} dN_p(d).$$

By the Möbius inverse formula,

$$nN_p(n) = \sum_{d|n} \mu(d)p^{\frac{n}{d}},$$

so there is (4.45), the Lemma holds.

**Corollary 4.4** *If d is a prime number, the degree in $\mathbb{F}_p[x]$ is d and the number of the first irreducible polynomial is $\frac{1}{d}(p^d - p)$, that is*

$$N_p(d) = \frac{1}{d}(p^d - p), \text{ if } d \text{ is a prime number.}$$

***Proof*** By (4.45),

$$N_p(d) = \frac{1}{d}\sum_{\delta|d} \mu(\delta)p^{\frac{d}{\delta}}$$
$$= \frac{1}{d}(p^d - p).$$

The Corollary holds.

Based on the above basic conclusions about finite fields, we introduce two methods for solving discrete logarithms. The first is the Silver–Pohlig–Hellman smoothing method, and the second is the so-called exponential integration method.

Silver–Pohlig–Hellman

Let $\mathbb{F}_q$ be a $q$-element finite field, $b$ is the generator, that is $\mathbb{F}_q^* =< b >$,

$$o(b) = |\mathbb{F}_q^*| = q - 1 = p_1^{\alpha_1} p_2^{\alpha_2} \cdots p_s^{\alpha_s}, \tag{4.46}$$

where $p_i$ is a different prime number. $p$ for each prime factor of $q - 1$, $p|q - 1$, if $p$ is relatively "small", the positive integer $q - 1$ is called a smooth positive integer. Under the condition that $q - 1$ is smooth, for each prime factor $p$, calculate all $p$-th unit roots $r_{p,j}$ in $\mathbb{F}_q^*$, where

$$r_{p,j} = b^{\frac{j(q-1)}{p}}, \ 1 \leq j \leq p. \tag{4.47}$$

Denote $R(p) = \{r_{p,j} | 1 \leq j \leq p\}$ is the root of $p$ $p$ subunits in $\mathbb{F}_q^*$, then in $\mathbb{F}_q^*$, we get a unit root table $R$.

$$\text{unit root table } R = \{R(p_1), R(p_2), \ldots, R(p_s)\}. \tag{4.48}$$

Now let's look at the calculation method of discrete logarithm in $\mathbb{F}_q^*$. Let $y \in \mathbb{F}_q^*$, the discrete logarithm of $y$ under base $b$ is $m$, that is $y = b^m$. When $y$ and $b$ are given, the value of $m$ is desired $(1 \le m \le q-1)$, by the prime factor decomposition of $q-1$ of formula (4.46), if for each $p_i^{\alpha_i}(1 \le i \le s)$, the minimum nonnegative residue of $m$ under mod $p_i^{\alpha_i}$ is $m_i = m \bmod p_i^{\alpha_i}$, according to the Chinese remainder theorem, there is a unique $m \bmod q-1$ such that

$$m \equiv m_i(\text{mod}(p_i^{\alpha_i})), \forall\, i, 1 \le i \le s.$$

Therefore, the discrete logarithm $m$ of $y$ is determined. Now the question is: let $p^\alpha || q-1$, we determine $m \bmod p^\alpha$. Let

$$m \bmod p^\alpha = m_0 + m_1 p + m_2 p^2 + \cdots + m_{\alpha-1} p^{\alpha-1}, 0 \le m_i < p$$

A is the minimum nonnegative residue of $m \bmod p^\alpha$, let's determine each $m_i$. First, we calculate $m_0$. Because $y = b^m$, so

$$y^{\frac{q-1}{p}} = b^{\frac{m(q-1)}{p}} = b^{\frac{m_0(q-1)}{p}}.$$

That is, $y^{\frac{q-1}{p}}$ is a unit root in $\mathbb{F}_q^*$, compare the unit root table $R$ in $\mathbb{F}_q^*$, then we have $m_0 = j, 1 \le j \le p$, which determines $m_0$. Next, calculate $m_1$, let $y_1 = \frac{y}{b^{m_0}} = b^{m-m_0}$, therefore, the discrete logarithm of $y_1$ is $m - m_0$, and

$$m - m_0 \equiv m_1 p + m_2 p^2 + \cdots + m_{\alpha-1} p^{\alpha-1}(\text{mod } p^\alpha),$$

so

$$y_1^{\frac{q-1}{p^2}} = b^{\frac{(m-m_0)(q-1)}{p^2}} = b^{\frac{m_1(q-1)}{p}}.$$

in other words, $y_1^{\frac{q-1}{p^2}}$ is a $p$ subunit root of $\mathbb{F}_q^*$, comparing the unit root table $R$, we can determine $m_1$. Continuing with this method, we can calculate $m_2, \ldots, m_{\alpha-1}$ in turn, so $m \bmod p^\alpha$ is calculated, then by the Chinese remainder theorem, the discrete logarithm $m$ of $y$ under $b$ is calculated.

Exponential integral method

Let $\mathbb{F}_q$ be the finite field of $q$ element, $q = p^n$, $p$ be a relatively small prime number, and $n$ be a large positive integer, so that the security of $q$ can meet certain requirements. Let $\mathbb{F}_p$ be the finite field of $p$ element, we can think of $\mathbb{F}_q$ as an $n$-th extension field of $\mathbb{F}_p$, according to the finite extension theory of the field, $\mathbb{F}_q$ equivalent to (isomorphism) a quotient ring of polynomial ring $\mathbb{F}_p[T]$ over $\mathbb{F}_p$. Let $f(T) \in \mathbb{F}_p[T]$ be the first irreducible polynomial of $n$ degree, then

$$\mathbb{F}_q = \mathbb{F}_p[T]/_{<f(T)>} = \{a_0 + a_1 T + \cdots + a_0 T^{n-1} | \forall\, a_i \in \mathbb{F}_p\}. \tag{4.49}$$

Therefore, any element $a$ in $\mathbb{F}_q$ is equivalent to a polynomial $a(T)$ on $\mathbb{F}_p$, where $\deg a(T) \leq n-1$. Let $b \in \mathbb{F}_q$ be the generator of $\mathbb{F}_q$, $b = b(T)$, if $a_0 \in \mathbb{F}_p$ is a constant polynomial, $a_0$ is called a constant in $\mathbb{F}_q$.

By Lemma 4.8, the discrete logarithm of the constant in $\mathbb{F}_q$ can be easily determined. Let $b' \in \mathbb{F}_p$ be the generator of $\mathbb{F}_p^*$, if $m'$ is the discrete logarithm of constant $a_0 \in \mathbb{F}_p$ to base $b'$, then by Lemma 4.8, $m = m'\frac{q-1}{p-1}$ is the discrete logarithm of $a_0 \in \mathbb{F}_q$ under base $b$. Take $m'(a_0)$ as the discrete logarithm of $a_0$ under base $b'$, since $p$ is small, we can easily calculate and list the discrete logarithms of all constants in $\mathbb{F}_q$:

$$L_0 = \{m'(a_0)\frac{q-1}{p-1} | a_0 \in \mathbb{F}_p\}. \tag{4.50}$$

Next, we determine the discrete logarithm of a nonconstant polynomial under base $b(T)$. Let $1 < m < n$, define

$$L_m = \{p(x) \in \mathbb{F}_p[x] | p(x) \text{ is monic irreducible polynomial, } \deg p(x) \leq m\}, \tag{4.51}$$

The number of irreducible polynomials in $L_m$ is written as $h_m$, that is $|L_m| = h_m$. We first calculate the discrete exponent of irreducible polynomials in $L_m$.

Let $b = b(T)$ be the generator of $\mathbb{F}_q^*$, $b(T) \in \mathbb{F}_p[T]$, $\deg b(T) \leq n-1$, obviously, when $t$ runs through all positive integers from 1 to $q-1$, $b^t(T)$ runs through all nonzero polynomials in Eq. (4.49). Appropriate choice $t$, let

$$b^t(T) \equiv c(T) (\text{mod } f(T)), \quad \deg c(T) \leq n-1.$$

Such that

$$c(T) = c_0 \prod_{p(T) \in L_m} p(T)^{\alpha_{c,p}},$$

denote the discrete logarithm of $a(T)$ under $b(T)$ with $\text{ind}(a(T))$, which can be obtained from the above formula,

$$\text{ind}(c(T)) - \text{ind}(c_0) \equiv \sum_{p(T) \in L_m} \alpha_{c,p} \text{ind}(p(T)) \ (\text{mod } q-1).$$

Because of $\text{ind}(c(T)) = t$, thus,

$$t - \text{ind}(c_0) \equiv \sum_{p(T) \in L_m} \alpha_{c,p} \text{ind}(p(T)) \ (\text{mod } q-1). \tag{4.52}$$

By (4.50), $\text{ind}(c_0)$ is known, therefore, the above formula is a linear equation with $h_m$ variables $\text{ind}(p(T))$. By continuously selecting the appropriate $t$, we can obtain $h_m$ independent linear equations, that is, the $h_m \times h_m$-order matrix formed by the coefficients of $h_m$ variables and $h_m$ linear equations is reversible under mod $q-1$, by Lemma 4.2, as long as its determinant and $q-1$ are coprime. From the knowledge of linear algebra, we can calculate all $\text{ind}(p(T))$ by solving the above linear equations, the following exponential integral table $B_m$ is obtained,

$$B_m = \{\text{ind}(p(T))|p(T) \in L_m\}. \tag{4.53}$$

With exponential integral table $B_m$, the discrete logarithm of any element $a(T) \in \mathbb{F}_q^*$ can be easily calculated. Let $a_1(T) = a(T)b(T)^t$, select the appropriate $t$ such that

$$a_1(T) \equiv a_0 \prod_{p(T)\in L_m} p(T)^{\alpha_a} \pmod{f(T)}.$$

Once the decomposition is established, there are

$$\text{ind}(a_1(T)) = \text{ind}(a_0) + \sum_{p(T)\in L_m} \alpha_a \text{ind}(p(T)).$$

Thus

$$\text{ind}(a(T)) = \text{ind}(a_1(T)) - t.$$

The discrete logarithm of $a(T)$ is obtained.

***Remark 4.1*** The key to the above calculation is to select an appropriate $m$ to obtain the exponential integral table $B_m$. This $m$ cannot be too large, because $h_m$ increases exponentially with $m$, for example, if $m$ is a prime number, then by Corollary 4.4,

$$h_m = |L_m| = \frac{1}{m}(p^m - p).$$

When $h_m$ is too large and calculating the exponential integral table $B_m$, a matrix of order $h_m \times h_m$ will be solved, and its computational complexity is exponential. Obviously, $m$ cannot be too small, the selection of $m$ depends on $p$ and $n$, when $p = 2, n = 127$, $m$'s best choice is $m = 17$. Select finite field $\mathbb{F}_q, q = 2^{127}$, because $q - 1 = 2^{127} - 1$ is a Mersenne prime. This is a popular option at present.

ElGamal cryptosystem

Using the computational complexity of discrete logarithm to design asymmetric cryptosystem is the basic idea of ElGamal cryptosystem. Each user randomly selects a finite field $\mathbb{F}_q$, $q = p^n$, $p$ is a sufficiently large prime number, and then calculates the generator $g$ of $\mathbb{F}_q^*$, select the positive integer $x$ randomly, $1 < x < q - 1$, and calculate $y = g^x$, to get the public key $P_e = (y, g, q)$, own private key $P_d = (x, g, q)$.

Encryption algorithm: To send an encrypted message to user $A$, user $B$ first corresponds each plaintext unit of plaintext space $P$ to an element in $\mathbb{F}_q^*$, and then encrypts each plaintext unit. Let $m \in \mathbb{F}_q^*$ be a plaintext unit, and user $B$ randomly selects an integer $k$, $1 < k < q - 1$, then, the public key $(y, g, q)$ of user $A$ is used to encrypt $m$, and the encryption algorithm $f$ is

$$f(m) = c', \text{ where} \begin{cases} c' = my^k, \\ c = g^k. \end{cases} \quad (4.54)$$

Get cryptosystemtext $(c, c')$.

Decryption algorithm: After receiving the cryptosystemtext $(c, c')$ sent by user $B$, user $A$ decrypts $(c, c')$ with its own private key $(x, g, q)$, decryption algorithm $f^{-1}$ is

$$f^{-1}(c') = c'c^{-x}. \quad (4.55)$$

**Lemma 4.11** *The encryption algorithm $f$ defined by Eq. (4.54) is a 1–1 correspondence of $\mathbb{F}_q^* \longrightarrow \mathbb{F}_q^*$, the inverse mapping $f^{-1}$ of $f$ is given by equation (4.55).*

***Proof*** By (4.54), $c = g^k$, $c' = my^k$, then

$$c'c^{-x} = my^k g^{-kx} = mg^{xk} g^{-xk} = m.$$

That is to say $f^{-1}(f(m)) = m$, conversely,

$$c'c^{-x} y^k = c' g^{-xk} g^{xk} = c'.$$

that is $f(f^{-1}(c')) = c'$, therefore, $f$ is the 1–1 correspondence of $\mathbb{F}_q^* \longrightarrow \mathbb{F}_q^*$ and the inverse mapping of $f$ is $f^{-1}$. The Lemma holds.

Finally, we discuss the computational complexity over finite fields.

**Lemma 4.12** *$\mathbb{F}_q$ is a finite field, $q = p^n$, $\alpha, \beta \in \mathbb{F}_q^*$, $k \geq 1$ is a positive integer, then*

$$Time(\alpha\beta) = O(\log^3 q),$$

$$Time(\frac{\alpha}{\beta}) = O(\log^3 q),$$

$$Time(\alpha^k) = O(\log k \log^3 q).$$

***Proof*** Let $f(x) \in \mathbb{F}_p[x]$, $\deg f(x) = n$, $f(x)$ is a monic irreducible polynomial, then

$$\mathbb{F}_q = \mathbb{F}_p[x]/_{<f(x)>} = \{a_0 + a_1 x + \cdots + a_{n-1}x^{n-1} | \forall\, a_i \in \mathbb{F}_p\}.$$

Let $\alpha, \beta \in \mathbb{F}_q^*$, then

$$\alpha = a_0 + a_1 x + \cdots + a_{n-1}x^{n-1}, \ \beta = b_0 + b_1 x + \cdots + b_{n-1}x^{n-1}.$$

The multiplication of two polynomials requires $n^2$ times of mod $p$ operation, and the bit operation times of each mod $p$ operation is $O(\log^2 p)$, so $\alpha \cdot \beta$ needs $O(n^2 \log^2 p) = O(\log^2 q)$- bit operation to get a polynomial on $\mathbb{F}_p[x]$. The resulting polynomial is divided by $f(x)$ to obtain a polynomial of degree $\leq n - 1$, that is, the final result of $\alpha \cdot \beta$, the number of bit operations required for this operation is $O(n \log^3 p)$. Therefore,

$$\text{Time}(\alpha\beta) = O(\log^3 q + n \log^3 p) = O(\log^3 q),$$

the same can be estimated $\text{Time}(\frac{\alpha}{\beta})$ and $\text{Time}(\alpha^k)$. The Lemma holds.

### *4.7.4 Knapsack Problem*

Given a pile of items with different weights, can you put all or several of these items into a backpack to make it equal to a given weight? This is a knapsack problem arising from real life. Abstract into mathematical problems: Suppose $A = \{a_0, a_1, \dots, a_{n-1}\}$ are $n$ sets of positive integers, $N$ is a positive integer. Is $N$ the sum of the elements of a subset in $A$? Using binary system, the knapsack problem in mathematics can be expressed as follows:

Knapsack problem: When $N$ and $A = \{a_0, a_1, \dots, a_{n-1}\}$ given, where each $a_i \geq 1$ is a positive integer, whether there is a binary integer $e = (e_{n-1}e_{n-2}\cdots e_1e_0)_2$ makes the following formula true,

$$\sum_{i=0}^{n-1} e_i a_i = N, \text{ where } e_i = 0 \text{ or } e_i = 1.$$

If $e$ exists, it is called knapsack problem $(A, N)$ solvable, denote as $\psi(A, N) = e$. If $N = 0$, then $\psi(A, 0) = 0$ (each $e_i = 0$) is called a trivial solution. Therefore, $N \geq 1$ is assumed to be a positive integer.

The above knapsack problem may have solutions, no solutions or multiple solutions. It is very difficult to solve the general knapsack problem $(A, N)$, which belongs to the "NP complete" problem. If the conjecture of "$P \neq NP$" holds, there is no general algorithm, and its computational complexity is polynomial of $n$ and $\log N$. However, under some special conditions, such as the so-called super-increasing sequence, the solution of the problem will be very easy. Next, we introduce the polynomial solution method on the premise of super-increasing sequence.

**Definition 4.11** A positive integer sequence $\{a_i\}_{i\geq 0}$ is called a super-increasing sequence, if each $a_i (i \geq 1)$ is greater than the sum of the previous $i$ positive integers, that is

$$a_i > \sum_{j=0}^{i-1} a_j, \ 1 \leq i < \infty. \tag{4.56}$$

The knapsack problem of super-increasing sequence is actually to find a monotonically decreasing index sequence $\{i_k\}_{k\geq 0}$, where $i_k > i_{k+1}$, $0 \leq i_k \leq n-1$, $\forall k \geq 0$. First, $i_0$ is defined as

$$i_0 = \max\{i|a_i \leq N\}. \tag{4.57}$$

Then consider $N - a_{i_0} = 0$, then the algorithm is completed, that $N = a_{i_0}$. If $N - a_{i_0} > 0$, then define

$$i_1 = \max\{i|a_i \leq N - a_{i_0}\}.$$

For any $k \geq 1$, define

$$i_k = \max\{i|a_i \leq N - a_{i_0} - \cdots - a_{i_{k-1}}\}. \tag{4.58}$$

If the equal sign in Eq. (4.58) holds, that is $a_{i_k} = N - a_{i_0} - \cdots - a_{i_{k-1}}$, then the algorithm completes and obtains the solution $N = a_{i_0} + a_{i_1} \cdots + a_{i_k}$ of $(A, N)$. If $i_k$ does not exist, that is

$$N - a_{i_0} - \cdots - a_{i_{k-1}} < a_i, \ \forall\, i \neq i_0, i_1, \ldots, i_{k-1},$$

call the algorithm terminated. Obvious indicators $i_0 > i_1 > \cdots > i_k > \cdots$. Let $I$ be a set of some indicators, and denote the above algorithm as $\psi$.

**Lemma 4.13** *Let $A = \{a_0, a_1, \ldots, a_{n-1}\}$ be a given set of positive integers, $a_i (i \geq 0)$ is a super-increasing sequence, $N$ is a positive integer. If there is a $k \geq 0$ that makes $\psi$ complete at $k$, that is $a_{i_k} = N - a_{i_0} - \cdots - a_{i_{k-1}}$, then the knapsack problem $(A, N)$ has a solution and the solution is*

$$\psi(A, N) = e = (e_{n-1}e_{n-2}\cdots e_1e_0)_2,$$

*where*

$$\begin{cases} e_i = 1, & \text{if } i \in I, \\ e_i = 0, & \text{if } i \notin I. \end{cases}$$

*If there is a $k \geq 0$, $\psi$ that terminates at $k$, i.e.,*

$$N - a_{i_0} - \cdots - a_{i_{k-1}} < a_i, \forall\, i \notin \{i_0, i_1, \ldots, i_{k-1}\}.$$

*Then the knapsack problem $(A, N)$ has no solution.*

***Proof*** If $\psi$ is completed at $k \geq 0$, then

$$N = a_{i_0} + a_{i_1} + \cdots + a_{i_k}, \ \ I = \{i_0, i_1, \ldots, i_k\},$$

Let $e_i = 1$, when $i \in I$; $e_i = 0$, when$i \notin I$, obviously,

$$\sum_{i=0}^{n-1} e_i a_i = N.$$

So $\psi(A, N) = e = (e_{n-1}e_{n-2}\cdots e_1e_0)_2$, if $k \geq 0$ exists so that $\psi$ terminates at $k$, that is

$$N - a_{i_0} - \cdots - a_{i_{k-1}} < a_i, \forall\, i \notin \{i_0, i_1, \ldots, i_{k-1}\}.$$

Then the knapsack problem $(A, N)$ has no solution. We can prove this conclusion by means of counter-evidence. If $(A, N)$ has a solution, you might as well make

$$N = a_{j_0} + a_{j_1} + \cdots + a_{j_t}.$$

Adjust the order, we can let $j_0 > j_1 > \cdots > j_t$. By the definition of $i_0$, and $a_{j_0} \leq N$, know $j_0 \leq i_0$, thus

$$N \geq a_{i_0} \geq a_{j_0} > \sum_{r=0}^{j_0-1} a_r \geq a_{j_0} + a_{j_1} + \cdots + a_{j_t}$$

contradict with $N = a_{j_0} + a_{j_1} + \cdots + a_{j_t}$, so $(A, N)$ has no solution. The Lemma holds.

MH knapsack public key encryption system

Merkle and Hellman first proposed an encryption method using knapsack problem in 1978, it is the first public key encryption password. Let $A = \{a_0, a_1, \ldots, a_{n-1}\}$ be a sequence of super-increasing positive integers, take $p, b$ as two prime numbers and satisfy

$$p > \sum_{i=0}^{n-1} a_i,\ \ 1 \leq b \leq p-1. \tag{4.59}$$

Calculate $t_i \equiv ba_i (\text{mod } p),\ \ 0 \leq i \leq n-1$, then the public key is $t = (t_0, t_1, \ldots, t_{n-1})$, private key are $A$ and $b$.

Encryption algorithm: The plaintext space $P = \mathbb{F}_2^n$, for each plaintext unit $m = (m_0m_1\cdots m_{n-1}) \in P$, encryption algorithm

$$c = f(m) \equiv \sum_{i=0}^{n-1} t_i m_i (\text{mod } p),\ \ 0 \leq c \leq p, \tag{4.60}$$

where $c$ is cryptosystemtext.

Decryption algorithm: First, use the private key $N \equiv b^{-1}c(\text{mod } p),\ \ 0 \leq N \leq p-1$. Then use the algorithm $\psi = f^{-1}$ of knapsack problem $(A, N)$ to solve

$$f^{-1}(N) = (m_0m_1\cdots m_{n-1}) \in \mathbb{F}_2^n, \tag{4.61}$$

to get plaintext $m = (m_0m_1\cdots m_{n-1})$.

The correctness of MH knapsack public key cryptography is attributed to the following Lemma.

**Lemma 4.14** *The encryption algorithm $f$ defined by Eq. (4.60) is a 1–1 correspondence of $\mathbb{F}_2^n \longrightarrow \mathbb{F}_p$, its inverse mapping $f^{-1}$ is given by equation (4.61).*

***Proof*** If $m = 0$ is the zero vector in $\mathbb{F}_2^n$, then $c = 0$, thus $N = 0$. Knapsack problem $(A, 0)$ has a unique trivial solution $\psi(A, 0) = 0 \in \mathbb{F}_2^n$ is a zero vector. Therefore, the zero vector in $\mathbb{F}_2^n$ is a 1–1 correspondence of the zero element in $\mathbb{F}_p$. Let $m \neq 0$, if

$$N \equiv b^{-1}c(\text{mod } p), \ \ c \equiv \sum_{i=0}^{n-1} t_i m_i (\text{mod } p).$$

Then

$$N \equiv \sum_{i=0}^{n-1} m_i b^{-1} t_i \equiv \sum_{i=0}^{n-1} m_i a_i (\text{mod } p).$$

By (4.59) and $0 \leq N < p$, to obtain

$$N = \sum_{i=0}^{n-1} m_i a_i, \Longrightarrow \psi(A, N) = m = m_0 m_1 \cdots m_{n-1}.$$

So we have

$$f^{-1}(f(m)) = m, \ \forall\, m \in \mathbb{F}_2^n.$$

Conversely, if

$$N = \sum_{i=0}^{n-1} m_i a_i,$$

then

$$bN \equiv \sum_{i=0}^{n-1} m_i a_i b \equiv \sum_{i=0}^{n-1} m_i t_i (\text{mod } p).$$

So there is $N \equiv b^{-1}c(\text{mod } p)$, that is

$$f(f^{-1}(c)) = c, \ \forall\, c \in \mathbb{F}_p.$$

It can be seen that $f$ is a 1–1 correspondence of $\mathbb{F}_2^n \longrightarrow \mathbb{F}_p$ and the inverse mapping is $f^{-1} = \psi$. The Lemma holds.

It can be seen from the above discussion that if $A = \{a_0, a_1, \ldots, a_{n-1}\}$ is not a super-increasing sequence, the decryption algorithm $f^{-1}$ is a difficult problem of "NP complete class", so the encryption and decryption algorithm defined by MH knapsack cryptosystem is the most typical trapdoor single function. Because of this, people believe that MH knapsack public key cryptography is very secure for a long time. However, in 1982, Shamir proved that a class of nonsuper-increasing sequences can be transformed into super-increasing sequences by a simple transformation $x \longrightarrow ax \bmod m$, which can be solved by polynomial algorithm. Although this kind of convertible nonsuper-increasing sequence knapsack problem is quite special, it is enough to shake people's confidence in the security of knapsack problem public key cryptosystem. It is now generally accepted that knapsack public key cryptography is no longer secure.

Shamir transform

Let $A_1 = \{\alpha_0, \alpha_1, \ldots, \alpha_{n-1}\}$ is a super-increasing sequence of positive integers. Randomly select four positive integers $m_1, a_1, m_2, a_2$, where

$$m_1 > \sum_{i=0}^{n-1} \alpha_i, \ m_2 > nm_1, \ (a_1, m_1) = (a_2, m_2) = 1. \tag{4.62}$$

A new positive integer sequence is defined by $m_1$ and $a_1$,

$$A_2 = \{\omega_0, \omega_1, \ldots, \omega_{n-1}\}, \text{ where } \omega_i = a_1\alpha_i \bmod m_1.$$

Where $a_1\alpha_i \bmod m_1$ represents the minimum nonnegative residue of $a_1\alpha_i \bmod m_1$, that is

$$0 \le \omega_i < m_1, \text{ and } \omega_i \equiv a_1\alpha_i (\bmod m_1). \tag{4.63}$$

By the third sequence of positive integers is defined by $m_2$ and $a_2$,

$$A_3 = \{u_0, u_1, \ldots, u_{n-1}\}, \ u_i = a_2\omega_i \bmod m_2,$$

that is

$$0 \le u_i < m_2, \ u_i \equiv a_2\omega_i (\bmod m_2). \tag{4.64}$$

Because $\{u_i\}$ is not a super-increasing sequence, if $A_3$ is used for encryption, it seems to be a general knapsack problem. Its difficulty will be NP complete, but Shamir transform will prove that its decryption algorithm is polynomial.

Let $x = (e_{n-1}e_{n-2}\cdots e_1e_0)_2 \in \mathbb{F}_2^n$ be clear text and encrypt with $A_3$,

$$c = f(x) = \sum_{i=0}^{n-1} e_i u_i, \tag{4.65}$$

get cryptosystemtext $c$. If decryption is required after receiving cryptosystemtext $c$, it is a general knapsack problem, but the problem of using private key $(b_1, m_1, b_2, m_2)$ will become quite simple, where

$$\begin{cases} 0 \le b_1 < m_1, & a_1b_1 \equiv 1 (\bmod m_1) \\ 0 \le b_2 < m_2, & a_2b_2 \equiv 1 (\bmod m_2). \end{cases}$$

First, note the minimum nonnegative residue of $b_2c$ under mod $m_2$,

$$N_0 = b_2c \bmod m_2 = \sum_{i=0}^{n-1} e_i\omega_i. \tag{4.66}$$

Because by (4.65),

$$b_2c \equiv \sum_{i=0}^{n-1} e_i b_2 u_i \equiv \sum_{i=0}^{n-1} e_i\omega_i (\bmod m_2).$$

By the assumption $m_2 > nm_1$ of formula (4.62), and (4.63), there is

$$0 \leq \sum_{i=0}^{n-1} e_i\omega_i < m_2.$$

So (4.66) holds. Then consider the minimum nonnegative residue $N = b_1 N_0 \bmod m_1 (0 \leq N < m_1)$ of $b_1 N_0 \bmod m_1$, by (4.63),

$$N = b_1 N_0 \equiv \sum_{i=0}^{n-1} e_i b_1 \omega_i \equiv \sum_{i=0}^{n-1} e_i \alpha_i (\bmod m_1).$$

So there is

$$N = \sum_{i=0}^{n-1} e_i\alpha_i,\ \alpha_i \in A_1.$$

Since $A_1$ is a super-increasing sequence, the algorithm of polynomial (see Lemma 4.13), we have

$$\psi(A_1, N) = (e_{n-1}e_{n-2}\cdots e_1 e_0)_2 = x.$$

To get plaintext $x$.

Therefore, Shamir uses simple transformation to transform the general knapsack problem into super-incremental knapsack problem. Although $A_3$ is very special, we have reason to doubt that the public key cryptography based on the general knapsack problem solving algorithm is not as secure as people think.

**Exercise 4**

1. Explain the following terms. (1) One secret at a time, (2) Completely confidential system, (3) Unique solution distance, (4) Improve the certification system.
2. Short answer:

   (1) What are the advantages and disadvantages of symmetric cryptosystem and asymmetric cryptosystem?
   (2) The goal of perfecting the certification system.

3. It is known that the plaintext is "Friday", and the cryptosystemtext obtained after encryption with $m = 2$'s Hill password is "POCFKU", find the key of Hill password.
4. Find the inverse matrix (mod $N$) of the following matrix:

$$A = \begin{bmatrix} 1 & 3 \\ 4 & 3 \end{bmatrix} \bmod 5, \ A = \begin{bmatrix} 1 & 3 \\ 4 & 3 \end{bmatrix} \bmod 29,$$

$$A = \begin{bmatrix} 15 & 17 \\ 4 & 9 \end{bmatrix} \bmod 26, \ A = \begin{bmatrix} 197 & 62 \\ 603 & 271 \end{bmatrix} \bmod 841.$$

5. In number theory, Fibonacci number is defined as $a_1 = 1, a_2 = 1, a_3 = 2$, when $n > 1, a_{n+1} = a_n + a_{n-1}$. Prove

$$\begin{bmatrix} a_{n+1} & a_n \\ a_n & a_{n-1} \end{bmatrix} = \begin{bmatrix} 1 & 1 \\ 1 & 0 \end{bmatrix}^n,$$

and $a_n$ is even if and only if $3|n$. More generally, find the law of $d|a_n$.

6. Suppose $N = mn$, and $(n, m) = 1$. A second-order matrix $A \in M_2(\mathbb{Z}_N)$ on n$\mathbb{Z}_N$, can consider $A \in M_2(\mathbb{Z}_m)$ and $A \in M_2(\mathbb{Z}_n)$, let $A_1$ and $A_2$ represent the elements of $A$ in $M_2(\mathbb{Z}_m)$ and $M_2(\mathbb{Z}_n)$, then prove

   (i) Mapping $A \xrightarrow{\sigma} (A_1, A_2)$ is a 1–1 correspondence between $M_2(\mathbb{Z}_N) \xrightarrow{\sigma} M_2(\mathbb{Z}_m) \times M_2(\mathbb{Z}_n)$.
   (ii) In the corresponding $\sigma$, $A$ is the invertible matrix (mod $N$) if and only if $A_1$ is the invertible matrix (mod $m$) and $A_2$ are the invertible matrix (mod $n$).

7. Let $p$ be a prime, $\alpha \geq 1$, then $A \in M_2(\mathbb{Z}_{p^\alpha})$ is a reversible square matrix if and only if $A \in M_2(\mathbb{Z}_p)$ is a reversible square matrix. By calculate, for $\forall\, \alpha \geq 1$, find the number of reversible matrices in $M_2(\mathbb{Z}_{p^\alpha})$.
8. Let $\varphi(N)$ be Euler function, $\varphi_2(N)$ is the number of invertible matrices in $M_2(\mathbb{Z}_N)$, calculation formula for $\varphi_2(N)$: that is, write a formula for $\varphi_2(N)$ similar to $\varphi(N)$. Known $\varphi(N) = N \prod_{p|N}(1 - \frac{1}{p})$, solve $\varphi_2(N) =$?
9. Let $\varphi_k(N)$ be the number of $k$-order reversible matrices in $M_k(\mathbb{Z}_N)$ and give the calculation formula of $\varphi_k(N)$.
10. According to exercise 8 and exercise 9, find the order of $k$-dimensional affine transformation group $G = (A, b)$ on $\mathbb{Z}_N$.
11. RSA is used for encryption, the alphabet of plaintext and cryptosystemtext is $\{0, 1, 2, \ldots, 39\}$ 40 numbers, of which $\{0, 1, 2, \ldots, 25\}$ 26 numbers are equivalent to English 26 letters. Blank $= 26$, $\bullet = 27$, $? = 28$, $\$ = 29$, number $\{0, 1, 2, \ldots, 9\} = \{30, 31, \ldots, 39\}$. Suppose all public keys $n_A$ satisfy $40^2 < n_A < 40^3$. Plaintext unit $m = m_1 m_2 \in \mathbb{Z}_{40}^2$, cryptosystemtext unit $c = c_1 c_2 c_3 \in \mathbb{Z}_{40}^3$. For any plaintext unit, $m = m_1 m_2$ corresponds to a number $m_2 40 + m_1$ of $\mathbb{Z}_{n_A}$, any cryptosystemtext $c = c_3 40^2 + c_2 40 + c_1 \in \mathbb{Z}_{n_A}$.

(i) Encrypting plaintext "$SEND\$7500$" with public key $(n_A, e_A)=(2047, 179)$.
(ii) Factor $n_A = 2047$ to find the private key $(n_A, d_A) = ?$
(iii) A password attacker can quickly find the private key $d_A$ without factoring 2047, so $n_A = 2047$ is a pretty bad choice. Why?

12. The computer attacks the public key $(n_A, e_A) = (536813567, 3602561)$ and finds the private key $d_A$. It shows that 29-bit $n_A$ is not safe in RSA system.
13. Assuming that the plaintext alphabet is $\{0, 1, \ldots, 26\}$, and the first 26 numbers are 26 letters in English, blank $= 26$. Cryptosystemtext alphabet adds "|" $= 27$ to the plaintext alphabet, a total of 28 numbers. If the plaintext unit is $m = m_1m_2m_3 \in \mathbb{Z}_{27}^3$, Cryptosystemtext unit is $c = c_1c_2c_3 \in \mathbb{Z}_{28}^3$. Then in the corresponding number of $Z_{n_A}$ (see exercise 11), we need $n_A$ to meet

$$19683 = 27^3 < n_A < 28^3 = 21952,$$

(i) If your decryption key is $(n_A, d_A) = (21583, 20787)$, decrypt cryptosystemtext is "$YSNAUOZHXXH$" (blank at the end).
(ii) If you know the Euler function $\varphi(n) = 21280$, calculate $e = d^{-1} \bmod \varphi(n)$ and factorize $n$.

14. Prove: In RSA, the 35 bit integer $n = 23360947609$ is a particularly bad choice. (Hint: $n = p \cdot q$ factorization, the size difference between $p$ and $q$ remains unchanged, and Fermat factorization can be used to attack.)
15. Let $n$ be a square free number, and $de \equiv 1(\bmod\ \varphi(n))$. It is proved that there is congruence

$$a^{de} \equiv a(\bmod\ n)$$

for all integers $a$.
16. The multiplication group $\mathbb{F}_{181}^*$ of finite field $\mathbb{F}_{181}$ is generated by $g = 2$, the discrete logarithm of 153 pairs of basis 2 is calculated by smoothing factor method.
17. In the knapsack problem, determine whether the following sequence is an over increasing sequence, whether the knapsack problem is solvable for a given $N$, and how many solutions there are:

(i) $A = \{2, 3, 7, 20, 35, 69\}$, $N = 45$;
(ii) $A = \{1, 2, 5, 9, 20, 49\}$, $N = 73$;
(iiii) $A = \{1, 3, 7, 12, 22, 45\}$, $N = 67$;
(iv) $A = \{2, 3, 6, 11, 21, 40\}$, $N = 39$;
(v) $A = \{4, 5, 10, 30, 50, 101\}$, $N = 186$.

18. If $A = \{a_i | i = 0, 1, 2, \cdots\}$ is an over increasing sequence and $a_0 = 1$, $a_i$ is the smallest positive integer satisfies $a_i \geq \sum_{j=0}^{i-1} a_j$, then $a_i = 2^i$ holds for $\forall\, i \geq 1$.

19. Let $A = \{a_0, a_1, \ldots, a_i, \ldots\}$ be a super-increasing sequence, where $a_i = 2^i (i \geq 1)$, then for any positive integer $N$, Knapsack problem $(A, N)$ has a unique solution.
20. Let $A = \{a_0, a_1, \ldots, a_i, \ldots\}$ be a super-increasing sequence, if for any positive integer $N$, knapsack problem $(A, N)$ always has a solution, prove $a_i = 2^i (i \geq 1)$.

## References


Adelman, L. M., Rivest, R. L., & Shamir, A. (1978). A method for obtaining digital signatures and public-key crypto system. *Communication of ACM, 21,* 120–126.

Adleman, L. M. (1979). A subexponential algorithm for the discrete logarithm problem with application to cryptography. In *Proceedings of the 20th Annual Symposium on the Foundations of Computer Science*, pp. 55–60.

Blum, M. (2022). *Coin-flipping by telephone–A protocol for solving impossible problems* (pp. 133–137). Spring-Compcan: IEEE Proceeding.

Coppersmith, D. (1984). Fast evaluation of logarithms in fields of characteristic two. *IEEE Transactions in Information Theory, IT-30*, 587–594.

Cover, T. M. (2003). *Fundamentals of information theory*. Tsinghua University Press (in Chinese).

Diffie, W., & Hellman, M. E. (1976). New direction in crytography. *IEEE Transactions in Information Theory, IT-22*, 644–654.

ElGamal, T. (1985). A public key cryptosystem and a signature scheme based on discrete logarithms. *IEEE Transactions in Information Theory, IT,314*, 469–472.

Fait, A. & Shamir, A. (2022). How to prove yourself: Practical solutions to identifications and signature problems. In *A advance in Crypology-CRYPTO'86* (Vol. 263, pp. 186–194). Springer-Verlag, LVCS.

Garey, M. R., & Johnson, D. S. (1979). *Computers and intractability: A guide to the theory of NP-completeness*. Freeman.

Goldreich, O. (2001). *Foundation of cryptography* Cambridge University Press.

Gordon, J. A. (1985). Strong prime are easy to find, advance in cryptology. In *Proceedings of Euro Crypt84* (pp. 216–223). Springer.

Hellman, M. E., & Merkle, R. C. (1978). Hiding information and signatures in trap door knapascks. *IEEE Transactions in Information Theory, IT-24*, 525–530

Hellman, M. E. (1979). The mathematics of public-key cryptography. *Scientific America, 241,* 146–157.

Hill, L. S. (1931). Concerning certain linear transformation apparatus of cryptography. *American Math Monthly, 38,* 135–154.

Kahn, D. (1967). *The codebreakers, the story of secret writing*. Macmillan.

Knuth, D. E. (1973). *The art of computer programming*. Addision-Wesley.

Koblitz, N. (1994). *A course in number theory and cryptograph*. Springer-Verlag.

Kranakis, E. (1986). *Primality and cryptogaphy*. John Wiley-Sons.

Massey, J. L. (1983). Logarithms in finite cyclic group-Cryptographic issues. In *Proceedings of the 4th Benelux Symposium on Information's Theory*, pp. 17–25.

Odlyzko, A. M. (1985). Discrete logarithms in finite fields and their cryptographic significance. In: *Advance in Cryptology, Proceedings of Eurocrypt 84*, pp. 224–314. Springer.

Rivest, R. L. (1985). RSA chips(past, present, and future). *Advances in Cryptology, Proceedings of Eurocrypt, 84,* 159–165.

Ruggiu, G. (1985). Cryptology and complexity theories, advances in cryptology. In *Proceedings of Eurocrypt* (Vol. 84, pp. 3–9), Springer

Schneier, B. (1996). *Applied cryptography*, John Wiley 8-sous.

Shamir, A. (1982). A polynomial time algorithm for breaking the basic Markle-Hellman Cryptosystem. In *Proceedings of the 23rd Annual Symposium on the Foundations of Computer Science*, pp. 145–152.
Shannon, C. E. (1949). Communication theory of secrecy system. *The Bell System Technical Journal, 28*, 656–715.
Stinson, D. R. (2003). *Principles and practice of cryptography, translated by Guodeng Feng*. Electronic Industry Press (in Chinese).
Trappe, W., & Washington, L. C. (2008). Cryptography and coding theory, translated by Quanlong Wang et al., people's Posts and Telecommunications Publishing House (in Chinese).
Wah, P., & Wang, M. Z. (1984). Realization and application of Massey-Omura lock. In *Proceedings of the International, Zürich Seminar*(1984),175-182.

# Chapter 5
# Prime Test

In the RSA algorithm in the previous chapter, we see that the decomposition of large prime factors constitutes the basis of RSA cryptosystem security. Theoretically, this security should not be questioned, because there is only the definition of prime in mathematics, and there is no general method to detect prime. The main purpose of this chapter is to introduce some basic prime test methods, including Fermat test, Euler test, Monte Carlo method, continued fraction method, etc., understanding the content of this chapter requires some special number theory knowledge.

## 5.1 Fermat Test

According to Fermat's congruence theorem (commonly known as Fermat's small theorem, which is a special case of Euler congruence theorem), if $n$ is a prime number, the following congruence formula holds for all integers $b$, $(b, n) = 1$,

$$b^{n-1} \equiv 1 (\mathrm{mod}\, n). \tag{5.1}$$

The above formula is an important characteristic of prime numbers. Although $n$ satisfying the above formula is not necessarily prime, it can be used as an important basis for detecting prime numbers, because we can conclude that $n$ not satisfying the above formula is definitely not a prime number. Using Formula (5.1) as the standard to detect prime numbers is called Fermat test.

**Definition 5.1** An odd number $n$, assuming that $n$ is a compound number (not a prime number) and there is a positive integer $b$, $(b, n) = 1$, satisfying

$$b^{n-1} \equiv 1 (\mathrm{mod}\, n),$$



Z. Zheng, *Modern Cryptography Volume 1*, Financial Mathematics and Fintech,
https://doi.org/10.1007/978-981-19-0920-7_5

the compound number $n$ is called a Fermat pseudo prime under base $b$.

The basic properties of pseudo prime numbers are discussed. Our working platform is a finite Abel group $\mathbb{Z}_n^*$, define as

$$\mathbb{Z}_n^* = \{\bar{a} | 1 \le a \le n, (a, n) = 1\}, \ n > 1, \tag{5.2}$$

where $\bar{a}$ is a congruence class of mod $n$ represented by $a$. The multiplication of two congruence classes is defined as $\bar{a} \cdot \bar{b} = \bar{ab}$; obviously, $\mathbb{Z}_n^*$ forms an Abel group of order $\varphi(n)$ under multiplication, in a finite group $G$, the order of a group element $g \in G$ is defined as

$$o(g) = \min\{m : g^m = 1, 1 \le m \le |G|\}.$$

$o(g) = 1$ if and only if $g$ is the unit element of group $G$. By the definition of $o(g)$, obviously,

$$g^t = 1 \Leftrightarrow o(g) | t. \tag{5.3}$$

The following two lemmas are the basic conclusions about the order of group element $g$.

**Lemma 5.1** *$G$ is a finite group, $g \in G$, $k \in \mathbb{Z}$ is an integer, then*

$$o(g^k) = \frac{o(g)}{(k, o(g))}, \tag{5.4}$$

*where the denominator is the greatest common divisor of $k$ and $o(g)$.*

***Proof*** Let $o(g) = m$, $o(g^k) = t$, obviously, $(g^k)^m = 1$, in particular,

$$g^{\frac{k \cdot m}{(k,m)}} = 1, \Longrightarrow t \left| \frac{m}{(k, m)} \right. .$$

On the other hand, by $g^{kt} = 1$, there is $m | kt$, thus

$$\frac{m}{(k, m)} \left| \frac{k}{(k, m)} t, \Longrightarrow \frac{m}{(k, m)} \right| t.$$

So we have $t = \frac{m}{(k,m)}$, the Lemma holds.

**Lemma 5.2** *Suppose $G$ is a finite Abel group, $a, b \in G$, $(o(a), o(b)) = 1$, then*

$$o(ab) = o(a)o(b).$$

***Proof*** Let $o(a) = m_1$, $o(b) = m_2$, then $(m_1, m_2) = 1$. Let $o(ab) = t$, by $(ab)^{m_1 m_2} = a^{m_1 m_2} b^{m_1 m_2} = 1$, there is $t | m_1 m_2$, on the other hand, $(ab)^t = 1$, then $(ab)^{t m_1} = 1$, thus

$b^{tm_1} = 1, m_2|m_1t, m_2|t$. By the same reason, there is $m_1|t$, thus $m_1m_2|t, t = m_1m_2$. The Lemma holds.

Back to the finite group $\mathbb{Z}_n^*$, any integer $a \in \mathbb{Z}$, $(a, n) = 1$, then $\bar{a} \in \mathbb{Z}_n^*$, we denote $o(\bar{a})$ with $o(a)$, $a$ is called the order mod $n$, obviously, $o(a) = o(b)$, if $a \equiv b(\bmod n)$. A basic problem in number theory is the existence of primitive roots of mod $n$. equivalently, is $\mathbb{Z}_n^*$ a cyclic group? If there is a positive integer $a$, $(a, n) = 1$, $o(\bar{a}) = |\mathbb{Z}_n^*| = \varphi(n)$, then $\mathbb{Z}_n^*$ is a cyclic group of order $\varphi(n)$, so that the primitive root of mod $n$ exists and $a$ is the primitive root of mod $n$.

**Lemma 5.3** (Existence of primitive root) *If and only if* $n = 2, 4, p^\alpha (\alpha \geq 1)$ *and* $a = 2p^\alpha (\alpha \geq 1)$ *four cases, the primitive root of* mod $n$ *exists, where* $p > 2$ *is an odd prime.*

***Proof*** If $n = 2, 4$, then the lemma holds. If $n = p$, then $\mathbb{Z}_n = \mathbb{F}_p$, $\mathbb{Z}_n^* = \mathbb{F}_p^*$, by Lemma 4.7 of Chap. 4, it can be seen that $\mathbb{F}_p^*$ is a cyclic group of order $(p - 1)$, so mod $p$ has primitive roots. Now, we need to prove for all positive integer $\alpha$, the primitive root of mod $p^\alpha$ also exists. Therefore, let $a$ be a primitive root of mod $p$, that is, the order of $a$ mod $p$ is $p - 1$. If the order of $a$ mod $p^\alpha$ is denoted by $o(a)$, then

$$a^{o(a)} \equiv 1(\bmod p^\alpha), \Longrightarrow a^{o(a)} \equiv 1(\bmod p),$$

so there is $p - 1|o(a)$. And the number of elements of $\mathbb{Z}_{p^\alpha}^*$ is $\varphi(p^\alpha) = p^{\alpha-1}(p - 1)$, obviously, $o(a)|p^{\alpha-1}(p - 1)$, thus, $o(a) = p^i(p - 1), 0 \leq i \leq \alpha - 1$.

We might as well let $o(a) = p - 1$, if $o(a) = p^i(p - 1)$,$1 \leq i$, then replace $a$ with $a^{p^i}$. By Lemma 5.1,

$$o(a^{p^i}) = \frac{p^i(p - 1)}{(p^i, p^i(p - 1))}.$$

Therefore, without losing generality, let $o(a) = p - 1$, then by Sylow theorem, when $\alpha > 1$, $p^{\alpha-a}|\varphi(p^\alpha)$, there is an integer $b$, $(b, n) = 1$, $b$ is $o(n) = p^{\alpha-1}$ in the order of mod $p^\alpha$, because of $(o(a), o(b)) = 1$, then by Lemma 5.2, there is

$$o(ab) = o(a)o(b) = p^{\alpha-1}(p - 1) = \varphi(p^\alpha),$$

So the primitive root of mod $p^\alpha$ exists.

When $n = 2p^\alpha$, $p > 2$ is odd prime, then $\varphi(n) = \varphi(p^\alpha)$. Thus, the primitive root $a$ of mod $p^\alpha$ is also an primitive root of mod $2p^\alpha$. The Lemma holds.

**Lemma 5.4** *Let* $n$ *be an odd compound number, then*

*(i)* $b \geq 1$ *is a positive integer,* $(b, n) = 1$, $n$ *is Fermat pseudo prime under base* $b$ *if and only if* $o(b)|n - 1$.
*(ii)* $n$ *is Fermat pseudo prime under bases* $b_1$ *and* $b_2$*, then it is Fermat pseudo prime under bases* $b_1b_2$ *and* $b_1b_2^{-1}$*, where* $b_2^{-1}$ *is the multiplicative inverse of* $b_2$ mod $n$.

(iii) *If exist one* $b \in \mathbb{Z}_n^*$ *does not satisfy Eq. (5.1), at least half of* $a, b \in \mathbb{Z}_n^*$ *do not satisfy Eq. (5.1).*

***Proof*** (i) and (ii) are trivial. (i) can be obtained by (5.3). And $b_1, b_2 \in \mathbb{Z}_n^*$,

$$\begin{cases} b_1^{n-1} \equiv 1(\bmod n), b_2^{n-1} \equiv 1(\bmod n). \Longrightarrow (b_1 b_2)^{n-1} \equiv 1(\bmod n). \\ b^{n-1} \equiv 1(\bmod n), \Longrightarrow (b^{-1})^{n-1} \equiv 1(\bmod n). \end{cases}$$

So there is (ii). To prove (iii). Let $n$ not be Fermat pseudo prime to base $b$, if $n$ is Fermat pseudo prime to base $a$, then $n$ is not Fermat pseudo prime to base $ab$. By (ii), therefore, if there is a base to make $n$ a Fermat pseudo prime number, there must be a base to make $n$ not a Fermat pseudo prime number, so more than half of the base $b$ must make $n$ not a Fermat pseudo prime number. The Lemma holds.

By Lemma 5.3, if there is a base $b$ so that $n$ is not Fermat pseudo prime, detect $a$, $1 \leq a \leq n$, $(a, n) = 1$ in sequence, whether $a^{n-1} \equiv 1(\bmod n)$; that is, there is more than 50% chance that find the exact $b$ such that $b^{n-1} \not\equiv 1(\bmod n)$, this proves that $n$ is not a prime number. Is it possible that all $a$, $1 \leq a \leq n$, $(a, n) = 1$, $n$ is Fermat pseudo prime to base $a$ The answer is yes, such a number $n$ is called Carmichael number.

**Definition 5.2** A Carmichael number $n$ is an odd compound number, and for $\forall\, b \in \mathbb{Z}_n^*$, there is

$$b^{n-1} \equiv 1(\bmod n).$$

For Carmichael number, we have the following engraving.

**Theorem 5.1** *Let n be a compound number, then*

(i) *If there is an integer* $a > 1$, $a^2 | n$, *then n is not a Carmichael number.*
(ii) *Assuming that n is a square free number, then n is a Carmichael number* $\Leftrightarrow$ *for all prime* $p$, $p | n$, *there is* $p - 1 | n - 1$.
(iii) *A Carmichael number is the product of at least three different prime numbers.*

***Proof*** Let's prove (i) first. Let $p^2 | n$, $p$ be a prime number, by Lemma 5.3, mod $p^2$ has primitive roots. Let $g$ be an original root of mod $p^2$, that is $o(g) = p(p-1)$, let

$$n' = \prod_{p' | n, p' \neq p} p', \ \ p' \text{ is a prime number.}$$

According to the Chinese remainder theorem, there is a positive integer $b$ such that

$$\begin{cases} b \equiv g(\bmod p^2), \\ b \equiv 1(\bmod n'). \end{cases}$$

Then $b$ is an primitive root of mod $p^2$, and $(b, n) = 1$. We assert that $n$ to base $b$ is not a Fermat pseudo prime. If $n$ to base $b$ is a Fermat pseudo prime, then

$$b^{n-1} \equiv 1(\bmod n), \Longrightarrow b^{n-1} \equiv 1(\bmod p^2), \Longrightarrow o(b)|n-1.$$

That is $p(p-1)|n-1$, but $p|n$ is contradict with $p|n-1$. So $b^{n-1} \not\equiv 1(\bmod n)$, $n$ is not Carmichael number, (i) holds.

Now to prove (ii). If $\forall\ p,\ p|n$, there is $p-1|n-1$, then $\forall\ b \in \mathbb{Z}_n^*$,

$$b^{n-1} = (b^{\frac{n-1}{p-1}})^{p-1} \equiv 1(\bmod p), \forall\ p|n.$$

Because $n$ is a square free number, so

$$b^{n-1} \equiv 1(\bmod n),\ \ \forall\ b \in \mathbb{Z}_n^*.$$

Therefore, $n$ is the Carmichael number. Conversely, if there is a prime number $p$, $p|n$, but $p-1 \nmid n-1$, Let $g$ be a primitive root of mod $p$, which is given by the Chinese remainder theorem,

$$\begin{cases} b \equiv g(\bmod p), \\ b \equiv 1\left(\bmod \dfrac{n}{p}\right). \end{cases}$$

Then $(b, n) = 1$, and

$$b^{p-1} \equiv g^{p-1} \equiv 1(\bmod p).$$

By $p-1 \nmid n-1$, then $g^{n-1} \not\equiv 1(\bmod p)$, so there is $b^{n-1} \not\equiv 1(\bmod n)$, this contradicts with the assumption that $n$ is the Carmichael number. So (ii) holds.

To prove (iii), we just need to exclude that $n$ is the product of two prime numbers. By (ii), let $n = pq$, $p < q$, if $n$ is a Carmichael number, then $q-1 \mid n-1$, but $n-1 = p(q-1+1)-1 = p(q-1)+p-1$, then

$$n-1 \equiv p-1(\bmod q-1),$$

this contradicts with $n-1 \equiv 0(\bmod q-1)$, so $n = pq$ must not be a Carmichael number, the Theorem holds.

Below we give some examples of Carmichael numbers, from property (ii) in Theorem 5.1, we can easily verify whether a square free number is Carmichael number.

***Example 5.1*** The following positive integers $n$ are Carmichael numbers,

$$n = 1105 = 5 \cdot 13 \cdot 7, n = 1729 = 1 \cdot 13 \cdot 19, n = 2465 = 5 \cdot 17 \cdot 29,$$
$$n = 2821 = 7 \cdot 13 \cdot 31, n = 6601 = 7 \cdot 23 \cdot 41.$$

***Example 5.2*** The positive integer $561 = 3 \cdot 11 \cdot 17$ is the smallest Carmichael number.

***Proof*** Defined by, the Carmichael number is odd and compound, so the minimum Carmichael number is

$$n = 3 \cdot p \cdot q, \text{ where } p - 1|n - 1, q - 1|n - 1, p < q \text{ is a prime.}$$

Let $p = 5$, $p = 7$, the congruence equation

$$3 \cdot p \cdot q \equiv 1(\bmod q - 1), q > p$$

has no prime solution $q$, when $p = 11$, the above formula has a minimum solution $q = 17$, so $n = 3 \cdot 11 \cdot 17$ is the smallest Carmichael number.

***Example 5.3*** For given prime number $r \geq 3$, then the congruence equations

$$\begin{cases} rpq \equiv 1(\bmod p - 1) \\ rpq \equiv 1(\bmod q - 1) \end{cases}$$

has only finite different prime solutions $p, q$. Let's leave this conclusion for reflection.

## 5.2 Euler Test

Let $p > 2$ be an odd prime, Euler test uses the Euler criterion in the quadratic residue of mod $p$ to detect whether a positive integer $n$ is prime. Like Fermat's test, it is obvious that the $n$ that passes the test cannot be determined as prime, but the $n$ that fails the test is certainly not prime. We know that when the positive integers $a$ and $n$ are given ($n > 1$), the solution of the quadratic congruence equation $x^2 \equiv a(\bmod n)$ is a famous "NP complete" problem. We can't find a general solution in an effective time. However, in the special case where $n = p > 2$ is an odd prime number, we have rich theoretical knowledge to discuss the quadratic residue of mod $p$, these knowledge include the famous Gauss quadratic reciprocal law and Euler criterion, which constitute the core knowledge system of elementary number theory. First, we introduce Legendre sign and let $p > 2$ be a given odd prime number.

$\mathbb{Z}_p^*$ is a $(p - 1)$-order cyclic group, $a \in \mathbb{Z}_p^*$ (i.e., $(a, p) = 1$), we define the Legendre symbolic function as

$$\left(\frac{a}{p}\right) = \begin{cases} 1, & \text{when } x^2 \equiv a(\bmod p) \text{ is solvable} \\ -1, & \text{when } x^2 \equiv a(\bmod p) \text{ is unsolvable} \end{cases}$$

If $(a, p) > 1$, that is $p \mid a$, we let $(\frac{a}{p}) = 0$, for $\forall\, a \in \mathbb{Z}$, Legendre symbolic function $(\frac{a}{p})$ is all defined, and it is a completely integral function of $\mathbb{Z} \to \{1, -1, 0\}$.

$$\left(\frac{ab}{p}\right)=\left(\frac{a}{p}\right)\left(\frac{b}{p}\right),\ \forall\, a,b\in\mathbb{Z}$$

and

$$\left(\frac{a}{p}\right)=\left(\frac{b}{p}\right),\quad \text{if } a\equiv b\ (\text{mod } p)\,.$$

If $(\frac{a}{p})=1$, then $x^2\equiv a(\text{mod } p)$ is solvable, $a$ is called a quadratic residue of mod $p$, if $(\frac{a}{p})=-1$, then $x^2\equiv a(\text{mod } p)$ is unsolvable, $a$ is called a quadratic nonresidue of mod $p$.

**Lemma 5.5** *$a\in\mathbb{Z}$, $p\nmid a$, then the necessary and sufficient condition for $a$ to be the quadratic residue of* mod $p$ *is*

$$a^{\frac{p-1}{2}}\equiv 1(\text{mod } p).$$

***Proof*** $\mathbb{Z}_p^*$ is a $p-1$-order cyclic group, let $g$ be a primitive root of mod $p$, that is $\bar{g}$ is the generator of $\mathbb{Z}_p^*$, that is $\forall a\in\mathbb{Z}$, $(a,p)=1$, we have

$$a\equiv g^t(\text{mod } p),\ \text{ where } 1\le t\le p-1.$$

Obviously, $a$ is the quadratic residue of mod $p \Leftrightarrow t$ is even. Therefore, if $t$ is even, then

$$a^{\frac{p-1}{2}}\equiv g^{\frac{t(p-1)}{2}}\equiv (g^{\frac{t}{2}})^{p-1}\equiv 1(\text{mod } p).$$

Conversely, if $a^{\frac{p-1}{2}}\equiv 1(\text{mod } p)$, then $o(a)\mid\frac{p-1}{2}$, and by Lemma 5.1, can calculate

$$o(a)=o(g^t)=\frac{p-1}{(t,p-1)}.$$

So

$$o(a)\mid\frac{p-1}{2}\Leftrightarrow 2|(t,p-1)\Leftrightarrow 2|t,$$

that is $t$ is even, thus, $a$ is a quadratic residue of mod $p$, the Lemma holds.

**Lemma 5.6** (Euler criterion). *For $\forall\, a\in\mathbb{Z}$, we have*

$$a^{\frac{p-1}{2}}\equiv\left(\frac{a}{p}\right)(\text{mod } p). \tag{5.5}$$

***Proof*** If $(a,p)>1$, that is $p|a$, the above formula holds. Might as well let $p\nmid a$. By Fermat congruence theorem $a^{p-1}\equiv 1(\text{mod } p)$, there is

$$(a^{\frac{p-1}{2}}+1)(a^{\frac{p-1}{2}}-1)\equiv 0(\text{mod } p).$$

Thus

$$a^{\frac{p-1}{2}} \equiv \pm 1 (\mathrm{mod}\ p).$$

If $a^{\frac{p-1}{2}} \equiv 1 (\mathrm{mod}\ p)$, by Lemma 5.5, then $(\frac{a}{p}) = 1$. If $a^{\frac{p-1}{2}} \equiv -1 (\mathrm{mod}\ p)$, then $(\frac{a}{p}) = -1$. So (5.5) holds.

**Definition 5.3** Suppose $n$ is an odd compound number, if there is an integer $b$, $(b, n) = 1$, it satisfies

$$b^{\frac{n-1}{2}} \equiv \left(\frac{b}{n}\right) (\mathrm{mod}\ n), \tag{5.6}$$

Call $n$ an Euler pseudo prime under base $b$. Where $(\frac{b}{n})$ is Jacobi symbol, define as

$$\left(\frac{b}{n}\right) = \left(\frac{b}{p_1}\right)^{\alpha_1} \left(\frac{b}{p_2}\right)^{\alpha_2} \cdots \left(\frac{b}{p_s}\right)^{\alpha_s}, \text{ if } n = p_1^{\alpha_1} \cdots p_s^{\alpha_s}. \tag{5.7}$$

From the definition, we obviously have a corollary: if $n$ is Euler pseudo prime under basis $b$, then $n$ is Fermat pseudo prime under basis $b$. This conclusion can be proved by squaring both sides of Eq. (5.6) at the same time.

The following example shows that the inverse of inference is not tenable; that is, if $n$ is Fermat pseudo prime under basis $b$, but not Euler pseudo prime.

***Example 5.4*** $n = 91$ is Fermat pseudo prime under basis $b = 3$, but not Euler pseudo prime. In fact, it's easy to calculate $3^6 \equiv 1 (\mathrm{mod}\ 91)$, thus $3^{90} \equiv 1 (\mathrm{mod}\ 91)$. From $3^6 \equiv 1 (\mathrm{mod}\ 91)$, we have

$$3^{42} \equiv 1 (\mathrm{mod}\ 91), \Longrightarrow 3^{45} \equiv 9 (\mathrm{mod}\ 91).$$

So 91 to base 3 is not an Euler pseudo prime.

***Example 5.5*** $n = 91$ to base $b = 10$ is an Euler pseudo prime. Because

$$10^{45} \equiv 10^3 \equiv -1 (\mathrm{mod}\ 91),$$

calculate Legendre symbols

$$\left(\frac{10}{91}\right) = \left(\frac{2}{91}\right) \cdot \left(\frac{5}{91}\right) = -1,$$

so $n = 91$ to base $b = 10$ is an Euler pseudo prime.

From the Euler criterion of Lemma 5.6, we can easily calculate the Legendre symbols of $-1$ and 2.

**Lemma 5.7** *Let $p > 2$ be an odd prime, then we have*

$$\left(\frac{-1}{p}\right) = (-1)^{\frac{p-1}{2}}, \quad \left(\frac{2}{p}\right) = (-1)^{\frac{1}{8}(p^2-1)}. \tag{5.8}$$

***Proof*** By Lemma 5.6,

$$(-1)^{\frac{p-1}{2}} \equiv \left(\frac{-1}{p}\right) (\text{mod } p),$$

Since both sides of the congruence are $\pm 1$, $p > 2$, there is $(\frac{-1}{p}) = (-1)^{\frac{p-1}{2}}$. To calculate the Legendre sign for 2, we notice that

$$\begin{cases} p - 1 \equiv (-1)^1 (\text{mod } p) \\ 2 \equiv 2 \cdot (-1)^2 (\text{mod } p) \\ p - 3 \equiv 3 \cdot (-1)^3 (\text{mod } p) \\ \vdots \\ r \equiv \dfrac{p-1}{2} \cdot (-1)^{\frac{p-1}{2}} (\text{mod } p), \end{cases}$$

where $r = \frac{p-1}{2}$, if $\frac{p-1}{2}$ is a even; $r = p - \frac{p-1}{2}$, if $\frac{p-1}{2}$ is an odd. There is

$$2 \cdot 4 \cdot 6 \cdots (p-1) \equiv \left(\frac{p-1}{2}\right)! (-1)^{\frac{1}{8}(p^2-1)} (\text{mod } p),$$

that is

$$2^{\frac{p-1}{2}} \equiv (-1)^{\frac{1}{8}(p^2-1)} (\text{mod } p),$$

by Lemma 5.6,

$$\left(\frac{2}{p}\right) \equiv (-1)^{\frac{1}{8}(p^2-1)} (\text{mod } p),$$

there is

$$\left(\frac{2}{p}\right) = (-1)^{\frac{1}{8}(p^2-1)},$$

Lemma 5.7 holds.

Let $(\frac{a}{n})$ be a Jacobi symbol, defined by Eq. (5.6), then Lemma 5.7 can be extended to Jacobi symbol.

**Lemma 5.8** *Let $n$ be an odd, then we have*

$$\left(\frac{-1}{n}\right) = (-1)^{\frac{n-1}{2}}, \quad \left(\frac{2}{n}\right) = (-1)^{\frac{1}{8}(n^2-1)}. \tag{5.9}$$

***Proof*** The square of any odd number is congruent 1 under mod 8, that is $a^2 \equiv 1 \pmod 8$. Write $n = a^2 \cdot p_1 p_2 \cdots p_t$, where $p_i$ are different prime numbers, then

$$n \equiv p_1 p_2 \cdots p_t \pmod 8.$$

Similarly, for $\forall\, n \in \mathbb{Z}$, by (5.7),

$$\left(\frac{b}{n}\right) = \left(\frac{b}{p_1}\right)\left(\frac{b}{p_2}\right)\cdots\left(\frac{b}{p_t}\right), \tag{5.10}$$

thus

$$\left(\frac{-1}{n}\right) = \left(\frac{-1}{p_1}\right)\left(\frac{-1}{p_2}\right)\cdots\left(\frac{-1}{p_t}\right) = (-1)^{\frac{p_1-1}{2}+\frac{p_2-1}{2}+\cdots+\frac{p_t-1}{2}} = (-1)^{\frac{n-1}{2}}. \tag{5.11}$$

The same can be proved $\left(\frac{2}{n}\right)$, the Lemma holds.

**Corollary 5.1** *For all odd numbers n, they are Euler pseudo prime under the base* $\pm 1$.

***Proof*** It is trivial that $n$ to 1 is an Euler pseudo prime number, and $n$ to $-1$ is an Euler pseudo prime number, which is directly derived from Lemma 5.8.

**Lemma 5.9** (Gauss. ) *Let p and q be two different odd primes, then*

$$\left(\frac{q}{p}\right)\left(\frac{p}{q}\right) = (-1)^{\frac{1}{4}(p-1)(q-1)}.$$

***Proof*** According to incomplete statistics, there are currently more than 270 methods to prove Gauss quadratic reciprocal law. In order to save space, we leave the proof to the readers, hoping that everyone can find their favorite proof method.

Next, we discuss the computational complexity of Fermat test and Euler test.

**Lemma 5.10** *Let n be an odd,* $1 \le b < n$, $(b, n) = 1$, *then*

$$\begin{cases} \text{Time}(n \text{ to base } b\text{'s Fermat test}) = O(\log^3 n), \\ \text{Time}(n \text{ to base } b\text{'s Euler test}) = O(\log^4 n). \end{cases}$$

***Proof*** By (5.1), the Fermat test of $n$ to base $b$ is actually an operation of $b^{n-1}$ to mod $n$, by the Lemma 1.5 of Chap. 1, bit operations of $b^{n-1} \bmod n$,

$$\text{Time}(b^{n-1} \bmod n) = O(\log n \log^2 n) = O(\log^3 n).$$

Euler test of $n$ to base $b$, by (5.6), the number of bit operations on the left is $O(\log^3 n)$. Find Jacobi symbol $\left(\frac{b}{n}\right)$, from Eq. (5.7) and quadratic reciprocal law, the calculation can be transformed into the calculation of Legendre symbol. Each reciprocal law is actually a division, so we only consider the calculation of Legendre symbols. By Euler criterion,

$$\text{Time}\left(\text{calculate }\left(\frac{b}{p}\right)\right) = \text{Time}\left(b^{\frac{p-1}{2}} \bmod p\right) = O(\log^3 n).$$

The number of prime factors of each $n$ has an estimated $O(\log\log n)$, so

$$\text{Time}\left(\text{calculate Jacobi symbol }\left(\frac{b}{n}\right)\right) = O(\log\log n \cdot \log^3 n) = O(\log^4 n).$$

We have completed the calculation of Lemma 5.10.

Solovay and Strassen proposed a probabilistic method to detect prime numbers by Euler test in 1977. When $n > 1$ is an odd number, $k$ numbers are randomly selected, $b_1, b_2, \ldots, b_k$, where $1 < b_i < n$, $(b_i, n) = 1$. Use Eq. (5.6) to calculate both sides of each $b$ in turn, and the required bit operation is $O(\log^4 n)$, if both sides of Eq. (5.6) are not equal, then $n$ is not a prime number and the test is terminated. If $k$ $b$ pass the Euler test of Eq. (5.6), then $n$ is the probability $< \frac{1}{2^k}$ of compound number, that is

$$P\{n \text{ is not prime}\} \le 2^{-k}.$$

The above formula is directly derived from Lemma 5.3. Let's introduce a better Miller–Rabin method than Solovay–Strassen method in a sense.

**Definition 5.4** Let $n$ be an odd compound number, write $n - 1 = 2^t \cdot m$, where $t \ge 1$, $m$ is an odd. Let $b \in \mathbb{Z}_n^*$, if $n$ and $b$ satisfy one of the following conditions,

$$b^m \equiv 1 (\bmod n), \text{ or exists one } r, 0 \le r < t, \text{ such that } b^{2^r m} \equiv -1 (\bmod n). \quad (5.12)$$

Then $n$ is called a strong pseudo prime under base $b$.

**Lemma 5.11** *Suppose $n \equiv 3 (\bmod 4)$, then $n$ is a strong Pseudoprime under base $b$ if and only if $n$ is an Euler Pseudoprime under base $b$.*

***Proof*** Because $n \equiv 3 (\bmod 4)$, then $n - 1 = 2m$, that is $t = 1$, $m = \frac{1}{2}(n-1)$. By Definition 5.4, $n$ is a strong pseudo prime under base $b$ if and only if

$$b^m = b^{\frac{n-1}{2}} \equiv \pm 1 (\bmod n).$$

Therefore, if $n$ is an Euler pseudo prime number under base $b$, the above formula holds, so it is also a strong pseudo prime number for base $b$. Conversely, if the above formula holds, because of $n \equiv 3 (\bmod 4)$, then $\frac{1}{2}(n-1)$ is an odd number, so $(\frac{-1}{n}) = -1$, and

$$\left(\frac{b}{n}\right)=\left(\frac{b}{n}\right)^{\frac{n-1}{2}}\equiv\left(\frac{b^{\frac{n-1}{2}}}{n}\right)\equiv b^{\frac{n-1}{2}}(\bmod n).$$

Therefore, $n$ to base $b$ is Euler pseudo prime. The Lemma holds.

Below we give the main results of this section.

**Theorem 5.2** *Let $n$ be an odd number, $b \in \mathbb{Z}_n^*$, then*

(i) *If $n$ to base $b$ is a strong pseudo prime, then $n$ to base $b$ is an Euler pseudo prime.*
(ii) *Base $b$, which makes $n$ a strong pseudo prime number, accounts for 25% of $1 \le b < n$, $(b, n) = 1$ at most.*

Before proving Theorem 5.2, let's introduce Miller–Rabin's test method, in order to test whether a large odd number $n$ is a prime number, we write $n - 1 = 2^t \cdot m$, $m$ is an odd number, $t \ge 1$, select one $b$ at random, $1 \le b < n$, $(b, n) = 1$. We first calculate $b^m \bmod n$, if we get the result is $\pm 1$, then $n$ passes the strong pseudo prime test (5.12). If $b^m \bmod n \ne \pm 1$, then we square $b^m \bmod n$ and find the minimum nonnegative residue of the squared number under $\bmod n$ to see if we get the result of $-1$ and perform $r$ times. If we can't get $-1$, then $n$ to base $b$ fails to test Formula (5.12). Therefore, it is asserted that $n$ to base $b$ is not a strong pseudo prime number. If $-1$ is obtained by $r$ squared, then $n$ passes the test under base $b$.

In Miller–Rabin's test, if $n$ to base $b$ fails to pass the test Formula (5.12), then $n$ must not be a prime number, if $n$ to randomly selected $k$ $b = \{b_1, b_2, \ldots, b_k\}$ pass the test, by property (ii) of 5.2, each $b_i$ accounts for no more than 25

$$P\{n \text{ not prime}\} \le \frac{1}{4^k}. \tag{5.13}$$

Compared with the Solovay–Strassen method using Euler test, the Miller–Rabin method using strong pseudo prime test is more powerful.

To prove 5.2, we first prove the following two lemmas.

**Lemma 5.12** *Let $G = \langle g \rangle$ be a finite group of order $m$, that is $o(g) = m$, then equation $x^k = 1$ has exactly $d$ solutions in $G$, $d = (k, m)$.*

***Proof*** $x \in G$, write $x = g^t$, then $x^k = g^{kt} = 1 \Leftrightarrow m|kt$, that is $\frac{m}{d}|\frac{k}{d} \cdot t$, thus $\frac{m}{d}|t$, let $t = \frac{m}{d} \cdot s$, then when $s = 1, 2, \ldots, d$, $x = g^t$ has exactly $d$ solutions. The Lemma holds.

**Lemma 5.13** *Let $p$ be an odd prime number, $p - 1 = 2^t m'$, $t \ge 1$, $m'$ is prime, then*

$$x^{2^r m} \equiv -1(\bmod p), m \text{ is odd} \tag{5.14}$$

*The number of solutions $N$ in $\mathbb{Z}_p^*$ satisfies*

$$N = \begin{cases} 0, & \text{if } r \geq t; \\ 2^r(m, m'), & \text{if } r < t. \end{cases}$$

***Proof*** Let $g$ be a generator of $\mathbb{Z}_p^*$, write $x = g^j$, $1 \leq j \leq p-1$, because $o(g) = p-1$, so

$$g^{\frac{p-1}{2}} \equiv -1(\text{mod } p).$$

Thus

$$x^{2^r m} \equiv -1(\text{mod } p) \Leftrightarrow 2^r mj \equiv \frac{p-1}{2}(\text{mod } p-1).$$

Namely,

$$2^r mj \equiv 0(\text{mod } p-1).$$

Because $p - 1 = 2^t m'$, the above formula is equivalent to

$$2^r mj \equiv 2^{t-1}m'(\text{mod } 2^t m'). \quad (5.15)$$

If $r > t - 1$, then the congruence has no solution to $j$, because $m$ and $m'$ are odd numbers, so when $r \geq t$, (5.14) is unsolvable. If $r < t$, let $d = (m, m')$, then

$$(2^r m, 2^t m') = 2^r d,$$

then Eq. (5.15) has exactly $d$ solutions for $j$. Each $j$ corresponds to one $x = g^j$, then the number of solutions of Eq. (5.14) to $x$ is $N = 2^r d$, the Lemma holds.

With the above preparation, we now give the proof of Theorem 5.2.

***Proof*** (*The proof of Theorem* 5.2). Let's first prove that (i), that is, $n$ and $b$ satisfy Eq. (5.12), we want to prove that formula (5.6) is satisfied; that is, if $n$ to base $b$ is a strong pseudo prime number, then $n$ to base $b$ is an Euler pseudo prime number, write $n - 1 = 2^t m$, $m$ is prime, we prove the property (i) of Theorem 5.2 in three cases.

(1) $b^m \equiv 1(\text{mod } n)$. In this case, it is obvious that $b^{\frac{n-1}{2}} \equiv 1(\text{mod } n)$. Let's prove $(\frac{b}{n}) = 1$, in fact,

$$1 = \left(\frac{1}{p}\right) = \left(\frac{b^m}{p}\right) = \left(\frac{b}{p}\right)^m = 1.$$

There is

$$b^{\frac{n-1}{2}} \equiv \left(\frac{b}{n}\right) \equiv 1(\text{mod } n).$$

That is $n$ to base $b$ is an Euler pseudo prime number.

(2) $b^{\frac{n-1}{2}} \equiv -1(\text{mod } n)$. In this case, we have to prove $(\frac{b}{n}) = -1$, let $p|n$ be any prime factor of $n$, write $p - 1 = 2^{t_1} m_1$, where $t_1 \geq 1$, $m_1$ is an odd number.

Let's calculate the Legendre symbol $(\frac{b}{p})$, in fact, $t_1 \geq t$, and

$$\left(\frac{b}{p}\right) = \begin{cases} -1, & \text{if } t_1 = t; \\ 1, & \text{if } t_1 > t. \end{cases} \tag{5.16}$$

Because

$$b^{\frac{n-1}{2}} = b^{2^{t-1}m} \equiv -1(\bmod n), \Longrightarrow b^{2^{t-1}mm_1} \equiv -1(\bmod n),$$

by $p|n$, we have

$$b^{2^{t-1}mm_1} \equiv -1(\bmod p). \tag{5.17}$$

If $t_1 < t$, from the above formula, there is

$$b^{2^{t_1}m_1} \equiv -1(\bmod p), \Longrightarrow b^{p-1} \equiv -1(\bmod p).$$

This contradicts Fermat's congruence theorem, so we always have $t_1 \geq t$. If $t_1 = t$, by (5.17), then

$$\left(\frac{b}{p}\right) \equiv b^{\frac{p-1}{2}} = b^{2^{t-1}m} \equiv -1(\bmod p).$$

Because if the above formula is 1, both sides will be $m$ power at the same time, which will contradict Formula (5.17). If $t_1 > t$, put both sides of Eq. (5.17) to the power of $2^{t_1-t}$ at the same time, then $(\frac{b}{p}) = 1$, so we have (5.16).
We now complete the proof of case (2) under the conclusion of Eq. (5.16), write $n = \prod_{p|n} p$, $p$ does not require different, define the positive integer $k$ as

$$k = \#\{p \mid p|n, p-1 = 2^{t_1}m_1, m_1 \text{ is odd}, t_1 = t\}.$$

By (5.16), then

$$\left(\frac{b}{n}\right) = \prod \left(\frac{b}{p}\right) = (-1)^k. \tag{5.18}$$

Let's prove that $k$ is an odd number, because $t_1 \geq t$, $p-1 = 2^{t_1}m_1, n-1 = 2^t m$, under mod $2^{t+1}$, we have

$$p \equiv \begin{cases} 1(\bmod 2^{t+1}), & \text{if } t_1 > t; \\ 1 + 2^t(\bmod 2^{t+1}), & \text{if } t_1 = t. \end{cases}$$

Because $n = 1 + 2^t(\bmod 2^{t+1})$, so

$$n \equiv 1 + 2^t \equiv 1 + k \cdot 2^t(\bmod 2^{t+1}),$$

So $k$ must be odd, by (5.18), then $(\frac{b}{n}) = -1$. Case (2) is proved.

(3) $b^{2^{r-1}\cdot m} \equiv -1(\bmod n)$, where $1 \leq r \leq t, n-1 = 2^t \cdot m$.
In this case, we replace $r$ of Eq. (5.12) with $r-1$. Because $r-1 \leq t-1$, so $b^{\frac{n-1}{2}} \equiv 1(\bmod n)$. To prove property (i) of Theorem 5.2, we have to prove $(\frac{b}{n}) = 1$, as in case (2), we let $p|n$, write $p-1 = 2^{t_1} \cdot m_1$, $m_1$ is odd, then we have $t_1 \geq r$, and

$$\left(\frac{b}{p}\right) = \begin{cases} -1, & \text{if } t_1 = r; \\ 1, & \text{if } t_1 > r. \end{cases} \tag{5.19}$$

The proof of Formula (5.19) is the same as that of case (2), write $n = \prod p$, $p$ is not required to be a different prime, define positive integer $k_1$:

$$k_1 = \#\{p \mid p|n, p-1 = 2^{t_1}m_1, m_1 \text{ is odd}, t_1 = r\}.$$

as in case (2), we have $(\frac{b}{n}) = (-1)^{k_1}$, similarly, under mod $2^{r+1}$, it can be proved that $k_1$ must be even. Thus $(\frac{b}{n}) = 1$, we have completed all the proofs of property (i) in Theorem 5.2.

Next, we prove property (ii) in Theorem 5.2. It is also discussed in three cases.

(1) $n$ can be divided by a square number; that is, there is a prime number $p$, $p^\alpha || n$, $\alpha \geq 2$.
In this case, we prove that there are at least $\frac{1}{4}(n-1)$ $b$, $b \in \mathbb{Z}_n^*$, $n$ to base $b$ is not Fermat prime number, let alone a strong pseudo prime. First, suppose $b^{n-1} \equiv 1(\bmod n)$, then there is a prime $p$, $p^2|n$, thus $b^{n-1} \equiv 1(\bmod p^2)$. Because $\mathbb{Z}_{p^2}^*$ is a $p(p-1)$-order cyclic group (see Theorem 5.3), let $g$ be a generator of $\mathbb{Z}_{p^2}^*$, then

$$\mathbb{Z}_{p^2}^* = \{g, g^2, \ldots, g^{p(p-1)}\}.$$

By Lemma 5.12, the number of $b$ satisfying $b^{n-1} \equiv 1(\bmod p^2)$ is $d$,

$$d = (n-1, p(p-1)) = (n-1, p-1).$$

Because $p|n$, so $p \nmid n-1$, and $p \nmid d$; therefore, the maximum possibility of $d$ is $p-1$; therefore, the proportion of $b$ in $b^{n-1} \equiv 1(\bmod p^2)$ in $1 \leq b < n$ shall not exceed

$$\frac{p-1}{p^2-1} = \frac{1}{p+1} \leq \frac{1}{4}.$$

Therefore, there is at most $b$ in the proportion of $\frac{1}{4}$, so that $n$ to base $b$ is Fermat prime, in case (1), we prove the property (ii) of Theorem 5.2.

(2) $n = pq$ are two different prime numbers.
In this case, let $p-1 = 2^{t_1}m_1, q-1 = 2^{t_2}m_2$, $m_1, m_2$ to be odd. Without losing generality, you can let $t_1 \leq t_2$. Let $b \in \mathbb{Z}_n^*$, in order for $n$ to base $b$ to be a strong pseudo prime number, it is necessary to satisfy

$$b^m \equiv 1 (\bmod p),\ b^m \equiv 1 (\bmod q) \tag{5.20}$$

or

$$b^{2^r m} \equiv -1 (\bmod p),\ b^{2^r m} \equiv -1 (\bmod q), 0 \le r < t. \tag{5.21}$$

By Lemma 5.12, the number of $b$ satisfied (5.20) is $\le (m, m_1)(m, m_2) \le m_1 m_2$. By Lemma 5.13, for each $r$, $0 \le r < \min(t_1, t_2) = t_1$, the number of $b$ satisfying $b^{2^r m} \equiv -1 (\bmod n)$ is $2^r (m, m_1) \cdot 2^r (m, m_2) < 4^r m_1 m_2$. Because $n = pq$, then $\varphi(n) = (p-1)(q-1), \Longrightarrow n - 1 > \varphi(n) = 2^{t_1+t_2}$, therefore, the proportion of $b$ of the strong pseudo prime of $n$ to base $b$ does not exceed

$$\frac{m_1 m_2 + m_1 m_2 + 4 m_1 m_2 + \cdots + 4^{t_1 - 1} m_1 m_2}{2^{t_1+t_2} m_1 m_2} = 2^{-t_1-t_2}\left(1 + \frac{4^{t_1} - 1}{4 - 1}\right) \tag{5.22}$$

in $1 \le b < n$, $(b, n) = 1$.
If $t_1 < t_2$, then the above formula shall not exceed

$$2^{-2t_1-1}\left(\frac{2}{3} + \frac{4^{t_1}}{3}\right) \le 2^{-3} \cdot \frac{2}{3} + \frac{1}{6} = \frac{1}{4}.$$

If $t_1 = t_2$, then $m_1 \ne m_2$, so $(m, m_1) \le m_1$ and $(m, m_2) \le m_2$, one must be strictly less than. The reason is that if they are equal, then $m_1 | m$, $m_2 | m$, $n - 1 = 2^t m, \Longrightarrow n - 1 = 2^t m = pq - 1 \equiv q - 1 (\bmod m_1)$, thus $m_1 | n - 1, \Longrightarrow m_1 | q - 1 = 2^{t_2} m_2, \Longrightarrow m_1 | m_2$, this is a contradiction. So $(m, m_1) \le m_1$ and $(m, m_2) \le m_2$ must have a strict less than 0. We have

$$(m, m_1) \cdot (m, m_2) \le \frac{1}{3} m_1 m_2.$$

If $m_1 m_2$ is substituted for $\frac{1}{3} m_1 m_2$ in Eq. (5.22), the proportion of $n$ to $b$ whose base $b$ is a strong pseudo prime number does not exceed

$$\frac{1}{3} 2^{-2t_1}\left(\frac{2}{3} + \frac{4^{t_1}}{3}\right) \le \frac{1}{18} + \frac{1}{9} = \frac{1}{6} < \frac{1}{4}.$$

We complete the proof of property (ii) of Theorem 5.2 in case (2).

(3) Finally, suppose $n = p_1 p_2 \cdots p_k$, $k \ge 3$ is the product of different prime factors. In this case, write $p_i - 1 = 2^{t_i} m_i$, $m_i$ as an odd number. As in case (2), without losing generality, it can make $t_1 \le t_j (1 \le j \le k)$. Similarly to the proof of formula (5.22), the proportion of $b$ satisfying that $n$ is a strong pseudo prime number for base $b$ does not exceed

$$\begin{aligned}
2^{-t_1-t_2-\cdots-t_k}\left(1+\frac{2^{k+1}-1}{2^k-1}\right) &\le 2^{-kt_1}\left(\frac{2^k-2}{2^k-1}+\frac{2^{kt_1}}{2^k-1}\right)\\
&= 2^{-kt_1}\cdot\frac{2^k-2}{2^k-1}+\frac{1}{2^k-1}\\
&\le 2^{-k}\frac{2^k-2}{2^k-1}+\frac{1}{2^k-1}\\
&= 2^{1-k}\\
&\le \frac{1}{4},
\end{aligned}$$

because $k \ge 3$, in this way, we have completed all the proofs of Theorem 5.2.

Euler test and strong pseudo prime test require some complex quadratic residual techniques. We summarize the main conclusions of this section as follows:

(A) $n$ to base $b$ is a strong pseudo prime number $\Rightarrow$ $n$ to base $b$ is an Euler pseudo prime number $\Rightarrow$ $n$ to base $b$ is a Fermat pseudo prime number; therefore, the strong pseudo prime test is the best way to detect prime numbers.

(B) Although no test can successfully detect a prime number at present, the probability detection method of strong pseudo prime number test, that is, Miller–Rabin method, can obtain that the success probability (see (5.13)) of detecting whether any odd number n is a prime number can be infinitely close to 1. That is

$$P\{\text{detect whether odd } n \ \text{ is prime}\} > 1-\varepsilon, \forall\, \varepsilon > 0 \text{ given}.$$

Moreover, the computational complexity of the detection algorithm is polynomial.

## 5.3 Monte Carlo Method

Using all the prime number test methods introduced in the previous two sections, for a huge odd number $n$, even if we already know that $n$ is not a prime number, we cannot successfully decompose $n$, because the prime number test does not provide prime factor decomposition information, A more direct method—like the sieve method—verifies whether the prime factor of $n$ is for prime numbers not greater than $\sqrt{n}$, because a compound number $n$ must have a prime factor $p$, $p \le \sqrt{n}$. Selected $p \le \sqrt{n}$, the bit operation required to divide $n$ by $p$ is $O(\log n)$, there are $O(\frac{\sqrt{n}}{\log n})$ prime numbers $p \le \sqrt{n}$ in total, therefore, the bit operation required for such a verification is $O(\sqrt{n})$. A more effective method was proposed by J. M. Pollard in 1975. We call it Monte Carlo method, or "rho" method.

First, find a convenient mapping $f$ of $\mathbb{Z}_n \xrightarrow{f} \mathbb{Z}_n$; for example, $f(x)$ is an integer coefficient polynomial, such as $f(x) = x^2 + 1$; secondly, a prime number $x_0$ is randomly generated, let $x_1 = f(x_0)$, $x_2 = f(x_1)$, ..., $x_{j+1} = f(x_j)(j = 0, 1, 2, \cdots)$. In these $x_j$, we want to find two integers $x_j$ and $x_k$, which are different elements in $\mathbb{Z}_n$, but there are some factors $d$ of $n$, $d|n$, and $x_j$ and $x_k$ are the same elements in $\mathbb{Z}_d$, that is to say

$$x_j \not\equiv x_k (\text{mod } n), (x_j - x_k, n) > 1. \tag{5.23}$$

Once $x_j$ and $x_k$ are found, the algorithm is said to be completed.

**Theorem 5.3** *Let $S$ be a set of $r$ elements, let $f : S \to S$ is a mapping, $x_0 \in S$, define $x_{j+1} = f(x_j)(j = 0, 1, 2, \ldots)$. Suppose $\lambda$ is a positive real number, let $l = 1 + [\sqrt{2\lambda r}]$, then the condition $x_0, x_1, \ldots, x_l$ is the ratio $\leq e^{-\lambda}$ of the mapping $f$ of elements in different s to the initial value $x_0$, $(f, x_0)$, $f$ in all mappings $S$ and all $x_0 \in S$.*

***Proof*** The total number of mappings $f$ from $f : S \to S$ is $r^r$, because each $x \in S$, we can arrange $r$ images for it, that is, $f(x)$ has $r$ choices. The initial value $x_0$ has $r$ choices, so the total number of $(f, x_0)$ is $r^{r+1}$. The question is which of these $(f, x_0)$ choices can satisfy the condition that $x_0, x_1, \ldots, x_l$ is a different element in $S$. we want to prove that the proportion of $(f, x_0)$ satisfying the condition in $r^{r+1}$ $(f, x_0)$ is not greater than $\leq e^{-\lambda}$.

When $x_0 \in S$ given, there are $r$ $x_0$ choices, then $x_1 = f(x_0)$ has only $r - 1$ choices and $x_2 = f(x_1)$ has only $r - 2$ choices, this goes on until $x_l = f(x_{l-1})$, there are only $r - l$ options. The remaining $x \in S$ and $f$ can be selected arbitrarily; that is, there are $r^{r-l}$ choices. Therefore, when $x_0$ is given, there are $N$ $f$ to make $(f, x_0)$ meet the required conditions, where

$$N = r^{r-l} \prod_{j=0}^{l} (r - j).$$

Divide $N$ by $r^{r+1}$, and the proportion of $(f, x_0)$ satisfying the condition is

$$\frac{N}{r^{r+1}} = r^{-l} \prod_{j=1}^{l} (r - j) = \prod_{j=1}^{l} \left(1 - \frac{j}{r}\right), \tag{5.24}$$

We notice that the real number $x \in (0, 1)$, then $\log(1 - x) < -x$. Take the logarithm to the right of the above formula, then

$$\sum_{j=1}^{l} \log\left(1 - \frac{j}{r}\right) < -\sum_{j=1}^{l} \frac{j}{r} = \frac{-l(l+1)}{2r} < -\frac{l^2}{2r}.$$

Because of $l = 1 + [\sqrt{2\lambda r}] > \sqrt{2\lambda r}$, from the above formula,

$$\sum_{j=1}^{l} \log\left(1 - \frac{j}{r}\right) < -\lambda.$$

By (5.24), we have

$$\frac{N}{r^{r+1}} \le \mathrm{e}^{-\lambda}.$$

We complete the proof of Theorem 5.3.

Monte Carlo method uses a polynomial $f(x) \in \mathbb{Z}[x]$, so that $n$ is a positive integer, and the congruence equation of mod $n$ is invariant to polynomial $f(x)$, that is

$$a \equiv b(\bmod n), \Longrightarrow f(a) \equiv f(b)(\bmod n). \tag{5.25}$$

$x_0 \in \mathbb{Z}_n$ given, $x_{j+1} = f(x_j)(j = 0, 1, \ldots)$, if you find an $x_{k_0} \in \mathbb{Z}_n$ that satisfies $x_{k_0} \equiv x_{j_0}(\bmod r)$, where $r|n$, $r > 1$, $k_0 > j_0$. By (5.25),

$$f(x_{k_0}) \equiv f(x_{j_0})(\bmod r), \Longrightarrow x_{k_0+1} \equiv x_{j_0+1}(\bmod r).$$

Thus for any $k > j$, if $k - j = k_0 - j_0$, there is $x_k \equiv x_j(\bmod r)$, this proves that a polynomial mapping $\mathbb{Z}_n \xrightarrow{f} \mathbb{Z}_n$ produces $k_0$ different residue classes under mod $r(r|n)$,

$$\{x_0, x_1, \ldots, x_{k_0-1}\}.$$

Therefore, there is the following Lemma 5.14.

**Lemma 5.14** *$f(x) \in \mathbb{Z}[x]$ is a polynomial, $n > 1$ is an positive integer, let $x_0 \in \mathbb{Z}_n$, $x_j = f(x_{j-1})(j = 1, 2, \ldots)$, if $k$ is the first subscript, there is a $j$, $0 \le j < k$, such that*

$$(x_k - x_j, n) = r > 1.$$

*Then $\{x_0, x_1, \ldots, x_{k-1}\}$ is $k$ different residual classes under* mod $r$, *so it is also $k$ different residual classes under* mod $n$. *Moreover, Monte Carlo calculation defined by $f$ can only produce $k$ different residual classes.*

We call the polynomial $f$ and the initial value $x_0$ described in Lemma 5.14 an average mapping. When the first subscript $k$ is very large, the amount of calculation is very large. Here we give an improved Monte Carlo algorithm.

$f(x) \in \mathbb{Z}[x]$ given, Monte Carlo algorithm needs to continuously calculate $x_k(k = 1, 2, \ldots)$. Let $2^h \le k < 2^{k+1}(h \ge 0)$, $j = 2^h - 1$; that is, $k$ is an $(h + 1)$-bit number, $j$ is the maximum $h$-bit number, compare $x_k$ with $x_j$ and calculate $(x_k - x_j, n)$, if $(x_k - x_j, n) > 1$, then the calculation is terminated, otherwise consider $k + 1$. The improved Monte Carlo algorithm only needs to calculate $(x_k - x_j, n)$ once for each $k$ , $j = 2^h - 1$. There is no need to verify every $j, 0 \le j < k$, when $k$ is very large, it reduces a lot of computation, but there is a disadvantage. It may miss the smallest subscript $k$ satisfying the condition, but the error is controllable. In fact, we have the following error estimation.

**Lemma 5.15** *$f(x) \in \mathbb{Z}[x]$, $n \geq 1$ given, $x_0 \in \mathbb{Z}_n$, $x_j = f(x_{j-1})(j = 1, 2, \ldots)$, let $k_0$ be the smallest subscript and satisfy $(x_{k_0} - x_{j_0}, n) > 1$, where $0 \leq j_0 < k_0$, assuming that k is the smallest positive integer satisfying $(x_k - x_j, n) > 1$ in the improved Monte Carlo algorithm, we have $k \leq 4k_0$.*

***Proof*** Suppose $k_0$ has $(h + 1)$ bits. Let $j = 2^{h+1} - 1, k = j + (k_0 - j_0)$. By Lemma 5.14, then

$$(x_{k_0} - x_{j_0}, n) > 1, \Longrightarrow (x_k - x_j, n) > 1.$$

Obviously, $j$ is the maximum number of $(h + 1)$ bits and $k$ is the number of $(h + 2)$ bits, so $k$ is the required subscript calculated by the improved Monte Carlo algorithm. Obviously,

$$k = j + (k_0 - j_0) \leq 2^{h+1} - 1 + 2^{h+1} < 4 \cdot 2^h \leq 4k_0.$$

Lemma 5.15 holds.

***Example 5.6*** Let $n = 91$, $f(x) = x^2 + 1$, $x_0 = 1$. By Monte Carlo algorithm, then $x_1 = 2$, $x_2 = 5$, $x_3 = 26$ and $x_4 = 40$ (because$26^2 + 1 \equiv 40 (\bmod 91)$). By the improved Monte Carlo algorithm, only $(x_4 - x_3, 91)$ needs to be detected to obtain

$$(x_4 - x_3, 91) = (14, 91) = 7.$$

**Lemma 5.16** *Let n be an odd number and a compound number, and r be a factor of n, $r|n$, $1 < r < \sqrt{n}$. Let $f(x) \in \mathbb{Z}[x]$, $x_0 \in \mathbb{Z}_n$ given, then the computational complexity of finding r by Monte Carlo algorithm $(f, x_0)$ is*

$$\text{Time}((f, x_0)) = O(\sqrt{n} \log^3 n) \text{ bits}. \tag{5.26}$$

*Further, there is a normal number C, so that for any positive real number $\lambda$, the success probability of Monte Carlo algorithm $(f, x_0)$ to find a nontrivial factor r of n is greater than $1 - e^{-\lambda}$, that is*

$$P\{(f, x_0) \text{find out } r|n, r > 1\} \geq 1 - e^{-\lambda}. \tag{5.27}$$

*The number of bit calculation operations required by the algorithm that depends on parameter $\lambda$ (to ensure the success rate of the algorithm) is $O(\sqrt{\lambda}\sqrt[4]{n}\log^3 n)$.*

***Proof*** From the discussion of computational complexity in Chap. 1, finding the maximum common divisor of two integers and the addition, subtraction, multiplication and division in mod $n$ are polynomial. Let $C_1$ satisfies

$$\text{Time}((y - z, n)) \leq C_1 \log^3 n, \text{ where } y, z \leq n.$$

$C_2$ satisfies

$$\text{Time}(f(x) \bmod n) \le C_2 \log^3 n, x \in \mathbb{Z}_n.$$

If $k_0$ is $(f, x_0)$, the first subscript in the calculation satisfies $(x_{k_0} - x_{j_0}, n) > 1$, by the improved Monte Carlo algorithm, we have $(x_k - x_j, n) > 1$, where $j = 2^h - 1$, $2^h \le k < 2^{h+1}$. By Lemma 5.15, $k \le 4k$. Thus

$$\text{Time}(\text{found by } (f, x_0)\ k) \le 4k_0(C_1 \log^3 n + C_2 \log^3 n). \tag{5.28}$$

Let $(x_{k_0} - x_{j_0}, n) = r > 1, r < \sqrt{n}$, by Lemma 5.14, $k_0 \le r$, so

$$\text{Time}(\text{find } r, r|n, r < \sqrt{n}) \le 4\sqrt{n}(C_1 \log^3 n, C_2 \log^3 n).$$

Equation (5.26) proved. In the sense of probability, that is, on the premise of allowing certain errors, Eq. (5.26) can be further improved.

Let $\lambda > 0$ be any given real number, by Lemma 5.3, ratio of $k_0 \ge 1 + \sqrt{2\lambda r}$ $< \mathrm{e}^{-\lambda}$, in other words, the probability of successfully finding $r, r|n, r \le \sqrt{n}$ is

$$P\{\text{find out } r, r|n, r < \sqrt{n}\} \ge 1 - \mathrm{e}^{-\lambda}.$$

In order to ensure the success rate, then $k_0 \le 1 + \sqrt{2\lambda r}$. By (5.28), the number of bit operations required shall not be greater than

$$4(1 + \sqrt{2\lambda r})(C_1 \log^3 n + C_2 \log^3 n) = O(\sqrt{\lambda}\sqrt[4]{n} \log^3 n).$$

We have completed the proof of Lemma.

***Remark 5.1*** A basic assumption of Monte Carlo method is that the integer coefficient polynomial $f$ can be used as an average mapping (see Lemma 5.14); this has not yet been proved.

## 5.4 Fermat Decomposition and Factor Basis Method

**Lemma 5.17** *Suppose $n$ is an odd number, there is a 1-1 correspondence between factorization $n = a \cdot b (a \ge b > 0)$ of $n$ and expression $n = t^2 - s^2$ ($t$ and $s$ are nonnegative integers) of $n$. The corresponding $\sigma : (a, b) \to (t, s)$ can be written as $\sigma((a, b)) = (t, s)$, where*

$$\sigma((a, b)) = \left(\frac{a+b}{2}, \frac{a-b}{2}\right).$$

*Inverse mapping is*

$$\sigma^{-1}((t, s)) = (t + s, t - s).$$

***Proof*** If $n = ab$, because both $a$ and $b$ are odd, then $n = (\frac{a+b}{2})^2 - (\frac{a-b}{2})^2$, so define

$$\sigma((a,b)) = \left(\frac{a+b}{2}, \frac{a-b}{2}\right).$$

Conversely, if $n = t^2 - s^2$, then $n = (t+s)(t-s)$. So define $\sigma^{-1}((t,s)) = (t+s, t-s)$, we prove $\sigma^{-1}\sigma = 1, \sigma\sigma^{-1} = 1$. By the definition,

$$\begin{cases} \sigma^{-1}\sigma((a,b)) = \sigma^{-1}\left(\frac{a+b}{2}, \frac{a-b}{2}\right) = (a,b), \\ \sigma(\sigma^{-1}((t,s))) = \sigma(t+s, t-s) = (t,s). \end{cases}$$

So $\sigma$ is a 1-1 correspondence between the two decomposition $n = ab = t^2 - s^2$, the Lemma holds.

The above simple lemma provides us with a method of factor decomposition, called Fermat factor decomposition: if $n = ab$, $a$ is very close to $b$, then $n = (\frac{a+b}{2})^2 + (\frac{a-b}{2})^2 = t^2 - s^2$, where $s$ is very small and $t$ is only a little larger than $\sqrt{n}$. Therefore, starting from $t = [\sqrt{n}] + 1$, we successively detect whether $t^2 - n$ is a complete square number. If not, we change it to $t = [\sqrt{n}] + 2$ for detection. In this way, until $t^2 - n = s^2$, we get $n = (t+s)(t-s)$ through Fermat factorization. This method is effective when $n = ab$, $a$ and $b$ are very close.

Fermat factor decomposition can be further expanded into a factor-based method to become a more effective factor decomposition method. Its basic idea is: in Fermat factorization, $t^2 - n^2$ is required to be a complete square, which is difficult to appear in practice, but $t^2 \equiv s^2 (\bmod n)$, $t \not\equiv \pm s (\bmod n)$ is easy to appear. Calculate the maximum common divisor $(t+s, n)$ and $(t-s, n)$, then we have factorization

$$n = (t+s, n)(t-s, n).$$

**Definition 5.5** Let B be $h$ different primes (maybe $p_1 = -1$), $B$ is called a factor base. An integer $b$ is called a $B$-number, if the minimum nonnegative residue of $b^2$ under $\bmod n$ can be expressed as the product of prime numbers in $B$, where $n$ is the given positive integer.

***Example 5.7*** Let $n = 4633$, $B = \{-1, 2, 3\}$, then 67, 68, 69 are all $B$-number, because $67^2 \equiv -144 (\bmod 4633)$, $68^2 \equiv -9 (\bmod 4633)$, $69^2 \equiv 128 (\bmod 4633)$.

If $b$ is a $B$-number, $b^2 \bmod n$ represents the minimum nonnegative residue of $b^2$ under $\bmod n$, by the definition,

$$b^2 \bmod n = \prod_{i=1}^{h} p_i^{\alpha_i}, \alpha_i \geq 0.$$

Let $e = \{e_1, e_2, \ldots, e_h\} \in \mathbb{F}_2^h$ be an $h$-dimensional binary vector, define

$$e_j = \begin{cases} 0, & \text{if } \alpha^j \text{ is even;} \\ 1, & \text{if } \alpha^j \text{ is odd.} \end{cases} \quad 1 \leq j \leq h.$$

$e$ is called the binary vector corresponding to $b$ if $\{b_i\} = A$ is a set of $B$-numbers. The binary vector corresponding to each $b_i$ is denoted as $e_i = \{e_{i_1}, e_{i_2}, \ldots, e_{i_h}\}$, denote $b_i^2 \bmod n$ with $a_i$. We have

$$\prod_{i\in A} a_i = \prod_{j=1}^{h} p_j^{\sum_{i\in A}\alpha_{ij}}, \text{ where } a_i = \prod_{j=1}^{h} p_j^{\alpha_{ij}},$$

Suppose $\sum_{i\in A} e_i = (0, 0, \ldots, 0)$ is the zero vector in $\mathbb{F}_2^h$, then

$$\sum_{i\in A} \alpha_{ij} \equiv 0 (\bmod 2), \forall\, 1 \leq j \leq h.$$

That is, $\prod a_i$ is a square number. Let $r_j = \frac{1}{2}\sum_{i\in A}\alpha_{ij}$, then

$$\prod_{i\in A} a_i = \left(\prod_{j=1}^{h} p_j^{r_j}\right)^2, \text{ define } c = \prod_{j=1}^{h} p_j^{r_j}, \tag{5.29}$$

On the other hand, $b_i \bmod n$ represents the minimum nonnegative residue of $b_i$ under $\bmod n$, let

$$b = \prod_{i\in A}(b_i \bmod n) = \prod_{i\in A}\delta_i, \tag{5.30}$$

where $\delta_i = b_i \bmod n$, that is $0 \leq \delta_i < n$, and $b_i \equiv \delta_i (\bmod n)$, thus

$$\prod_{i\in A} b_i \equiv b (\bmod n).$$

Because of $a_i = b_i^2 \bmod n$, that is $0 \leq a_i < n$, and $b_i^2 \equiv a_i (\bmod n)$. There is

$$\prod_{i\in A} b_i^2 = b^2 \equiv \prod_{i\in A} a_i = c^2 (\bmod n).$$

Two different integers $b$ and $c$ defined by Eqs. (5.29) and (5.30) satisfy $b^2 \equiv c^2 (\bmod n)$, We write the above analysis as the following lemma.

**Lemma 5.18** *Let $A = \{b_1, b_2, \ldots, b_i, \ldots\}$ be a finite set of some $B$-numbers, let $e_i = (e_{i_1}, e_{i_2}, \ldots, e_{i_h}) \in \mathbb{F}_2^h$ be the binary vector corresponding to $b_i$, $a_i = b_i^2 \bmod n$, $\delta_i = b_i \bmod n$. If $\sum_{i\in A} e_i = 0$ is the zero vector in $\mathbb{F}_2^h$, then $\prod_{i\in A} a_i$ is a square number. Write*

$$a_i = \prod_{j=1}^{h} p_j^{\alpha_{ij}}, \quad \prod_{i \in A} a_i = \prod_{j=1}^{h} p_j^{\sum_{i \in A} \alpha_{ij}} = c^2.$$

*where*

$$c = \prod_{j=1}^{h} p_j^{\frac{1}{2}\sum_{i \in A} \alpha_{ij}},$$

*Further let* $b = \delta_1\delta_2 \cdots$, *we have* $b^2 \equiv c^2 (\bmod n)$.

From the above lemma, if $b^2 \equiv c^2 (\bmod n)$, $b \not\equiv \pm c (\bmod n)$. Then we will find a nontrivial factor $d = (b + c, n)$ of $n$. Now the question is, if $b^2 \equiv c^2 (\bmod n)$, how likely is $b \not\equiv \pm c (\bmod n)$? Might as well make $(b, n) = (c, n) = 1$, otherwise both sides are divided by $(b, n)^2$: by $b^2 \equiv c^2 (\bmod n)$, $\Longrightarrow (bc^{-1})^2 \equiv 1 (\bmod n)$. The problem is transformed into how many solutions $x$ are in $x^2 \equiv 1 (\bmod n)$, $1 \leq x < n$.

**Lemma 5.19** *Let $n$ be an odd number, then the number of solutions of $x^2 \equiv 1 (\bmod n)$ is $2^r$, where $r$ is the number of different prime factors of $n$.*

***Proof*** If $r = 1$, then $n = p^\alpha (\alpha \geq 1)$, $p$ is an odd prime, now $x^2 \equiv 1 (\bmod p^\alpha)$ has two solutions $x = \pm 1$, because let $g$ be the original root of mod $p^\alpha$, then $x = g^t (1 \leq t \leq p^{\alpha-1}(p-1))$, $x^2 = 1 \Leftrightarrow p^{\alpha-1}(p-1) | 2t$. So there are only two solutions $t = \frac{1}{2} p^{\alpha-1}(p-1)$ and $t = p^{\alpha-1}(p-1)$. So $x \equiv \pm 1 (\bmod p^\alpha)$. If $n = p_1^{\alpha_1} \cdots p_r^{\alpha_r}$, then the number of solutions of $x^2 \equiv 1 (\bmod n)$ deduced from the Chinese remainder theorem is $2^r$. The Lemma holds!

**Lemma 5.20** *$n$ is an odd number and is the product of the power of more than two different primes, $B = \{p_1, p_2, \ldots, p_h\}$ is a factor base. Randomly select two $B$-numbers $b$ and $c$, then $b^2 \equiv c^2 (\bmod n)$, $\Longrightarrow b \equiv \pm c (\bmod n)$'s rate is $\leq \frac{1}{2}$.*

***Proof*** $x^2 \equiv 1 (\bmod n)$ has $2^r$ different solutions $(\bmod n)$, $r \geq 2$. The two solutions corresponding to $x \equiv \pm 1 (\bmod n)$ correspond to $b \equiv \pm c (\bmod n)$. Thus

$$b^2 \equiv c^2 (\bmod n), \Longrightarrow b \equiv \pm c (\bmod n)\text{'s rate } \leq \frac{2}{2^r} \leq \frac{1}{2},$$

Lemma 5.20 holds.

According to Lemma 5.20, $b$ and $c$ are selected by using factor basis, if $b \equiv \pm c (\bmod n)$, then select failure, and the probability of failure is $\leq \frac{1}{2}$. If the selection fails, select another $b_1$ and $c_1$, in this way, we randomly select $k$ $b$ and $c$ equally almost independently, and the probability of success of $b \not\equiv \pm c (\bmod n)$ is

$$P\{b^2 \equiv c^2 (\bmod n), b \not\equiv \pm c (\bmod n)\} \geq 1 - \frac{1}{2^k}. \tag{5.31}$$

In other words, the probability of finding a nontrivial factor $d = (b + c, n)$ of $n$ by using the factor base can be infinitely close to 1. Below, we systematically summarize the factor base decomposition method as follows:

Factor-based method

Let $n$ be a large odd number and $y$ be an appropriately selected integer (e.g., $y \le n^{\frac{1}{10}}$), let the factor base be

$$B = \{-1, p \mid p \text{ is prime}, \ p \le y\}.$$

Select a certain number of $B$-number at random, $A_1 = \{b_1, b_2, \ldots, b_N\}$, usually $N \le \pi(y) + 2$ will meet the needs. Each $b_i$ is expressed as the product of prime numbers in $B$. Calculate the corresponding binary vector $e_i$, select a subset $A \subset A_1$ in $A_1$, such that $\sum_{i \in A} e_i = 0$, $b_i$ corresponding to binary vector $e_i$, denote as $A = \{b_1, b_2, \ldots, b_i, \ldots\}$. Let

$$b = \prod_{i \in A}(b_i \bmod n) = \prod_{i \in A} \delta_i, \quad \text{where } \delta_i = b_i \bmod n$$

and

$$c = \prod_{j \in B} p_j^{r_j} \bmod n, \ r_j = \frac{1}{2}\sum_{i \in A} \alpha_{ij}.$$

We have $b^2 \equiv c^2 (\bmod n)$, if $b \equiv \pm c (\bmod n)$, then reselect the subset $A$, Until finally $b \not\equiv \pm c (\bmod n)$, in this way, we find a nontrivial factor $d|n$ of $n$, $d = (b + c, n)$. Therefore, there is factorization $n = d \cdot \frac{n}{d}$.

Factor decomposition using factor-based method cannot guarantee the success rate of 100% because $b \not\equiv \pm c (\bmod n)$ cannot be deduced from $b^2 \equiv c^2 (\bmod n)$, however, the success probability of factorization for large odd $n$ can be infinitely close to 1. Under the condition of success probability $\ge 1 - \frac{1}{2^k}$ ($k$ is a given normal number), the computational complexity of factorization $n$ of by factor-based method can be estimated as

$$\text{Time(factor-based method to } n \text{ factorization)} = O(\mathrm{e}^{c\sqrt{\log n \log \log n}}). \tag{5.32}$$

The proof of Formula (5.32) is relatively complex. No detailed proof is given here. Interested readers can refer to pages 136–141 of (Pomerance, 1982a) in reference 5. The exact value of $C$ in (5.32) is unknown. It is generally guessed that $C = 1 + \varepsilon$, where $\varepsilon > 0$ is any small positive real number.

Let $k$ be the number of bits of $n$, and the estimate on the right of (5.32) can be written as $O(\mathrm{e}^{c\sqrt{k \log k}})$. Therefore, the computational complexity of the factor-based method is sub-exponential. Compared with the Monte Carlo method introduced in the previous section (see (5.31)), its computational complexity is exponential, because

$$O(\sqrt{n}) = O(\mathrm{e}^{c_1 k}), \ \text{where } c_1 = \frac{1}{2}\log 2.$$

As we all know, the security of RSA public key cryptography is based on the prime factorization $n = pq$ of $n$. Although there is no general method to factorize any large odd $n$, although Monte Carlo method and factor-based method are probability calculation methods, the probability of successful factorization is very large, The disadvantage is that their computational complexity is exponential and sub exponential, which is the reason for choosing huge prime numbers $p$ and $q$ in RSA.

## 5.5 Continued Fraction Method

In the factor-based method introduced in the previous section, $b^2 \bmod n$ can be the residual of the minimum absolute value of $b^2$ under $\bmod n$, that is

$$b^2 \equiv b^2 \bmod n (\bmod n), \quad |b^2 \bmod n| \leq \frac{n}{2}.$$

In this way, $b^2 \bmod n$ can be decomposed into the product of some smaller prime numbers. The continued fraction method is the best method at present. How to find the integer $b$, so that $|b^2 \bmod n| < 2\sqrt{n}$, $b^2 \bmod n$ is more likely to be decomposed into the product of some small prime numbers. First, we introduce what is continued fraction and some basic properties.

Suppose $x \in \mathbb{R}$ is a real number, $[x]$ is the integer part of $x$, and $\{x\}$ is the decimal part of $x$. Let $a_0 = [x]$, if $\{x\} \neq 0$, and let $a_1 = [\frac{1}{\{x\}}]$, because of $x = [x] + \{x\}$, there is

$$x = a_0 + \frac{1}{\{x\}} = a_0 + \frac{1}{a_1 + \{\{x\}^{-1}\}}.$$

If $\{\{x\}^{-1}\} \neq 0$, write

$$a_2 = [\{\{x\}^{-1}\}^{-1}],$$

consider

$$\{\{\{x\}^{-1}\}^{-1}\}^{-1},$$

So we got

$$x = a_0 + \frac{1}{a_1 + \frac{1}{a_2 + \frac{1}{a_3 + \cdots}}}.$$

The above formula is called the continued fraction expansion of real number $x$. To save space, write $x = [a_0, a_1, \ldots, a_n, \ldots]$, if and only if $x$ is a rational number, the continued fraction of $x$ is expanded to be finite, denote as

$$x = [a_0, a_1, \ldots, a_n], \text{ where } a_n > 1.$$

It is called the standard expansion of rational number $x$.

**Definition 5.6** $x = [a_0, a_1, \ldots, a_n, \ldots]$ is the continued fraction expansion of $x$, for $i \geq 0$, call $\frac{b_i}{c_i} = [a_0, a_1, \ldots, a_i]$ the $i$th asymptotic fraction of $x$, specially,

$$\frac{b_0}{c_0} = \frac{a_0}{1}, \ \frac{b_1}{c_1} = \frac{a_1 a_0 + 1}{a_1}.$$

The progressive fraction $\frac{b_i}{c_i}$ of the real number $x$ is a reduced fraction, that is $(b_i, c_i) = 1$, and has the following properties.

**Lemma 5.21** *$x = [a_0, a_1, \ldots, a_n, \cdots]$ is the continued fraction expansion of $x$, $\frac{b_i}{c_i}$ is the asymptotic fraction, then*

*(i) when $i \geq 2$,*

$$\frac{b_i}{c_i} = \frac{a_i b_{i-1} + b_{i-2}}{a_i c_{i-1} + c_{i-2}}. \tag{5.33}$$

*(ii) If $i \geq 1$, then*

$$b_i c_{i-1} - b_{i-1} c_i = (-1)^{i-1}. \tag{5.34}$$

***Proof*** We prove that (i) by induction. Obviously, the proposition of $i = 2$ holds, that is

$$\frac{b_2}{c_2} = \frac{a_2 b_1 + b_0}{a_2 c_1 + c_0} = \frac{a_2(a_1 a_0 + 1) + a_0}{a_2 a_1 + 1}.$$

If the proposition holds for $i$, that is

$$\frac{b_i}{c_i} = \frac{a_i b_{i-1} + b_{i-2}}{a_i c_{i-1} + c_{i-2}}.$$

Then write $[a_0, a_1, \ldots, a_i, a_{i+1}] = [a_0, a_1, \ldots, a_i + \frac{1}{a_{i+1}}]$,

$$\frac{b_{i+1}}{c_{i+1}} = \frac{\left(a_i + \frac{1}{a_{i+1}}\right) b_{i-1} + b_{i-2}}{\left(a_i + \frac{1}{a_{i+1}}\right) c_{i-1} + c_{i-2}} = \frac{a_{i+1} b_i + b_{i-1}}{a_{i+1} c_i + c_{i-1}}.$$

So (i) holds.

We prove Formula (5.34) by induction, when $i = 1$,

$$b_1 c_0 - b_0 c_1 = a_1 a_0 + 1 - a_1 a_0 = 1 = (-1)^0.$$

So when $i = 1$, the proposition holds, and when $i$, the proposition holds, that is

$$b_i c_{i-1} - b_{i-1} c_i = (-1)^{i-1}.$$

Then

$$\begin{aligned} b_{i+1}c_i - b_ic_{i+1} &= (a_{i+1}b_i + b_{i-1})c_i - b_i(a_{i+1}c_i + c_{i-1}) \\ &= b_{i-1}c_i - b_ic_{i-1} \\ &= (-1)^i. \end{aligned}$$

Lemma 5.21 holds.

Continued fractions have many important applications in numbers, such as rational approximation of real numbers and rational approximation of algebraic numbers. Periodic continued fractions are an important special case in rational approximation of algebraic numbers. $x = [a_0, a_1, \ldots, a_n, \ldots]$. If these $a_i$ occur in cycles of a certain length, they are called periodic continued fractions. The famous Lagrange theorem shows that the necessary and sufficient condition for the expansion of the continued fraction of $x$ into a periodic continued fraction is that $x$ is a quadratic real algebraic number. Here we do not discuss some profound properties of continued fractions, but only prove some properties we need.

**Lemma 5.22** *Let $x > 1$ be a real number, $\frac{b_i}{c_i} (i \geq 0)$ is the asymptotic fraction of $x$, then*

$$|{b_i}^2 - x^2c_i^2| < 2x, \forall\, i \geq 0.$$

***Proof*** Because $x$ is between progressive scores $\frac{b_i}{c_i}$ and $\frac{b_{i+1}}{c_{i+1}}$, by property (ii) of Lemma 5.21, there is

$$\left|\frac{b_{i+1}}{c_{i+1}} - \frac{b_i}{c_i}\right| = \frac{1}{c_ic_{i+1}}, i \geq 0.$$

Thus

$$\begin{aligned} |{b_i}^2 - x^2c_i^2| &= c_i^2\left|x - \frac{b_i}{c_i}\right|\left|x + \frac{b_i}{c_i}\right| \\ &< c_i^2 \cdot \frac{1}{c_ic_{i+1}}\left(x + \left(x + \frac{1}{c_ic_{i+1}}\right)\right). \end{aligned}$$

So

$$\begin{aligned} |{b_i}^2 - x^2c_i^2| - 2x &< 2x\left(-1 + \frac{c_i}{c_{i+1}} + \frac{1}{2xc_{i+1}^2}\right) \\ &< 2x\left(-1 + \frac{c_i}{c_{i+1}} + \frac{1}{c_{i+1}}\right) \\ &< 2x\left(-1 + \frac{c_{i+1}}{c_{i+1}}\right) = 0. \end{aligned}$$

The Lemma holds.

**Lemma 5.23** *Let $n$ be a positive integer and $n$ not a complete square. Let $\{\frac{b_i}{c_i}\}_{i\geq 0}$ be the asymptotic fraction of the continued fraction expansion of $\sqrt{n}$, and $b_i^2$* mod $n$ *be the residue of the minimum absolute value of $b_i^2$ under* mod $n$, *then we have*

$$b_i^2 \bmod n < 2\sqrt{n}, \ \forall\, i \geq 0.$$

***Proof*** By Lemma 5.22, let $x = \sqrt{n}$, then

$$b_i^2 \equiv b_i^2 - nc_i^2 (\bmod n).$$

Because

$$|b_i^2 - nc_i^2| < 2\sqrt{n}, \Longrightarrow b_i^2 \bmod n < 2\sqrt{n}, \ \forall\, i \geq 0.$$

The Lemma holds.

Combining the above Lemma 5.23 with the factorization method, we obtain the continued fraction decomposition method.

Continued fraction decomposition method:

The operations of $\bmod n$ involved in this algorithm, except that it is specially pointed out, are the minimum nonnegative residue of $\bmod n$. If $n$ is a large odd number, it is also a compound number, first let $b_{-1} = b$, $b_0 = a_0 = [\sqrt{n}]$, and $x_0 = \sqrt{n} - a_0 = \{\sqrt{n}\}$, calculate $b_0^2 \bmod n$, in fact, $b_0^2 \bmod n = b_0^2 - n$. Second, consider $i = 1, 2, \ldots$. To determine $b_i$, we proceed in several steps:

1. Let $a_i = [\frac{1}{x_{i-1}}]$, and $x_i = \frac{1}{x_{i-1}} - a_i (i \geq 1)$.
2. Let $b_i = a_i b_{i-1} + b_{i-2}$, the minimum nonnegative residual $b_i \bmod n$ of $b_i$ under $\bmod n$ is still recorded as $b_i$.
3. calculate $b_i^2 \bmod n$.

By Lemma 5.23, $b_i^2 \bmod n < 2\sqrt{n}$, it can be decomposed into the product of some small prime numbers. If a prime number $p$ appears in the decomposition of two or more $b_i^2 \bmod n$, or in the decomposition of an $b_i^2 \bmod n$, $p$ appears to an even power, $p$ is called a standard prime number, in other words, a standard prime $p$ is

$$p | b_i^2 \bmod n, \ \ p | b_j^2 \bmod n, i \neq j.$$

Or

$$p^\alpha \| b_i^2 \bmod n, \ \ \alpha \text{ is even}.$$

We choose factor base $B$ as

$$B = \{-1, \ \text{standard prime}\}.$$

In this way, all $b_i^2 \bmod n$ are $B$-numbers, and the corresponding binary vector is $e_i$. Select a subset $A = \{b_i\}, \Longrightarrow \sum_{i \in A} e_i = 0$. Let

$$b = \prod_{i \in A} (b_i \bmod n) = \prod_{i \in A} \delta_i$$

and $c = \prod_{j \in B} p_j^{r_j}$, where

$$r_j = \frac{1}{2}\sum_{i\in A}\alpha_{ij}, \forall j \in B.$$

If $b \not\equiv \pm c(\bmod n)$, then $(b+c, n)$ is a nontrivial factor of $n$, and we obtain the factorization of $n$. If $b \equiv \pm c(\bmod n)$, then another subset $A$ is selected and repeated to complete the continued fraction factorization method.

***Example 5.8*** The continued fraction method is used to factorize $n = 9073$.

Solution: We calculate $a_i, b_i$ and $b_i^2 \bmod n$ in turn, where $b_i = (a_i b_{i-1} + b_{i-2}) \bmod n$, the table is as follows:

| $i$ | 0 | 1 | 2 | 3 | 4 |
|---|---|---|---|---|---|
| $a_i$ | 95 | 3 | 1 | 26 | 2 |
| $b_i$ | 95 | 286 | 381 | 1119 | 2619 |
| $b_i^2 \bmod n$ | $-48$ | 139 | $-7$ | 87 | $-27$ |

From the value of $b_i^2 \bmod n$, we can choose the factor base $B$ as $B = \{-1, 2, 3, 7\}$. Then $b_i^2 \bmod n$ is the number of $B$-number, when $i = 0, 2, 4, \ldots$. The corresponding binary vector is

$$e_0 = (1, 4, 1, 0), e_2 = (1, 0, 0, 1), \text{ and } e_4 = (1, 0, 3, 0).$$

Easy to calculate $e_0 + e_4 = (0, 0, 0, 0)$. Therefore, we choose

$$\begin{cases} b = 95 \cdot 2619 \equiv 3834 \bmod 9073; \\ c = 2^2 \cdot 3^2 = 36. \end{cases}$$

Because $b^2 \equiv c^2(\bmod 9073)$, that is $3834^2 = 36^2(\bmod 9073)$, but $3834 \not\equiv \pm 36$ $(\bmod 9073)$, so we get a nontrivial factor of $n = 9073$, $d = (3834 + 36, 9073) = 43$. Thus $9073 = 43 \cdot 211$, the factorization of 9073 is obtained.

**Exercise 5**

1. $p$ is a prime, if and only if $b^{p-1} \equiv 1(\bmod p^2)$, $p^2$ to base $b$ is a Fermat pseudo prime.
2. What is the minimum pseudo prime number with Fermat pseudo prime for base 5? What is the minimum Fermat pseudo prime number for base 2?
3. $n = pq, p \neq q$ are two primes, let $d = (p-1, q-1)$, it is proved that $n$ to base $b$ is Fermat pseudo prime number, if and only if $b^d \equiv 1(\bmod n)$, and calculate the number of bases $b$.
4. If $b \in \mathbb{Z}_n^*$, $n$ to base $b$ is Fermat pseudo prime, then $n$ to base $-b$ and $b$ are Fermat pseudo prime numbers.
5. If $n$ to base 2 is Fermat pseudo prime, then $N = 2^n - 1$ is also Fermat pseudo prime.
6. If $n$ to base $b$ is Fermat pseudo prime, and $(b-1, n) = 1$, then $N = \frac{b^n-1}{b-1}$ to base $b$ is also Fermat pseudo prime.

7. Prove that the following integers are Carmichael numbers:

$$1105 = 5 \cdot 13 \cdot 17, 1729 = 7 \cdot 13 \cdot 19, 2465 = 5 \cdot 17 \cdot 29, 2821 = 7 \cdot 13 \cdot 31,$$

$$6601 = 7 \cdot 23 \cdot 41, 29{,}341 = 13 \cdot 37 \cdot 61, 172{,}081 = 7 \cdot 13 \cdot 31 \cdot 61, 278{,}545 = 5 \cdot 17 \cdot 29 \cdot 113.$$

8. Find all Carmichael numbers of form $3pq$ and all Carmichael numbers of form $5pq$.
9. Prove that 561 is the minimum Carmichael number.
10. If $n$ to base 2 is a Fermat pseudo prime, prove $N = 2^n - 1$ is a strong pseudo prime.
11. There are infinite Euler pseudo primes and strong pseudo primes for base 2.
12. If $n$ to base $b$ is a strong pseudo prime, then $n$ to base $b^k$ is also a strong pseudo prime for any integer $k$.
13. The Fermat factorization method is used to decompose the positive integer as follows:

$$n = 8633, n = 809{,}009, n = 92{,}296{,}873, n = 88{,}169{,}891.$$

14. The Fermat factorization method is used to decompose the positive integer as follows:

$$n = 68{,}987, n = 29{,}895{,}581, n = 19{,}578{,}079, n = 17{,}018{,}759.$$

15. Expand the rational number $x = \frac{45}{89}, x = \frac{55}{89}, x = 1.13$ into continued fractions.
16. Let $a$ be a positive integer, $x = [a, a, a, \cdots]$, calculate $x =?$

## References


Adelman, L. M., Pomerance, C., & Rumely, R. S. (1983). On distinguishing prime number from composite numbers. *Annals of Mathematics, 117,* 173–206.

Berent, R. P., & Pollared, J. M. (1981). Factorization of the eighth Fermat number. *Mathematics of Computation, 36,* 627–630.

Blair, W. D., Lacampague, C. B., & Selfridge, J. L. (1986). Factoring large numbers on a pocket calculator. *The American Mathematical Monthly, 93,* 802–808.

Brent, R. P. (1980). An improved Monte Carlo factorization algorithm. *BIT, 20,* 176–184.

Cohen, H., & Lenstra, H. W. (1984). Primality testing and Jacobi sums. *Mathematics of Computation, 142,* 297–330.

Dawonport, H. (1982). *The higher arithmetic*. Cambridge University Press.

Dickson, L. E. (1952). *History of the theory of number* (Vol. 1). Chelsea.

Dixon, J. D. (1984). Factorization and primality tests. *The American Mathematical Monthly, 91,* 333–352.

Guy, R. K. (1975). How to factor a number. In *Proceedings of the 5th Manitoba Conference on Numerical Mathematics* (pp. 49–89).

Kranakis, E. (1986). *Primality and cryptography*. Wiley.

Lehman, R. S. (1974). Factoring large number. *Mathematics of Computation, 28*, 637–646.
Lehmer, D. H., & Powers, R. E. (1931). On factoring large number. *Bulletin of the American Mathematical Society, 37*, 770–776.
Miller, G. L. Riemann's hypothesis and tests for primality. In *Proceedings of the 7th Annual ACM Symposium on the Theory of Computing* (pp. 234–239).
Morrison, M. A., & Brillhart, J. (1975). A method of factoring and the factorization of $\mathbb{F}_7$. *Mathematics of Computation, 29*, 183–205.
Pollard, J. M. (1975). A Monte Carlo method for factorization. *BIT, 15*, 331–334.
Pomerance, C. (1981). Recent development in primality testing. *The Mathematical Intelligencer,3*, 97–105.
Pomerance, C. (1982a). Analysis and comparison of some integer factoring algorithms. *Computation Methods in Number Theory, Part I*.
Pomerance, C. (1982b). The search for prime number. *Scientific American,427*, 136–147.
Pomerance, C., & Wagstaff, S. S. (1983). Implementation of the continued fraction integer factoring algorithm. In *Proceedings of the 12th Winnipeg Conference on Numerical Methods and computing*.
Rabin, M. O. (1980). Probabilities algorithms for testing Primality. *Journal of Number Theory, 12*, 128–138.
Solovag, R., & Strassen, V. (1977). A fast Munte Carlo test for primality. *SIAM Journal for Computing, 6*, 84–85.
Wagon, S. (1986). Primality testing. *The Mathematical Intelligence, 8*, 58–61.
Wunderlich, M. C. (1979). A running time and analysis of Brillhart's continued fraction factoring method. *Number Theory, Carbondale, Springer Lecture Notes, 175*, 328–342.
Wunderlich, M. C. (1985). Implementing the continued fraction factoring algorithm on parallel machines. *Mathematics of Computation, 44*, 251–260.

# Chapter 6
# Elliptic Curve

In 1985, mathematician v. Miller introduced elliptic curve into cryptography for the first time. In 1987, mathematician N. Koblitz further improved and perfected Miller's work and formed the famous elliptic curve public key cryptosystem. Elliptic curve public key cryptosystem, RSA public key cryptosystem and ElGamal public key cryptosystem based on discrete logarithm are recognized as the three major public key cryptosystems, which occupy the most prominent position in modern cryptography. Compared with RSA cryptography, elliptic curve cryptography can provide the same or higher level of security with a shorter key; compared with ElGamal cryptosystem, they are based on the same mathematical principle and are essentially based on discrete logarithm cryptosystem. ElGamal cryptosystem is based on the discrete logarithm of multiplication group over finite field, and elliptic curve cryptosystem is based on the discrete logarithm of Mordell group of elliptic curve over finite field, but choosing elliptic curve has more flexibility than choosing finite field, so elliptic curve cryptosystem has attracted more attention This paper systematically and comprehensively introduces elliptic curve cryptography from the three aspects of cryptography mechanism and factorization, in order to make readers better understand and master this public key cryptography mechanism.

## 6.1 Basic Theory

The working platform of this chapter is a field $E$, especially $E = \mathbb{R}$(real number field), $E = \mathbb{C}$(complex field), $E = \mathbb{Q}$(rational number field) or $E = \mathbb{F}_q$(Finite field of $q$ elements) four common fields. The characteristic $\chi(E)$ of a field $E$ is the order of the multiplicative unit element $e$ of $E$ in the additive group. That is, $\chi(E) = o(e)$ is a prime number or $\infty$, specifically,



Z. Zheng, *Modern Cryptography Volume 1*, Financial Mathematics and Fintech,
https://doi.org/10.1007/978-981-19-0920-7_6

$$\chi(E) = \begin{cases} \infty, & \text{if } E = \mathbb{C}, \mathbb{R}, \mathbb{Q}, \\ p, & \text{if } E = \mathbb{F}_q, q = p^r. \end{cases}$$

**Definition 6.1** (i) Suppose $E$ is a field, the character of $E$ $\chi(E) \neq 2, 3$, $f(x) = x^3 + ax + b \in E[x]$ is a cubic polynomial and has no multiple roots in the split field. An elliptic curve in field $E$ refers to the set of finite points $(x, y) \in E^2$ plus infinity on the "plane," where the finite point $(x, y)$ satisfies

$$y^2 = x^3 + ax + b, \text{ where } a \in E,\ b \in E \text{ given.}$$

$C_E$ represents the elliptic curve, and "O" represents the infinity point, i.e.,

$$C_E = \{(x, y) \in E^2 | y^2 = x^3 + ax + b\} \cup \{O\}. \tag{6.1}$$

(ii) If $\chi(E) = 2$, then an elliptic curve $C_E$ on the field $E$ with the characteristic of 2 is defined as

$$C_E = \{(x, y) \in E^2 | y^2 + y = x^3 + ax + b\} \cup \{O\}. \tag{6.2}$$

(iii) If $\chi(E) = 3$, $x^3 + ax^2 + bx + c \in E[x]$ has no multiple roots in the split field, then an elliptic curve $C_E$ on $E$ is defined as

$$C_E = \{(x, y) \in E^2 | y^2 = x^3 + ax^2 + bx + c\} \cup \{O\}. \tag{6.3}$$

Let $F(x, y) \in E[x, y]$ be a bivariate polynomial, then $F(x, y) = 0$ defines an algebraic curve $C$ on $E$. $(x_0, y_0) \in C$ is called a nonsingular point on $C$, if at least one of the partial derivatives $\frac{\partial F}{\partial x}$ and $\frac{\partial F}{\partial y}$ at $(x_0, y_0)$ is not 0. If $\chi(E) \neq 2, 3$, let $f(x) = x^3 + bx + c$, then the finite points of an elliptic curve $F(x, y) = y^2 - f(x) = 0$ on $E$ are nonsingular points, which is the same as that in $\chi(E) = 2$, $\chi(E) = 3$. Therefore, an elliptic curve is also called a nonsingular cubic curve.

Among many profound arithmetic properties of elliptic curves, Mordell group on elliptic curves is the most beautiful and important basic property. Firstly, we introduce Mordell group when $E = \mathbb{R}$ is familiar with real number field and then extend it to finite field.

Elliptic curve over real number field

**Definition 6.2** Let $E = \mathbb{R}$ be real number field, $C_E$ is an elliptic curve, $P$ and $Q$ are two points on $C_E$, that is $P \in C_E$, $Q \in C_E$, we define addition according to the following rules:

(1) If $P = O$ is infinity, define that $P + P = O$ is still infinity; that is, infinity is the unit element of addition, and the negative element of $P$ is $-P = P = O$.

(2) If $P = (x, y) \in C_E$ is a finite point. Define $-P = (x, -y)$, obviously, $-P \in C_E$, is the specular reflection point of point $P$ on the $xy$−plane.

(3) If $P, Q \in C_E$ are two finite points, they have different $x$-coordinates (i.e., $P = (x_1, y_1)$, $Q = (x_2, y_2)$, $x_1 \neq x_2$), then there is exactly a point $R$ on the connecting

line between $P$ and $Q$ on the $xy-$plane, which is the intersection of the connecting line and the elliptic curve, define $P + Q = -R$, is the specular reflection point of $R$. If $Q$ is infinity. Then define $P + O = P$.

(4) If $Q = -P$, that is, $P$ and $Q$ have the same $x$-coordinate, and $P + Q = O$ is defined as infinity.

(5) If $P = Q$ is a finite point on $C_E$. Then the tangent of $C_E$ at $P$ has exactly an intersection $R$ with $C_E$, define $P + P = -R$.

We use the geometric construction method to define the addition on elliptic curve $C_E$, for the connection of finite points with different $x$-coordinates and why the tangent at the finite point has only a unique intersection with $C_E$, it needs strict mathematical proof. We attribute it to the following lemma.

**Lemma 6.1** *Let $P = (x_1, y_1)$, $Q = (x_2, y_2)$ be two finite points on elliptic curve $C_E$, and $x_1 \neq x_2$, then*

*(i) The line between $P$ and $Q$ has only a unique intersection $R = (x_3, y_3)$ with $C_E$, satisfies $R \neq P$, $R \neq Q$, where*

$$\begin{cases} x_3 = (\frac{y_2-y_1}{x_2-x_1})^2 - x_1 - x_2, \\ y_3 = -y_1 + (\frac{y_2-y_1}{x_2-x_1})(x_1 - x_3). \end{cases} \tag{6.4}$$

*(ii) Let $\alpha$ be the value of derivative $\frac{dy}{dx}$ at point $P$, then the tangent of point $P$ and $C_E$ only have a unique intersection $R = (x_3, y_3)$, $R \neq P$, where*

$$\begin{cases} x_3 = (\frac{3x_1^2+a}{2y_1})^2 - 2x_1, \\ y_3 = -y_1 + (\frac{3x_1^2+a}{2y_1})(x_1 - x_3). \end{cases} \tag{6.5}$$

***Proof*** Let the functional equation of the connecting line between $P$ and $Q$ be $y = \alpha x + \beta$ on the $xy-$plane, where

$$\alpha = \frac{y_2 - y_1}{x_2 - x_1}, \quad \beta = y_1 - \alpha x_1.$$

A point $(x, \alpha x + \beta)$ on line $y = \alpha x + \beta$ is on elliptic curve $C_E$ if and only if

$$(\alpha x + \beta)^2 = x^3 + ax + b. \tag{6.6}$$

Therefore, the three solutions of $x^3 - (\alpha x + \beta)^2 + ax + b = 0$ are $x$, and each solution will produce an intersection. But we assume that $P$ and $Q$ are at the intersection, so there is only the third intersection $R = (x, \alpha x + \beta) = (x_3, \alpha x_3 + \beta)$. Because the three solutions $x_1, x_2, x_3$ of equation (6.6) satisfy the following relationship

$$x_1 + x_2 + x_3 = \alpha^2.$$

There is

$$\begin{cases} x_3 = (\frac{y_2-y_1}{x_2-x_1})^2 - x_1 - x_2, \\ y_3 = \alpha x_3 + \beta = -y_1 + (\frac{y_2-y_1}{x_2-x_1})(x_1 - x_3). \end{cases}$$

Thus, (6.4) holds. If point $Q$ is infinitely close to point $P$, the connecting line becomes the tangent of curve $C_E$ at point $P$, now

$$\alpha = \frac{dy}{dx}|_{(x_1,y_1)} = \frac{3x_1^2 + a}{2y_1}.$$

So the tangent has a unique intersection with $C_E$, $R \neq P$, $R = (x_3, \alpha x_3 + \beta)$, where

$$\begin{cases} x_3 = \alpha^2 - 2x_1 = (\frac{3x_1^2+a}{2y_1})^2 - 2x_1, \\ y_3 = -y_1 + (\frac{3x_1^2+a}{2y_1})(x_1 - x_3). \end{cases}$$

(6.5) holds, So as to complete the proof of Lemma (Fig. 6.1).

***Example 6.1*** On the real plane, we give a specific example $y^2 = x^3 - x$ to illustrate the addition rule on this elliptic curve:

The point of $C_E$ in the left half plane is called the torsion point of $C_E$, and the point of C in the right half plane is called the free point of $C_E$.

***Remark 6.1*** In Lemma 6.1, if $P = (x_1, 0)$, that is $y_1 = 0$, then the only intersection of the tangent of point $P$ and $C_E$ is defined as the infinity point "$O$" .

From Definition 6.1 and Lemma 6.1, we have the following important corollaries.

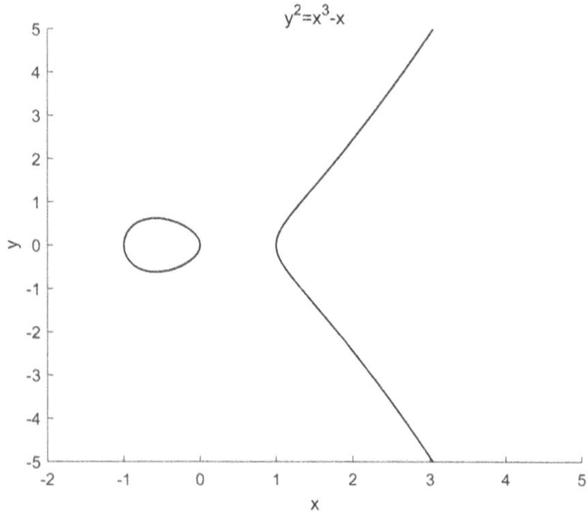


**Fig. 6.1** Elliptic Curve

**Corollary 6.1** *(i) All points of elliptic curve $C_E$ form an Abel group under addition, in which the infinity point "O" is the zero element of the group. This group is called Mordell group.*

*(ii) If $P = (x_1, y_1)$, $Q = (x_2, y_2)$ is a rational point, that is, $x_1, y_1, x_2, y_2$ is a rational number, then another unique intersection $R$ between the line between $P$ and $Q$ and $C_E$ is also a rational point.*

***Proof*** $(i)$ is directly given by Definition 6.1. Conclusion $(ii)$ is directly derived from Formula (6.4) and Formula (6.5) of Lemma 6.1.

Elliptic curves over rational fields

Let $E = \mathbb{Q}$, then $a, b, c$ in Definition 6.1 are rational numbers. Elliptic curves over rational number fields are one of the most important research topics in modern number theory. There are many important conclusions and famous number theory problems related to them, such as the famous "BSD" conjecture, the ancient congruence problem and so on. Mordell theorem is the most basic conclusion of elliptic curves over rational fields. Since cryptography only cares about elliptic curves over finite fields, here we briefly introduce some important results without proof.

Let $C_E$ be an elliptic curve in the field of rational numbers. From Corollary 6.1, all points of $C_E$ form an Abel group $G$. In algebra, an Abel group is equivalent to a module over an integer ring, so an Abel group is also called $\mathbb{Z}-$module. The Mordell group on elliptic curve $C_E$ is regarded as a $\mathbb{Z}-$module $G$, according to the decomposition theorem of modules on the principal ideal ring, a $\mathbb{Z}-$module can be decomposed into the direct sum of a twisted module and a free module. Therefore, the Mordell group $G$ on $C_E$ has

$$G = Tor(G) \oplus Free(G).$$

Mordell first proved the following important conclusions. Mordell Theorem: The Abel group $G$ on elliptic curve $C_E$ ($E = \mathbb{Q}$ is a rational number field) is finitely generated; in other words, $G$ is a finitely generated $\mathbb{Z}$-module. Therefore, Mordell group $G$ can be decomposed into

$$G = Tor(G) \oplus \mathbb{Z}(\alpha_1, \alpha_2, \ldots, \alpha_r).$$

where $\alpha_1, \alpha_2, \ldots, \alpha_r$ is a set of bases of free module $Free(G)$ and $r$ is the rank of free module. The rank $r$ is only known to be finite, but how to calculate it is a famous number theory problem. The so-called BSD conjecture holds that $r$ can be given by the function value of $L$-function on elliptic curve, but it has not been fully proved at present.

Another problem related to elliptic curves is the ancient congruence problem, which can be traced back to Plato's time in ancient Greece.

The congruent number problem: if $n > 1$ is a positive integer, is there a right triangle with rational side length, and its area is exactly $n$?

This problem is equivalent to the rank $r > 0$ of elliptic curve $y^2 = x^3 - n^2x$, at present, this problem has not been completely solved. Chinese mathematicians prof. Tian Ye have made substantial progress in this problem.

Elliptic curves over finite fields

Let $E = \mathbb{F}_q$ be a $q$-element finite field, $q = p^r$, and $p$ be a prime number. Let

$$F(x, y) = \begin{cases} y^2 - f(x), & \text{if } p \neq 2, \\ y^2 + y - f(x), & \text{if } p = 2. \end{cases} \tag{6.7}$$

where

$$f(x) = \begin{cases} x^3 + ax + b, & \text{if } p \neq 2, \\ x^3 + ax^2 + bx + c, & \text{if } p = 2. \end{cases} \quad a, b, c \in \mathbb{F}_q. \tag{6.8}$$

Then an elliptic curve $C_E$ on $\mathbb{F}_q$ is defined as

$$C_E = \{(x, y) \in \mathbb{F}_q^2 | F(x, y) = 0\} \cup \{O\}. \tag{6.9}$$

where "$O''$ is the infinity point.

Obviously, the number of points in $C_E$ is limited, let $N_q = |C_E|$, be called the number of points of elliptic curve in $\mathbb{F}_q$. $N_q \leq 2q + 1$ is a trivial estimate, because each $x \in \mathbb{F}_q$ has at most two $y$ values, together with the infinity point. The more accurate estimation of $N_q$ depends on the Riemann hypothesis on the field of univariate algebraic functions proved by A. Weil, which is a very profound result in mathematics. A. Hasse proved the following results when $F(x, y)$ is an elliptic curve.

**Theorem 6.1** (Hasse Theorem) *Let $N_q$ be the number of elliptic curve $F(x, y) = 0$ at the midpoint of $\mathbb{F}_q$, then we have*

$$|N_q - (q + 1)| \leq 2\sqrt{q}.$$

***Proof*** Let $\chi$ be a quadratic real feature in $\mathbb{F}_q$, that is

$$\chi(a) = \begin{cases} 0, & \text{if } a = 0, \\ 1, & \text{if } a = b^2, b \in \mathbb{F}_q, \\ -1, & \text{otherwise}. \end{cases}$$

By definition, it is obvious that the number of solutions of $u^2 = a$ in $\mathbb{F}_q$ is $1 + \chi(a)$, so suppose $N_q$ is the number of solutions of elliptic curve $F(x, y) = 0$ in $\mathbb{F}_q$, where $F(x, y)$ is given by Eq. (6.7), then

$$\begin{aligned} N_q &= 1 + \sum_{x \in \mathbb{F}_q} (1 + \chi(x^3 + ax + b)) \\ &= q + 1 + \sum_{x \in \mathbb{F}_q} \chi(x^3 + ax + b). \end{aligned} \tag{6.10}$$

We use $\mathbb{F}_q(x)$ to represent the rational function field on $\mathbb{F}_q$, then the univariate algebraic function field defined by $y^2 = f(x)$ can be regarded as a quadratic finite extension field on $\mathbb{F}_q(x)$. The genus $d$ of this function field is $d = 3$. Hasse can prove that the Riemann hypothesis on this special algebraic function field is true; that is, all zeros of the corresponding Riemann $\xi$−function lie on the straight line of $s = \frac{1}{2} + it$. A special case of this conclusion is

$$|\sum_{x \in \mathbb{F}_q} \chi(x^3 + ax + b)| \leq (d - 1)\sqrt{q} = 2\sqrt{q}. \tag{6.11}$$

By (6.10),

$$|N_q - (q + 1)| \leq 2\sqrt{q}.$$

We have completed the proof.

***Remark 6.2*** (6.11) is called the characteristic sum over a finite field, so that $g(x) \in \mathbb{F}_q[x]$ is any polynomial and $\chi$ is any nontrivial multiplication characteristic over $\mathbb{F}_q$, according to A. Weil's famous theorem, we have the following general characteristics and estimates,

$$|\sum_{x \in \mathbb{F}_q} \chi(g(x))| \leq (\deg g - 1)\sqrt{q}.$$

Let's briefly introduce A. Weil's theorem. Let $\mathbb{F}_{q^n}$ be an $n$-th extension on $\mathbb{F}_q$, that is $n = [\mathbb{F}_{q^n} : \mathbb{F}_q]$. $N_{q^n}$ is the number of solutions of elliptic curve $F(x, y) = 0$ in extended field $\mathbb{F}_{q^n}$. Zeta function $Z(T, C_E)$ on elliptic curve $C_E$ is defined as the formal power series of $T$:

$$Z(T) = Z(T, C_E) = \exp(\sum_{n=1}^{+\infty} \frac{1}{n} N_{q^n} T^n). \tag{6.12}$$

where $\exp(a) = e^a$ is an exponential function. A. Weil proves that $Z(T)$ is a rational function, i.e.,

$$Z(T) = \frac{qT^2 - \alpha T + 1}{(1 - T)(1 - qT)}. \tag{6.13}$$

where $\alpha$ is an integer depends on elliptic curve $C_E$. In fact, the above formula is valid for general algebraic curves. Because of $N_q = q + 1 - \alpha$, and $\alpha^2 - 4q < 0$ (Hasse theorem). Therefore, zeta function $Z(T)$ has two complex roots, that is, the two solutions $\alpha_1$ and $\alpha_2$ of $qT^2 - \alpha T + 1 = 0$, and $|\frac{1}{\alpha_1}| = |\frac{1}{\alpha_2}| = \sqrt{q}$. This is the

Riemann hypothesis on elliptic curves. A. Weil proved it on general algebraic curves for the first time. See Chap. 5 of Silverman (1986 of reference 6) for the specific proof process.

From the above a. Weil theorem, take logarithms on both sides of Eq. (6.12) and compare the coefficients on both sides of the formal power series. Let $N_{q^n}$ be the number of points of the elliptic curve in $\mathbb{F}_{q^n}$, then

$$|N_{q^n} - (q^n + 1)| \leq 2q^{\frac{n}{2}} (n \geq 1). \tag{6.14}$$

The above formula can also be derived directly from Hasse theorem.

Now let's look at a specific elliptic curve in $\mathbb{F}_2$, $y^2 + y = x^3$; thus, we have a better understanding of A. Weil's theorem. Because $F(x, y) = y^2 + y - x^3 = 0$ has three points in $\mathbb{F}_2$, the zeta function on the elliptic curve,

$$\begin{aligned} Z(T) &= \exp\left(\sum_{n=1}^{+\infty} \frac{N_n}{n} T^n\right) \\ &= \frac{2T^2 + 1}{(1 - T)(1 - 2T)}. \end{aligned}$$

Write $2T^2 + 1 = (1 - \alpha_1 T)(1 - \alpha_2 T)$, where $\alpha_1 = i\sqrt{2}$, $\alpha_2 = -i\sqrt{2}$. Take logarithms on both sides of the above formula and compare the coefficients of $T^n$ on both sides,

$$N_n = \begin{cases} 2^n + 1, & \text{if } n \text{ is odd}, \\ 2^n + 1 - 2(-2)^{\frac{n}{2}}, & \text{if } n \text{ is even}. \end{cases}$$

Where $N_n$ represents the number of points of elliptic curve $y^2 + y = x^3$ in $\mathbb{F}_{2^n}$.

Finally, the Mordell group of elliptic curve on $\mathbb{F}_q$ is a finite Abel group of order $N_q$; according to the classification theorem of finite Abel groups, this group can be expressed as the direct sum of two cyclic groups, which will be further explained when necessary.

## 6.2 Elliptic Curve Public Key Cryptosystem

An elliptic curve over a finite field $\mathbb{F}_q$ forms a finite Abel group $G$, which is similar to $\mathbb{F}_q^*$; therefore, the elliptic curve public key cryptosystem can be constructed by using discrete logarithm. Compared with other public key cryptosystems based on discrete logarithm (such as ElGamal cryptosystem), elliptic curve cryptosystem has greater flexibility, because when a huge $q$ is selected, the working platform of ElGamal cryptosystem has only one multiplication group $\mathbb{F}_q^*$, but multiple elliptic curves can be defined on $\mathbb{F}_q$, so there will be multiple Mordell groups to choose, and elliptic curve cryptosystem has greater concealment and security.

Before introducing elliptic curve cryptography, we first discuss the computational complexity on two species group. The computational complexity of multiplication over finite field $\mathbb{F}_q$ has been discussed in Chap. 4 Lemma 4.12, specially, $\alpha \in \mathbb{F}_q$, $k$ is an integer, then $Time(\alpha^k) = O(\log k \log^3 q)$. In the case of elliptic curves, the Mordell group $G$ is an addition operation, so that $P \in G$ is a point. $kP$ means that the points $P$ are added $k$ times continuously.

**Lemma 6.2** *Let $E = \mathbb{F}_q$ be a $q$-element finite field, $C_E$ be an elliptic curve on $\mathbb{F}_q$ defined by Weierstrass equations (6.7), (6.8) and (6.9), $P \in G$, $G$ be a Mordell group on $C_E$, then for any integer $k$,*

$$\begin{cases} Time(kP) = O(\log k \log^3 q), & \text{if } k \le N_q; \\ Time(kP) = O(\log^4 q), & \text{if } k > N_q. \end{cases}$$

*where $N_q$ is the number of points of curve $C_E$ and the order of Mordell group $G$.*

***Proof*** Let $P = (x, y)$, $y \neq 0$, then $P + P = (x', y')$, where $x'$ and $y'$ are determined by Equation (6.5), (6.5) (addition, subtraction, multiplication, division, etc.) involved in the formula shall not exceed 20 calculations, and the bit operation times of each calculation is $O(\log^3 q)$. By the "repeated square method," $kP$ can be transformed into $\log k$ steps, thus

$$Time(kP) = O(\log k \log^3 q).$$

If $y = 0$, defined by $P + P = O$ and "repeated square method," there is $Time(kP) = O(\log k)$.

If $k > N_q$, because $N_q \cdot P = 0$, let $k = s \cdot N_q + r$, $1 \le r \le N_q$, thus $kP = rP$. We can calculate $rP$. Thus

$$Time(kP) = O(\log N_q + \log N_q \log^3 q) = O(\log^4 q).$$

We use Hasse's theorem: $N_q = q + 1 + O(\sqrt{q})$, there is $N_q = O(q)$, thus

$$\log N_q = O(\log q).$$

Lemma 6.2 holds.

Secondly, we consider how to correspond a plaintext unit $m$ to a point on a given elliptic curve $C_E$, which is a necessary premise for encryption using elliptic curve. Unfortunately, there is no definite algorithm for polynomial bit operation, which can correspond any huge integer $m$ to a point on any elliptic curve. Instead, we can only choose the probability algorithm with sufficiently low error probability to realize the correspondence from number to point. The so-called probability algorithm does not guarantee 100% success rate (therefore, each operation depends on your luck), but the success probability should be large enough, that is

$$P\{\text{number to point correspondence}\} > 1 - \varepsilon, \ \varepsilon > 0 \text{ sufficient small.}$$

Next, we introduce a probabilistic algorithm to realize the correspondence from number to point, which makes theoretical preparation for the construction of elliptic curve cryptosystem.

Probabilistic algorithm

Treat each plaintext unit as an integer $m$, $0 \le m < M$, $k$ is an integer. Select a finite field $\mathbb{F}_q$, $q = p^r$ satisfies $q > kM$. We write the positive integer $n$ from 1 to $kM$ as follows,

$$1 \le n \le kM, n = mk + j, 0 \le m < M, 1 \le j \le k. \tag{6.15}$$

**Lemma 6.3** *There is a 1-1 correspondence $\tau$ between the set of integers $A = \{1, 2, \ldots, kM\}$ and a subset of finite field $\mathbb{F}_q (q > kM)$.*

***Proof*** Because $q = p^r$, let $g(x) \in \mathbb{F}_q[x]$ be a monic irreducible polynomial, and $\deg g(x) = r - 1$, from the finite extension theory of fields, $\mathbb{F}_q$ is isomorphic to a quotient ring of polynomial ring $\mathbb{F}_p[x]$ over subfield $\mathbb{F}_p$, that is

$$\mathbb{F}_q \cong \mathbb{F}_p[x]/_{<g(x)>} = \{a_0 + a_1 x + \cdots + a_{r-1}x^{r-1} | a_i \in \mathbb{F}_p\}.$$

Each element $\alpha \in \mathbb{F}_q$ uniquely corresponds to a polynomial $a_0 + a_1 x + \cdots + a_{r-1}x^{r-1}$, we write

$$\alpha = (a_{r-1}a_{r-2} \cdots a_1 a_0)_p,$$

is called the $p$-ary representation of $\alpha$.

For every $m$, $0 \le m < M$, each $j$, $1 \le j \le k$, then it uniquely corresponds to $n = mk + j$, express $n$ as a $p$-ary number, if the $p$-ary of $n$ is expressed as $(a_{r-1}a_{r-2} \cdots a_1 a_0)_p$, then let $\tau(n) = \alpha \in \mathbb{F}_q$. The uniqueness represented by $p$-ary, then $\tau$ is an injection.

$$A' = \{\tau(n) | 1 \le n \le kM\} \subset \mathbb{F}_q.$$

Therefore, we establish a 1-1 correspondence $\tau$ of $A \to A'$. The Lemma holds.

Next, for each $m (0 \le m < M)$, we establish a 1-1 correspondence $\sigma$ between $m$ and the point on elliptic curve $C_E$. Arbitrary choice $1 \le j \le k$, then $n = mk + j$ corresponds to an element in $\mathbb{F}_q$, that is $\tau(n) = x_j \in \mathbb{F}_q$. For each $x_j$, consider the solution of the following equation.

$$y^2 = f(x_j) = x_j^3 + ax_j + b. \tag{6.16}$$

If the above equation has a solution, let $y_1$ be one of the solutions, then $P_m = (x_j, y_1) \in C_E$, we let $\sigma(m) = P_m$, the inverse mapping $\sigma^{-1}(P_m)$ of $\sigma$ is

$$\sigma^{-1}(P_m) = [\frac{\tau^{-1}(x_j) - 1}{k}]. \tag{6.17}$$

where $\tau$ is the 1-1 correspondence in lemma 6.3. Because $\tau^{-1}(x_j) = mk + j$, so

$$[\frac{\tau^{-1}(x_j) - 1}{2}] = [m + \frac{j-1}{k}] = m.$$

So $\sigma^{-1}$ is exactly the inverse mapping of $\sigma$. From $\sigma$, a 1-1 correspondence between each $m$ and the point on the elliptic curve is established. $\sigma$ is called a probabilistic algorithm.

**Lemma 6.4** *Probability algorithm $\sigma$ can successfully achieve probability $\geq 1 - \frac{1}{2^k}$, that is*

$$P\{\textit{is generated by } \sigma \textit{ and the number } m \textit{ corresponds to 1-1 of the point}\} \geq 1 - \frac{1}{2^k}.$$

***Proof*** When $m, 0 \leq m < M$ given, $n = mk + j$, where $k$ is any given positive integer, $1 \leq j \leq k$. By Lemma 6.3, $\tau(n) = x_j \in \mathbb{F}_q$, then the probability that $f(x_j) = x_j^3 + ax_j + b$ is a square number is $\frac{1}{2}$, in other words, the probability that equation (6.16) has a solution in $\mathbb{F}_q$ is $\frac{1}{2}$; therefore, the probability of no solution is also $\frac{1}{2}$. We randomly and independently select $j$, $1 \leq j \leq k$, the error probability of each $j$ (no solution in Eq. (6.16)) is $\frac{1}{2}$; therefore, the error probability of $k$ $j$ is $\frac{1}{2^k}$. Once Equation (6.16) has a solution, then $P_m = (x_j, y) \in C_E$, we can establish the 1-1 correspondence $\sigma$ between $m$ and points on $C_E$, $\sigma(m) = P_m$. Thus

$$P\{\sigma \text{ Successfully implemented}\} \geq 1 - \frac{1}{2^k}.$$

We complete the proof of lemma.

***Remark 6.3*** $f(x_j) = x_j^3 + ax_j + b$ is a square number, that is, the probability that Equation (6.16) has a solution is exactly $N_q/2q$, where $N_q$ is the number of points of $C_E$. By Hasse's theorem, $N_q/2q$ is very close to $\frac{1}{2}$.

**Definition 6.3** Let $C_E$ be an elliptic curve over a finite field $\mathbb{F}_q$ and $B \in C_E$ be a point. For any point $P$ on $C_E$, if there is an integer $x$, such that $xB = P$, $x$ is called the discrete logarithm of $P$ to base $B$.

With the above preparation, we can establish elliptic curve public key cryptosystem.

Diffie–Hellman key conversion principle

Symmetric cryptosystem, also known as classical cryptosystem or traditional cryptosystem, is the mainstream cryptosystem before the advent of public key cryptosystem. It has high efficiency because its encryption and decryption share the same algorithm (such as DES, the data encryption standard algorithm launched by the American Bureau of standards in 1977). When Diffie and Hellman proposed asymmetric cryptosystem, they pointed out that symmetric cryptosystem and asymmetric cryptosystem are not completely separated. The two cryptosystems are interrelated and can even be used together. Diffie–Hellman key conversion principle is based on the following mathematical principles.

**Lemma 6.5** *Let $p$ be a prime number, $q = p^r$, $\mathbb{F}_q$ is a $q$-element finite field, $\mathbb{Z}_{p^r}$ is the residual class ring* mod $p^r$, *$\mathbb{Z}_p^r$ is an $n$-dimensional vector space on $\mathbb{F}_p$, then $\mathbb{F}_q$, $\mathbb{Z}_{p^r}$, $\mathbb{Z}_p^r$ have 1-1 correspondence with each other.*

***Proof*** $\mathbb{F}_q$ is an $r$-th finite extension on $\mathbb{F}_p$, so the additive group $\mathbb{F}_q^+$ of $\mathbb{F}_q$ is isomorphic with $\mathbb{F}_p^r$, that is $\mathbb{F}_q^+ \cong \mathbb{F}_p^r$, therefore, there is a 1-1 correspondence between $\mathbb{F}_q$ and $\mathbb{F}_p^r$. Each $a = (a_0, a_1, \ldots, a_{r-1}) \in \mathbb{F}_p^r$, define

$$\sigma(a) = a_0 + a_1 p + \cdots + a_{r-1} p^{r-1} \in \mathbb{Z}_p^r.$$

Then $\sigma$ is a surjection and a injection of $\mathbb{F}_p^r \to \mathbb{Z}_{p^r}$, so $\sigma$ is a 1-1 correspondence of $\mathbb{F}_p^r \to \mathbb{Z}_{p^r}$. Since there is a 1-1 correspondence between $\mathbb{F}_q$ and $\mathbb{F}_p^r$ and a 1-1 correspondence between $\mathbb{F}_p^r$ and $\mathbb{Z}_{p^r}$, there is also a 1-1 correspondence between $\mathbb{F}_q$ and $\mathbb{Z}_{p^r}$, the Lemma holds.

From the above lemma, we have the following conclusions.

**Lemma 6.6** *Let $N$ be a positive integer. $\mathbb{Z}_N$ is a residue class ring* mod $N$. *Then for any prime $p$, there is a finite field $\mathbb{F}_{p^r}$ such that there is an injection $\sigma$ of $\mathbb{Z}_N \to \mathbb{F}_{p^r}$, this injection is also called embedded mapping.*

***Proof*** When $N$ given, for any prime $p$, express $N$ as a $p$-ary number, then exists a positive integer $r \geq 1$, such that $p^{r-1} \leq N < p^r$. We write

$$\mathbb{Z}_N = \{0, 1, 2, \ldots, N-1\} \subset \{0, 1, 2, \ldots, N-1, N, \ldots, p^r - 1\} = \mathbb{Z}_{p^r}.$$

That is, $\mathbb{Z}_N$ is regarded as a subset of $\mathbb{Z}_{p^r}$. Let $\mathbb{Z}_{p^r} \xrightarrow{\sigma} \mathbb{F}_{p^r}$ be 1-1 correspond, so $\sigma$ gives that $\mathbb{Z}_N \to \mathbb{F}_{p^r}$ is an injection. The Lemma holds.

From the above conclusions, we can establish Diffie–Hellman's key conversion principle. Because symmetric cryptographic keys are related to the numbers of $\mathbb{Z}_N$, each number in $\mathbb{Z}_N$ can be embedded into a finite field $\mathbb{F}_q$ by Lemma 6.6. Therefore, the discrete logarithm on $\mathbb{F}_q$ can encrypt each embedded number asymmetrically, so that the two cryptosystems can be combined with each other.

Taking the affine cryptosystem introduced in Chap. 4 as an example, $A$ is a $k \times k$-order reversible square matrix in $\mathbb{Z}_N$, $b = (b_1, b_2, \ldots, b_k) \in \mathbb{Z}_N^k$ is a given vector, affine transformation $f = (A, b)$ gives the encryption algorithm of each plaintext unit $m = m_1 m_2 \cdots m_k \in \mathbb{Z}_N^k$.

$$f(m) = c = A \begin{pmatrix} m_1 \\ \vdots \\ m_k \end{pmatrix} + \begin{pmatrix} b_1 \\ \vdots \\ b_k \end{pmatrix}.$$

Let $A = (a_{ij})_{k \times k}$, each $a_{ij} \in \mathbb{Z}_N$. By Lemma 6.6, we can embed $a_{ij}$ into a finite field $\mathbb{F}_q$. $a_{ij}$ is encrypted again by using the discrete logarithm algorithm on $\mathbb{F}_q$, so that the two cryptosystems can be effectively combined.

In the case of elliptic curve, we introduce the workflow of Diffie–Hellman elliptic curve cryptography. First, the user selects a public finite field $\mathbb{F}_q$, and an elliptic curve $C_E$ on $\mathbb{F}_q$, randomly select a point $P \in C_E$, let $P = (x, y)$, then $x \in \mathbb{F}_q$. By Lemma 6.5, $x$ corresponds to an $r$-dimensional vector $(a_0, a_1, \ldots, a_{r-1})$ in $\mathbb{F}_p^r$ space (where $q = p^r$), consider $(a_0, a_1, \ldots, a_{r-1})$ as a $p$-ary number, that is

$$(a_0, a_1, \ldots, a_{r-1}) \rightarrow a_0 + a_1 p + \cdots + a_{r-1} p^{r-1}.$$

Then $(a_0, a_1, \ldots, a_{r-1})$ can be used as the key of other cryptosystems, especially symmetric cryptosystems.

Secondly, the user selects a common point $B \in C_E$, like a finite field, as the basis of the discrete logarithm on the Mordell group. The difference from finite field is that the Mordell group on elliptic curve is not a cyclic group, so point $B$ is not the generator of Mordell group. However, we require order $o(B)$ of $B$ to be as large as possible $(o(B)|N_q)$. When point $B$ is selected, the working platform of elliptic curve cryptography is actually the subgroup $< B >$ generated by $B$.

In order to generate the key, each user can randomly select an integer $a$, whose order is roughly the same as $N_q$, as their own user's private key, $a$ should be strictly confidential. Calculate $aB = A \in C_E$. Point $A$ is the public key of each user. Now each user has its own public key $(A, B)$ and private key $(a, B)$.

Massey–Omura elliptic curve cryptography.

In order to encrypt and send a plaintext unit $m(0 \leq m < M)$, $m$ is corresponding to the only point $P_m \in C_E$ on elliptic curve $C_E$ by using the probability algorithm introduced earlier. Let $N = N_q = |C_E|$; that is, the order of Mordell group is known. Each user randomly selects an integer $e$ to satisfy

$$1 \leq e \leq N, \text{ and } (e, N) = 1.$$

$d = e^{-1} \bmod N$ is calculated by Euclidean rolling division method, that is

$$de \equiv 1 (\bmod N), \text{ and } 1 \leq d \leq N.$$

Suppose user $A$ wants to encrypt and send plaintext message $P_m$ to user $B$, so that $(e_A, d_A)$ and $(e_B, d_B)$ are the respective private keys of $A$ and $B$. First, $A$ sends a message $e_A P_m$ to $B$, and then $B$ returns the message $e_B e_A P_m$ to $A$, $A$ can calculate the message by using the private key $d_A$. Because $N P_m = 0$, $d_A e_A \equiv 1 (\bmod N)$, so

$$d_A e_B e_A P_m = e_B P_m.$$

Finally, user $A$ sends the calculation result $e_B P_m$ to $B$, and user $B$ can read the original real message $P_m$ of user $A$ by using the private key $d_B$, because $d_B e_B \equiv 1 (\bmod N)$, so

$$d_B e_B P_m = P_m.$$

It should be noted that even if user $B$ receives the message $e_A P_m$ sent by $A$ for the first time, $e_A P_m$ is given to user $B$ as a point $Q = e_A P_m$ on the elliptic curve. If $B$ does not calculate the discrete logarithm, $e_A$ and $d_A$ are not known. Although the last user $B$ already knows the plaintext $P_m$, the calculation of the discrete logarithm of $Q$ under base $P_m$ is very complex. Similarly, when user $A$ receives a reply from user $B$ and calculates $e_B P_m$, he cannot know $B$'s private key $(e_B, d_B)$.

ElGamal elliptic curve cryptography

ElGamal cryptosystem is another elliptic curve cryptosystem completely different from Massey–Omura cryptosystem. In this system, the order $N$ of Mordell group of elliptic curve does not need to be known. All users jointly select a fixed finite field $\mathbb{F}_q$, an elliptic curve $C_E$ on $\mathbb{F}_q$ and a fixed point $B \in C_E$ on $C_E$ as the basis of discrete logarithm. Each user randomly selects an integer $a(0 \leq a < N_q)$ as the private key, calculates $Q = aB \in C_E$ and discloses it. Its workflow is as follows:

If user $A$ wants to encrypt and send a plaintext unit $P_m$ to user $B$, the public key of $A$ is $Q_A = a_A \cdot B$, the private key is $a_A$, the public key of $B$ is $Q_B = a_B \cdot B$ and the private key is $a_B$. The encryption algorithm of $A \xrightarrow{f} B$ is

$$f(m) = f(P_m) = (kB, P_m + kQ_B) = c. \tag{6.18}$$

The decryption algorithm is that user $B$ multiplies the first number with private key $a_B$ and then subtracts the second number. That is,

$$f^{-1}(c) = P_m + kQ_B - a_B(kB). \tag{6.19}$$

Because $Q_B = a_B \cdot B$, there is

$$f^{-1}(c) = P_m + ka_B \cdot B - ka_B \cdot B = P_m.$$

Where $k$ is an integer randomly selected by user $A$. This integer $k$ does not appear in cryptosystemtext $c$ and is called a layer of " mask" added by user $A$ to protect plaintext $P_m$. In fact, the cryptosystemtext $c = (A_1, A_2)$ received by user $B$ is two points on elliptic curve $C_E$, where

$$A_1 = kB, A_2 = P_m + kQ_B = P_m + k(a_B \cdot B).$$

Even if the third user knows the private key $a_B$ of user $B$ (assuming that the private key of user $B$ is not secure), decryption with $A_2 - a_B \cdot B$ cannot obtain plaintext $P_m$, because

$$A_2 - a_B \cdot B = P_m + kQ_B - a_B B = P_m + k(a_B \cdot B) - a_B \cdot B \neq P_m,$$

if $k \neq 1$.

The two elliptic curve cryptosystems introduced above are based on the selected elliptic curve $C_E$ and a point $B$ on $C_E$ as the basis of discrete logarithm. How to randomly select $C_E$ and $B$ needs further research.

**Lemma 6.7** *Let $x^3 + ax + b \in \mathbb{F}_q[x]$ be a cubic polynomial, then $x^3 + ax + b = 0$ have no multiple roots in the split domain if and only if the discriminant $4a^3 + 27b^2 \neq 0$.*

***Proof*** This conclusion can be deduced directly from the root formula of cubic algebraic equation.

In order to randomly select an elliptic curve on $\mathbb{F}_q$, $C_E$ is determined by equation $y^2 = x^3 + ax + b$ at $\chi(\mathbb{F}_q) \neq 2, 3$. Randomly select three elements $(x_0, y_0, a)$ in $\mathbb{F}_q$, let

$$b = y_0^2 - (x_0^3 + ax_0).$$

Check whether $f(x) = x^3 + ax + b$ has multiple roots. From Lemma 6.7, just check whether discriminant $4a^3 + 27b^2$ is 0. If $f(x)$ has no multiple roots, then select the elliptic curve $y^2 = x^3 + ax + b$. Where $(x_0, y_0) \in C_E$ is a point on an elliptic curve. So let $B = (x_0, y_0)$ is the base of discrete logarithm. Similarly, for $q = 2^r$ or $q = 3^r$, we can also randomly draw an elliptic curve $C_E$ and determine the basis $B \in C_E$ of the discrete logarithm at the same time.

It should be noted that at present, no algorithm can calculate the number of points $N_q$ of any elliptic curve. Some special algorithms, such as schoof algorithm, are quite complex and lengthy in practical application, although the computational complexity is polynomial.

Now we introduce the second method of selecting elliptic curves, called mod $p$ method. An elliptic curve $C_E$, if $E$ is a number field, such as $E = \mathbb{R}, \mathbb{Q}, \mathbb{C}$, $C_E$ is called a global curve. We use the mod $p$ method to convert a global curve into a "local" curve. Firstly, a point $B \in C_E$ on a global curve $C_E$ and $C_E$ is selected, where $B$ is the group element of Mordell group, its addition order is $\infty$, where $E = \mathbb{Q}$ is the rational number field.

$$C_E : y^2 = x^3 + ax + b, \ a, b \in \mathbb{Q}.$$

Let $p$ be a prime number and coprime with the integers in the denominators of $a$ and $b$, then we obtain an elliptic curve on $\mathbb{F}_p$,

$$C_E \bmod p : y^2 \equiv x^3 + ax + b (\bmod p), a, b \in \mathbb{F}_p.$$

and a point $B \bmod p$ on $C_E \bmod p$, when localizing an elliptic curve, the choice of prime $p$ only needs to satisfy

$$p \nmid a \text{and } b \text{ 's denominator}, \ \text{and } 4a^3 + 27b^2 \not\equiv 0 (\bmod p).$$

In fact, we can ask further

$$N_p = |C_E \bmod p| = \text{prime}. \tag{6.20}$$

In this way, the Mordell group of $C_E$ mod $p$ is a cyclic group, and any finite point of $C_E$ mod $p$ will be the generator of the group. At present, there is no deterministic algorithm for selecting the prime number $p$ satisfying Formula (6.20), and it is generally speculated that a probabilistic algorithm with success probability $\geq O(\frac{1}{\log p})$ exists.

## 6.3 Elliptic Curve Factorization

In 1986, mathematician H.W. Lenstra used elliptic curve to find a new method of factor decomposition. Lenstra's method has greater advantages than the known old algorithms in many aspects, which is also one of the main reasons why elliptic curve has attracted more and more attention in the field of cryptography, We first introduce a classical factorization method called Pollard $(p-1)$ algorithm.

$(p-1)$ algorithm

Suppose $n$ is a compound number, and $p$ is a prime factor of $n$; of course, $p$ is unknown and needs to be further determined. If $p-1$ happens to have some small prime factors, or all prime factors of $p-1$ are not too large, the essence of $(p-1)$ method is to find the prime factor $p$ with this property of $n$. $(p-1)$ method can be completed in the following steps:

1. Let $B$ be a positive integer. Select a positive integer $k$ so that $k$ is a multiple of most positive integers smaller than $B$, for example, $k = B!$, or $k$ can be the least common multiple of all positive integers smaller than $B$.
2. Select a positive integer $a$ to satisfy $2 \leq a \leq n-2$, $(a, n) = 1$, such as $a = 2$, or $a = 3$, and any randomly selected positive integer.
3. Using the "repeated square method" to calculate the minimum nonnegative residual $a^k \bmod n$ of $a^k$ under mod $n$.
4. The maximum common divisor $d = (a^k - 1, n)$ of $a^k - 1$ and $n$ is calculated by Euclidean rolling division method.
5. If $d = 1$ or $d = n$, that is, if $d$ is the trivial factor of $n$, re select $a$, and then repeat steps 1–4 above.

In order to explain the working principle of $(p-1)$ algorithm, we further assume that $k$ is a multiple of all positive integers less than $B$, and $p|n$,

$$p - 1 = \prod p_i^{\alpha_i}, \text{ where } \forall\ p_i^{\alpha_i} \leq B. \tag{6.21}$$

There is $p-1|k$. By Fermat congruence theorem,

$$a^{p-1} \equiv 1 (\text{mod } p), \Longrightarrow a^k \equiv 1 (\text{mod } p).$$

So $p|d$, where $d = (a^k - 1, n)$.

**Definition 6.4** Suppose $n$ is a compound number, $p|n$. $B$ is a sufficiently large positive integer arbitrarily selected, and $p$ is called $B-$smooth prime, if Eq. (6.21) holds. That is, $p-1$ can be decomposed into the product of prime powers less than $B$.

**Lemma 6.8** *Suppose $n$ is a compound number and $B$ is a positive integer. If $n$ has a $B-$smoothing prime factor $p$, select $k$ and $a$ according to the algorithm steps* $1-4$, *then we have $d=(a^k-1,n)>1$, so we have factor decomposition $n=d\cdot\frac{n}{d}$.*

***Proof*** If $p$ is a smoothing prime factor of $n$, then we have $p|(a^k-1,n)$, thus $d>1$. The Lemma holds.

In the above algorithm, if $d=(a^k-1,n)=n$. That is $n|a^k-1$, if the algorithm fails, we must reselect $a$ and carry out a new round of testing.

***Example 6.2*** Factorization of $n=540143$, if $(p-1)$ method is used, then choose $B=8, k=840$, is the least common multiple of $1,2,\ldots,8$, let $a=2$, calculate the minimum nonnegative residue of $2^{840}$ under mod $n$,

$$2^{840}\equiv 53047(\bmod 540143).$$

Calculate $(2^{840}-1,n)$,

$$d=(2^{840}-1,n)=(53046,540143)=421.$$

So we have factorization $540143=421\times 1283$.

Pollard's $(p-1)$ method is essentially the multiplication group of $\mathbb{Z}_p$, the order of $\mathbb{Z}_p^*$ cannot be divided by a huge prime number; otherwise, this method will not work. Lenstra can overcome this disadvantage by using elliptic curves for factor decomposition, because there are many elliptic curves to choose from, we can always hope that the order of Mordell group on an elliptic curve is not divided by a huge prime number. Next, we introduce Lenstra's method in detail. First, we discuss the elliptic curve mod $n$.

The following general assumption is that $n$ is an odd number and a compound number, $p|n$ ($p$ is unknown) and $p>3$. Let $m$ be a positive integer, $x_1,x_2$ be two rational numbers, and the denominators of $x_1$ and $x_2$ are mutually prime with $m$, so that $x_1-x_2=\frac{c}{d}$ is a reduced fraction, then define

$$x_1\equiv x_2(\bmod m),\ \text{ if } m|c. \tag{6.22}$$

**Lemma 6.9** *Suppose $x_1\in\mathbb{Q}$ is a rational number, if its denominator and $m$ are mutually prime, there is a unique nonnegative integer $r$, such that $x_1\equiv r(\bmod m)$. $r$ is called the nonnegative residue of $x_1$ under* mod $m$, *denote as $r=x_1 \bmod m$.*

***Proof*** Write $x_1=\frac{b}{a}$, where $(a,m)=1$, $x_1-x=\frac{-ax+b}{a}$, because the congruence equation $-ax+b\equiv 0(\bmod m)$ has a unique solution $r$, $0\le r<m$. So there is a unique $r$ such that $x_1\equiv r(\bmod m)$. The Lemma holds.

In order to randomly generate an elliptic curve $C_E$ over the rational number field $\mathbb{Q}$, we randomly select three integers $a, x_0, y_0 \in \mathbb{Z}$, let $b = y_0^2 - x_0^3 - ax_0$ to satisfy

$$\Delta = 4a^2 + 27b^2 \neq 0, \text{ and } (\Delta, n) = 1. \tag{6.23}$$

We get an elliptic curve $C_E : y^2 = x^3 + ax + b$, where $(x_0, y_0) \in C_E$. Because $a, b \in \mathbb{Z}$, $\Delta = 4a^2 + 27b^2$ and $n$ are coprime, then for all prime $p$, $p|n, \Longrightarrow \Delta \not\equiv 0 (\text{mod } p)$. Therefore, as a cubic algebraic equation over a finite field $\mathbb{F}_p, x^3 + ax + b$ has no multiple roots, so we obtain a "local" elliptic curve $C_E$ mod $p$, where

$$C_E \bmod p : y^2 \equiv x^3 + ax + b (\text{mod } p). \tag{6.24}$$

And a point $(x_0 \bmod p, y_0 \bmod p) \in C_E \bmod p$ on $C_E \bmod p$, let's write this point on $C_E \bmod p$ with $P$, that is

$$P = (x_0 \bmod p, y_0 \bmod p) \in C_E \bmod p.$$

Next, we want to calculate $kP$, like the "continuous square method" of multiplication, and there is a similar continuous doubling method for addition.

**Lemma 6.10** *When k is a huge integer, the computational complexity of* $kP$ *is*

$$Time(kP) = \log k \cdot Time(P).$$

***Proof*** $k$ is expressed as a binary integer, i.e.,

$$k = a_0 + a_1 2 + a_2 2^2 + \cdots + a_{m-1} 2^{m-1}, \forall\, a_i = 0 \text{ or } 1.$$

We can double continuously, that is, $2^j P + 2^j P = 2 \cdot 2^j P (0 \leq j \leq m - 2)$, thus obtain $kP$, $m$ is the binary digit of $k$, $m = O(\log k)$, there is

$$Time(kP) = \log k \cdot Time(P).$$

The Lemma holds.

**Theorem 6.2** *Let* $C_E$ *be an elliptic curve over the rational field* $\mathbb{Q}$*, define the equation as* $y^2 = x^3 + ax + b$*, where* $a, b \in \mathbb{Z}$*, and* $(4a^3 + 27b^2, n) = 1$*. Let* $P_1$ *and* $P_2$ *be two points on* $C_E$*, and their denominators are coprime with n, and* $P_1 \neq -P_2$*,* $P_1 + P_2 \in C_E$*. Let* $P_1 + P_2 = (x, y)$*, then the necessary and sufficient condition for the denominator of x and y to be mutually prime with n is that there is no prime factor* $p|n$ *of n,* $P_1 \bmod p$ *and* $P_2 \bmod p$ *are two points on the local curve* $C_E \bmod p$*,*

$$P_1 \bmod p + P_2 \bmod p = 0.$$

***Proof*** Let $P_1 = (x_1, y_1)$, $P_2 = (x_2, y_2)$ is the two points on $C_E$. $P_1 + P_2 = (x, y)$. If the denominators of $x$ and $y$ are coprime with $n$, we have to prove

$$P_1 \bmod p + P_2 \bmod p \neq 0, \forall\, p|n. \tag{6.25}$$

If $x_1 \not\equiv x_2 (\bmod p)$, it is obvious that Formula (6.4) is true from Formula (6.25). Might as well make $x_1 \equiv x_2 (\bmod p)$. If $P_1 = P_2$, now $x_1 = x_2$, $y_1 = y_2$, we only need $p \nmid 2y_1$. If $p|2y_1$, because the coordinates of $2P_1 = (x, y)$ are determined by equation (6.5),

$$\begin{cases} x = (\frac{3x_1^2+\alpha}{2y_1})^2 - 2x_1; \\ y = y_1 - (\frac{3x_1^2+\alpha}{2y_1})(x_1 - x). \end{cases}$$

Where $\alpha = \frac{3x_1^2+a}{2y_1}$. By $p|2y_1$, $\Longrightarrow 3x_1^2 + \alpha \equiv 0 (\bmod p)$. Because $n$ is an odd number, so $p|y_1$, we have

$$\begin{cases} x_1^3 + ax_1 + b \equiv 0 (\bmod p); \\ 3x_1^2 + a \equiv 0 (\bmod p). \end{cases}$$

That is, $x_1$ is the root of $f(x) = x^3 + ax + b$ and derivative $f'(x) = 3x^2 + a (\bmod p)$. This is contradictory to $(4a^3 + 27b^2, n) = 1$. So you might as well let $P_1 \neq P_2$, now $x_1 \equiv x_2 (\bmod p)$, $x_1 \neq x_2$(because $P_1 \neq -P_2$), we can write

$$x_2 = x_1 + tp^r, r \geq 1.$$

The numerator and denominator of $t$ and $p$ are mutually prime, which can be deduced from Formula (6.4),

$$y_2 = y_1 + sp^r.$$

On the other hand, by $y_2^2 = x_2^3 + ax_2 + b$, there is

$$\begin{aligned} y_2^2 &= (x_1 + tp^r)^3 + a(x_1 + tp^r) + b \\ &\equiv x_1^3 + ax_1 + b + tp^r(3x_1^2 + a)(\bmod p) \\ &\equiv y_1^2 + tp^r(3x_1^2 + a)(\bmod p). \end{aligned} \tag{6.26}$$

But $x_1 \equiv x_2 (\bmod p)$, $y_1 \equiv y_2 (\bmod p)$, there is

$$P_1 \bmod p + P_2 \bmod p = 2P_1.$$

The above formula is infinite if and only if $y_1 \equiv y_2 \equiv 0 (\bmod p)$. If $y_1 \equiv y_2 \equiv 0 (\bmod p)$, then $y_2^2 - y_1^2 = (y_2 - y_1)(y_2 + y_1)$ will be divided by $p^{r+1}$. Therefore, Equation (6.26) contains $3x_1^2 + a \equiv 0 (\bmod p)$. It's impossible. Because $x^3 + ax + b (\bmod p)$ has no multiple roots, $x_1$ cannot be the roots of $x^3 + ax + b$ and derivative $3x^2$ under mod $p$. This proves that Formula (6.25) holds under the assumption.

Conversely, if Eq. (6.25) holds, we prove that the denominator of $P_1 + P_2$ and $n$ are coprime. Fixed $p|n$, if $x_1 \not\equiv x_2 (\bmod p)$, from equation (6.4), the denominator of $P_1 + P_2$ and $p$ are coprime. Might as well make $x_1 \equiv x_2 (\bmod p)$, then $y_2 \equiv$

$\pm y_1 (\text{mod } p)$. Because $P_1 \bmod p + P_2 \bmod p \neq 0$, we have $y_2 \equiv y_1 \not\equiv 0(\text{mod } p)$. First assume $P_2 = P_1$, then Equation (6.5) and the fact of $y_1 \not\equiv 0(\text{mod } p)$ prove that the denominator of $P_1 + P_2 = 2P_1$ and $p$ is coprime. Finally, let $P_2 \neq P_1$, we write $x_2 = x_1 + tp^r$, $(t, p) = 1$, using the congruence of Formula (6.26), there are

$$\frac{y_2^2 - y_1^2}{x_2 - x_1} \equiv 3x_1^2 + a(\text{mod } p).$$

Because $p \nmid y_2 + y_1 \equiv 2y_1 (\text{mod } p)$, so the denominator of

$$\frac{y_2^2 - y_1^2}{(y_2 + y_1)(x_2 - x_1)} = \frac{y_2 - y_1}{x_2 - x_1}$$

cannot be divided by $p$, by (6.4), the denominator of $P_1 + P_2$ cannot be divided by $p$. Since $p|n$ is arbitrary, we complete the proof of the whole theorem.

Lenstra algorithm.

Let $n$ be an odd compound number, we hope to find a nontrivial factor $d$ of $n$, $d|n$, $1 < d < n$, so there is factorization $n = d \cdot \frac{n}{d}$. Previously, we have introduced the random selection of an elliptic curve $C_E$ on rational number field $\mathbb{Q}$ and a point $P$ on $C_E$. Lenstra's algorithm hopes to factorize $n$ by $(C_E, P)$. There is no doubt that the Lenstra algorithm to be explained below is also a probability algorithm. If $(C_E, P)$ cannot be factorized successfully, as long as the probability of failure is $p < 1$, select another elliptic curve and a point above. If this continues, after randomly and independently selecting $n$ elliptic curves, the probability of successful factorization of $n$,

$$P\{n = d \cdot \frac{n}{d}\} \geq 1 - p^n (p < 1).$$

When $n$ is sufficiently large, the success probability of Lenstra algorithm can be infinitely close to 1. Therefore, the so-called Lenstra algorithm can be simply summarized as an algorithm that factorizes $n$ by using any rational elliptic curve $(C_E, P)$, and its failure probability is $p < 1$.

Let $(C_E, P)$ be a given rational elliptic curve, and $B$ and $C$ be the positive upper bound of selection. Let $k$ be divided by some small prime powers, to be exact,

$$k = \prod_{1 < l \leq B} l^{\alpha_l}, \tag{6.27}$$

where $\alpha_l$ is the largest index satisfying $l^{\alpha_l} \leq C$. Thus $\alpha_l = [\frac{\log C}{\log l}]$.

Next, we calculate $kP(\text{mod } n)$, by (6.4) and (6.5), if $x_2 - x_1$ and $2y_1$ have a rational number whose denominator and $n$ are not prime, for example $d = (x_2 - x_1, n)$, $1 < d < n$; Then we have factorization $n = d \cdot \frac{n}{d}$. If $d = n$, then re select point $P$ on rational elliptic curves $C_E$ and $C_E$. By Theorem 6.2, $d > 1$ appears in these rational numbers $x_2 - x_1$ and $y_1$ if and only if there is a $k_1$, such that

$$k_1 \cdot (P \bmod p) = 0, \forall\ p|n.$$

From the selection of equation $k$ in (6.27), there is a maximum probability $k_1|k$, thus

$$k \cdot (P \bmod p) = 0, \forall\ p|n.$$

Therefore, in Lenstra algorithm, by calculating the rational point $kP$, there is a great probability that there is a certain $p$, $p|n$ such that

$$k(P \bmod p) = 0,\ p|n,\ \text{is a prime number.} \tag{6.28}$$

By Theorem 6.2, let $P = (x_1, y_1)$, $(k-1)P = (x_2, y_2)$, thus $d = (x_2 - x_1, n) > 1$ or $(2y_1, n) = d' > 1$, we obtain the nontrivial factorization of $n$.

From the above Lenstra algorithm, the key problem is to calculate $k \cdot P$. Using the continuous doubling method given in Lemma 6.10, we only need to calculate $2P, 2(2P), 2(4P), \ldots, 2^{\alpha_2}P, 3(2^{\alpha_2}P), 3(3 \cdot 2^{\alpha_2}P) \cdots 3^{\alpha_3}2^{\alpha_2}P$, this continues until $(\prod_{1<l\leq B} l^{\alpha_l})P$, i.e., $kP$.

For the probability estimation and computational complexity of Lenstra algorithm, see 1986 of reference 6.

**Exercise 6**

1. Let $C_E = \{(x, y) \in \mathbb{C} \mid y^2 = x^3 + ax + b, a, b \in \mathbb{R}\}$ is a complex elliptic curve, then $C_E \bigcap \mathbb{R}^2$ is a subgroup of $C_E$, determine all subgroups of $C_E$ whose coordinates are real numbers.
2. The points of order $n$ on complex elliptic curve and real elliptic curve are determined.
3. Take an example of a rational elliptic curve $C_E$, there are exactly two points on $C_E$ with order 2. Another example is that there are exactly four points on $C_E$ with order 2.
4. Let $C_E$ is a real elliptic curve, $P \in C_E$ is a finite point, determine the geometric equivalence condition of $o(P) = 2$, $o(P) = 3$, $o(P) = 4$.
5. Calculate the order of points on the following rational elliptic curves:
   $(i)$ $P = (0, 16)$, $C_E : y^2 = x^3 + 256$;
   $(ii)$ $P = (\frac{1}{2}, \frac{1}{2})$, $C_E : y^2 = x^3 + \frac{x}{4}$;
   $(iii)$ $P = (3, 8)$, $C_E : y^2 = x^3 - 43x + 16$;
   $(iv)$ $P = (0, 0)$, $C_E : y^2 + y = x^3 - x^2$.
6. Proved that the following elliptic curve has exactly $q + 1$ points in $\mathbb{F}_q$:
   $(a)$ $y^2 = x^3 - x$, when $q \equiv 3 (\bmod 4)$;
   $(b)$ $y^2 = x^3 - 1$, when $q \equiv 2 (\bmod 3)$, $q$ is odd;
   $(c)$ $y^2 + y = x^3$, when $q \equiv 2 (\bmod 3)$.
7. Let $q = 2^r$, the elliptic curve $C_E$ on $\mathbb{F}_q$ be: $y^2 + y = x^3$; $P = (x, y) \in C_E$, calculate $2P$ and $-P$. If $q = 16$, prove that every point on $C_E$ has order 3.
8. Please give a probabilistic algorithm to find a nonsquare number in the finite field $\mathbb{F}_q$.

9. The deterministic algorithm can map the embedding of plaintext units to any $\mathbb{F}_q-$ elliptic curve. Please give the specific algorithm process for the following elliptic curves:
   (1) $C_E : y^2 = x^3 - x$, when $q \equiv 3 (\text{mod}\, 4)$,
   (2) $C_E : y^2 + y = x^3$, when $q \equiv 2 (\text{mod}\, 3)$.
10. Let $C_E$ be an elliptic curve on the finite field $\mathbb{F}_p$, and $N_r$ represents the number of midpoint of $C_E$ in the finite field $\mathbb{F}_{p^r}$, then
   $(i)$ If $p > 3$, when $r > 1$, $N_r$ is not prime.
   $(ii)$ When $p = 2, 3$, a counterexample is given to show that $N_r$ is a prime number.
11. Take an example of an elliptic curve $C_E$, which has only one point on $\mathbb{F}_4$, the infinity point. Take $N_r$ as the number of points of $C_E$ on $\mathbb{F}_{4^r}$, then $N_r$ is the square of Mersenne prime $2^r - 1$.
12. Decompose $n = 53467$ at $k = 840$, $a = 2$ using Pollard's $(p - 1)$ method.
13. Let $n_k = 2^{2^k} + 1$ be Fermat number, the following is Pepin's method to detect whether $n_k$ is a prime number:
   $(i)$ $n_k$ is a prime, if and only if there is an integer $a$, $a^{2^{2^k-1}} \equiv -1 (\text{mod}\, n_k)$.
   $(ii)$ If $n_k$ is a prime, then $a \in \mathbb{Z}_{n_k}^*$ over 50% has the congruence property of $(i)$.
   $(iii)$ When $k > 1$, we can always choose $a = 3, 5$, or $a = 7$.

## References


Fulton, W. (1969). *Algebraic curves*. Benjamin.

Gupta, R., & Murty, M. R. (1986). Primitive points on elliptic curves. *Composition Mathematics , 58*, 13–44.

Koblitz, N. (1984). *Introduction to elliptic curves and modular forms*. Springer.

Koblitz, N. (1987). *Elliptic curves cryptosystems, mathematics of computation* (Vol. 48).

Koblitz, N. *Primality of the number of points on an elliptic curve over finite field.*

Koblitz, N. (1982). Why study equations over Finite Fields. *Mathematics Magazine, 55,* 144–149.

Lang, S. (1978). *Elliptic curves: diophantine analysis*. Springer.

Lenstra Jr, H. W. (1986). *Elliptic curves and number-theoretic algorithms*. Report 86-19, Mathematics Institute University of Van Amsterdam.

Lenstra Jr, H. W. (1986). *Factoring integers with elliptic curves*. Report 86-18,Mathematic Institute University of Van Amsterdam.

Miller, V. (1985). *Use of elliptic curves in cryptography*. Abstracts for Crypto 85.

Odlyzko, A. M. (1985). Discrete logarithms in finite fields and their cryptographic significance. In *Advance in cryptography*, Proceeds of Eurocrypt (Vol. 84, pp. 224–314). Springer.

Pollard, J. M. (1974). Theorems on factorization and primality testing. In *Proceedings Cambridge Phil Soc, 76*, 521–528.

Schoof, H. (1985). Elliptic curves over finite fields and the computation of square roots mod $p$. *Mathematics of Computation, 44,* 483–494.

Silverman, J. (1986). *The arithmetic of elliptic curves*. Springer.

# Chapter 7
# Lattice-Based Cryptography

## 7.1 Geometry of Numbers

Let $\mathbb{R}^n$ be an $n$-dimensional Euclidean space and $x = (x_1, x_2, \ldots, x_n) \in \mathbb{R}^n$ be an $n$-dimensional vector, $x$ can be a row vector or a column vector, depending on the situation. If $x \in \mathbb{Z}^n$, then $x$ is called a integral point. $\mathbb{R}^{m\times n}$ is all $m \times n$-dimensional matrices on $\mathbb{R}$. $x = (x_1, x_2, \ldots, x_n) \in \mathbb{R}^n$, $y = (y_1, y_2, \ldots, y_n) \in \mathbb{R}^n$, define the inner product of $x$ and $y$ as

$$\langle x, y\rangle = \sum_{i=1}^{n} x_i y_i. \tag{7.1}$$

The length $|x|$ of vector $x$ is defined as

$$|x| = \sqrt{\langle x, x\rangle} = \sum_{i=1}^{n} x_i^2. \tag{7.2}$$

$\lambda \in \mathbb{R}$, then $\lambda \cdot x$ is defined as

$$\lambda x = (\lambda x_1, \lambda x_2, \ldots, \lambda x_n). \tag{7.3}$$

If the inner product $\langle x, y\rangle = 0$ of two vectors $x$ and $y$, $x$ and $y$ are said to be orthogonal, denote as $x \perp y$.

**Lemma 7.1** *Let $x, y \in \mathbb{R}^n$, $\lambda \in \mathbb{R}$ is any real number, then*

*(i)* $|x| \geq 0$, $|x| = 0$ *if and only if* $x = 0$ *is a zero vector;*
*(ii)* $|\lambda x| = |\lambda||x|$, $\forall\, x \in \mathbb{R}^n$, $\lambda \in \mathbb{R}$;
*(iii) (Trigonometric inequality)* $|x + y| \leq |x| + |y|$, *and* $|x - y| \geq ||x| - |y||$;



Z. Zheng, *Modern Cryptography Volume 1*, Financial Mathematics and Fintech,
https://doi.org/10.1007/978-981-19-0920-7_7

*(iv) (Pythagorean theorem) If and only if $x \perp y$, we have*

$$|x \pm y|^2 = |x|^2 + |y|^2.$$

***Proof*** (i) and (ii) can be derived directly from the definition. To prove (iii), let $x = (x_1, x_2, \dots, x_n)$, $y = (y_1, y_2, \dots, y_n) \in \mathbb{R}^n$, by Hölder inequality:

$$\left|\sum_{i=1}^{n} x_i y_i\right| \leq \left[\left(\sum_{i=1}^{n} x_i^2\right)\left(\sum_{i=1}^{n} y_i^2\right)\right]^{\frac{1}{2}}.$$

So there is

$$\begin{aligned} |x+y|^2 &= \sum_{i=1}^{n}(x_i+y_i)^2 = \sum_{i=1}^{n} x_i^2 + 2\sum_{i=1}^{n} x_i y_i + \sum_{i=1}^{n} y_i^2 \\ &\leq \sum_{i=1}^{n} x_i^2 + 2\left|\sum_{i=1}^{n} x_i y_i\right| + \sum_{i=1}^{n} y_i^2 \\ &\leq \left(\sqrt{\sum_{i=1}^{n} x_i^2} + \sqrt{\sum_{i=1}^{n} y_i^2}\right)^2 = (|x|+|y|)^2, \end{aligned}$$

so (iii) holds. Then, by the definition of inner product,

$$\langle x \pm y, x \pm y\rangle = \langle x, x\rangle \pm 2\langle x, y\rangle + \langle y, y\rangle,$$

if $x \perp y$, then

$$|x \pm y|^2 = |x|^2 + |y|^2.$$

Conversely, if $x$ is not orthogonal to $y$, then $\langle x, y\rangle \neq 0$, thus

$$|x \pm y|^2 \neq |x|^2 + |y|^2.$$

Lemma 7.1 holds.

From Pythagorean theorem, for orthogonal vector $x \perp y$, we have the following conclusion,

$$|x + y| = |x - y|, \text{ if } x \perp y. \tag{7.4}$$

**Definition 7.1** Let $\mathscr{R} \subset \mathbb{R}^n$ be a subset, $0 \in \mathscr{R}$, $\mathscr{R}$ is called a symmetric convex body of $\mathbb{R}^n$, if

(i) $x \in \mathscr{R}, \Rightarrow -x \in \mathscr{R}$ (Symmetry);
(ii) Let $x, y \in \mathscr{R}$, $\lambda \geq 0$, $\mu \geq 0$, and $\lambda + \mu = 1$, then $\lambda x + \mu y \in \mathscr{R}$ (Convexity).

The following example is a famous example of a symmetric convex body defined by a set of linear inequalities. Let $A \in \mathbb{R}^{m\times n}$ be an $m \times n$-order matrix, $c = (c_1, c_2, \ldots, c_n) \in \mathbb{R}^n$, and $\forall\ c_i \geq 0$, $\mathscr{R}(A, c)$ is defined as the set of solutions of $x = (x_1, x_2, \ldots, x_n) \in \mathbb{R}^n$ defined by the following $m$ linear inequalities, let $A = (a_{ij})_{m\times n}$,

$$\left|\sum_{j=1}^{n} a_{ij}x_j\right| \leq c_i,\ \ 1 \leq i \leq m. \tag{7.5}$$

We have

**Lemma 7.2** *For any $A \in \mathbb{R}^{m\times n}$, and any positive vector $c = (c_1, c_2, \ldots, c_n) \in \mathbb{R}^n$, then $\mathscr{R}(A, c)$ is a symmetric convex body in $\mathbb{R}^n$.*

***Proof*** Obviously zero vector $x = (0, 0, \ldots, 0) \in \mathscr{R}(A, c)$, and if $x \in \mathscr{R}(A, c) \Rightarrow -x \in \mathscr{R}(A, c)$. So we only prove the convexity of $\mathscr{R}(A, c)$. Suppose $x, y \in \mathscr{R}(A, c)$, let

$$z = \lambda x + \mu y, \lambda > 0, \mu > 0, \lambda + \mu = 1.$$

Then for any $1 \leq i \leq m$, we have

$$\begin{aligned}
&|a_{i1}z_1 + a_{i2}z_2 + \cdots + a_{in}z_n| \\
&\leq \lambda|a_{i1}x_1 + a_{i2}x_2 + \cdots + a_{in}x_n| + \mu|a_{i1}y_1 + a_{i2}y_2 + \cdots + a_{in}y_n| \\
&\leq \lambda c_i + \mu c_i = c_i.
\end{aligned}$$

So there is $z = \lambda x + \mu y \in \mathscr{R}(A, c)$. Thus, $\mathscr{R}(A, c)$ is a symmetrical convex body. Lemma 7.2 holds.

**Lemma 7.3** *Let $\mathscr{R} \subset \mathbb{R}^n$ be a symmetrical convex body, $x \in \mathscr{R}$, then when $|\lambda| \leq 1$, we have $\lambda x \in \mathscr{R}$.*

***Proof*** By convexity, let

$$\rho = \frac{1}{2}(1 + \lambda),\ \ \sigma = \frac{1}{2}(1 - \lambda).$$

Then $\rho \geq 0, \sigma \geq 0$, and $\rho + \sigma = 1$. So there is

$$\rho x + \sigma(-x) = \lambda x \in \mathscr{R}.$$

The Lemma holds.

**Lemma 7.4** *If $x, y \in \mathscr{R}$, then $\lambda x + \mu y \in \mathscr{R}$, where $\lambda, \mu$ are real numbers, and satisfies $|\lambda| + |\mu| \leq 1$.*

***Proof*** Let $\eta_1$ be the sign of $\lambda$ and $\eta_2$ be the sign of $\mu$, then by Lemma 7.3,

$$x' = \eta_1(|\lambda| + |\mu|)x \in \mathscr{R},$$
$$y' = \eta_2(|\lambda| + |\mu|)y \in \mathscr{R}.$$

Let $\rho = \frac{|\lambda|}{|\lambda|+|\mu|}$, $\sigma = \frac{|\mu|}{|\lambda|+|\mu|}$, then $\rho + \sigma = 1$. By definition, we have

$$\lambda x + \mu y = \rho x' + \sigma y' \in \mathscr{R},$$

thus the Lemma holds. And this result is not difficult to be extended to the case of $n$ variables.

**Lemma 7.5** (Blichfeldt) *Let $\mathscr{R} \subset \mathbb{R}^n$ be any region in $\mathbb{R}^n$ and $V$ be the volume of $\mathscr{R}$. If $V > 1$, then there are two different vectors $x \in \mathscr{R}$, $x' \in \mathscr{R}$ so that $x - x'$ is an integral point (thus a nonzero integral point).*

***Proof*** For $\forall x = (x_1, x_2, \ldots, x_n) \in \mathbb{R}^n$, we define

$$[[x]] = ([x_1], [x_2], \ldots, [x_n]) \in \mathbb{Z}^n \tag{7.6}$$

and

$$[x] = (\delta_1, \delta_2, \ldots, \delta_n) \in \mathbb{Z}^n, \tag{7.7}$$

where $[x_i]$ is the square bracket function of $x_i$ and $\delta_i$ is the nearest integer to $x_i$.

For each integral point $u \in \mathbb{Z}^n$, define

$$\mathscr{R}_u = \{x \in \mathscr{R} | [[x]] = u\}$$

and

$$D_u = \{x - u | x \in \mathscr{R}_u\}.$$

Because $\mathscr{R}_{u_1} \cap \mathscr{R}_{u_2} = \varnothing$, if $u_1 \neq u_2$. Thus by $\mathscr{R} = \bigcup_u \mathscr{R}_u$,

$$\Rightarrow V = \text{Vol}(\mathscr{R}) = \sum_u \text{Vol}(\mathscr{R}_u)$$
$$= \sum_u V_u > 1,$$

where $V_u = \text{Vol}(\mathscr{R}_u)$. Thus $V_u = \text{Vol}(D_u)$. If $D_u$ is disjoint, then

$$\sum_u V_u = \text{Vol}\left(\bigcup_u D_u\right) \subset [0, 1) \times \cdots \times [0, 1).$$

There is

$$\sum_u V_u \leq 1,$$

so there is a contradiction. Therefore, there must be two different integral points $u$ and $u'$ $(u \neq u') \Rightarrow D_u \cap D_{u'} \neq 0$, that is $x - u = x' - u' \Rightarrow x - x' = u - u' \in \mathbb{Z}^n$. The Lemma holds.

**Lemma 7.6** (Minkowski) *Let $\mathscr{R}$ be a symmetric convex body, and the volume of $\mathscr{R}$*

$$V = \mathrm{Vol}(\mathscr{R}) > 2^n,$$

*then $\mathscr{R}$ contains at least one nonzero integer point.*

***Proof*** Let

$$\frac{1}{2}\mathscr{R} = \left\{\frac{1}{2}x | x \in \mathscr{R}\right\}.$$

Thus

$$\mathrm{Vol}\left(\frac{1}{2}\mathscr{R}\right) = \frac{1}{2^n}V > 1,$$

by Lemma 7.5, there are integral points where $x', x'' \in \frac{1}{2}\mathscr{R} \Rightarrow x' - x'' = u$ is nonzero. We prove $u \in \mathscr{R}$. Write $x' = \frac{1}{2}y$, $x'' = \frac{1}{2}z$, where $y, z \in \mathscr{R}$. Then

$$u = \frac{1}{2}y - \frac{1}{2}z, \quad y \in \mathscr{R}, \; z \in \mathscr{R}.$$

By Lemma 7.4, then $u \in \mathscr{R}$. The Lemma holds.

***Remark 7.1*** The above Minkowski's conclusion cannot be improved, that is $V > 2^n$, it cannot be improved to $V \geq 2^n$. A counterexample is

$$\mathscr{R} = \{x \in \mathbb{R}^n | x = (x_1, x_2, \ldots, x_n), \forall\, |x_i| < 1\}.$$

Obviously $\mathrm{Vol}(\mathscr{R}) = 2^n$, but there is no nonzero integer point in ordinary $\mathscr{R}$.

When $\mathrm{Vol}(\mathscr{R}) = 2^n$, in order to make a symmetric convex body $\mathscr{R}$ still have nonzero integral points, we need to make some supplementary constraints on $\mathscr{R}$, first, we consider the bounded region. Let $\mathscr{R} \subset \mathbb{R}^n$, call $\mathscr{R}$ bounded, if

$$\mathscr{R} = \{x = (x_1, x_2, \ldots, x_n) \in \mathbb{R}^n || x_i| \leq B, 1 \leq i \leq n\},$$

where $B$ is a bounded constant.

**Lemma 7.7** *Let $A \in \mathbb{R}^{n \times n}$ be a reversible matrix, $d = |\det(A)| > 0$, $c = (c_1, c_2, \ldots, c_n) \in \mathbb{R}^n$ is a positive vector, that is $\forall\, c_i > 0$, then the symmetric convex body $\mathscr{R}(A, c)$ defined by Eq. (7.5) is bounded and its volume*

$$\mathrm{Vol}(\mathscr{R}(A, c)) = 2^n d^{-1} c_1 c_2 \cdots c_n.$$

***Proof*** Let $A = (a_{ij})_{n\times n}$. Write $Ax = y$, then $x = A^{-1}y$. And let $A^{-1} = (b_{ij})_{n\times n}$, then for any $x_i$, there is

$$|x_i| = \left|\sum_{j=1}^{n} b_{ij}y_j\right| \leq \sum_{j=1}^{n} |b_{ij}| \cdot c_j \leq B,$$

where $B$ is a bounded constant. Therefore, $\mathscr{R}(A, c)$ is a bounded set. Obviously

$$\text{Vol}(\mathscr{R}(A, c)) = \int\cdots\int_{x=(x_1,x_2,\ldots,x_n)\in\mathscr{R}(A,c)} \mathrm{d}x_1\mathrm{d}x_2\cdots\mathrm{d}x_n,$$

do variable replacement $Ax = y$, then

$$\mathrm{d}x = \mathrm{d}x_1\cdots\mathrm{d}x_n = \frac{1}{|\det(A)|}\mathrm{d}y_1\mathrm{d}y_2\cdots\mathrm{d}y_n.$$

Thus

$$\begin{aligned}\text{Vol}(\mathscr{R}(A, c)) &= \frac{1}{|\det(A)|}\int_{-c_1}^{c_1}\cdots\int_{-c_n}^{c_n} \mathrm{d}y_1\mathrm{d}y_2\cdots\mathrm{d}y_n \\ &= 2^n d^{-1}\prod_{i=1}^{n} c_i,\end{aligned}$$

Lemma 7.7 holds.

***Remark 7.2*** In (7.5), "$\leq$" is changed to "$<$" to define $\mathscr{R}(A, c)$, and the above lemma is still holds.

Now consider the general situation, let $A = (a_{ij})_{m\times n}$. If $m > n$, and $\text{rank}(A) \geq n$, then $\mathscr{R}(A, c)$ defined by Eq. (7.5) is still a bounded region. Obviously if $m < n$, or $m = n$, $\text{rank}(A) < n$, then $\mathscr{R}(A, c)$ is an unbounded region, and $V = \infty$. Therefore, we have the following Corollary.

**Corollary 7.1** *Let* $A = (a_{ij})_{m\times n}$, $m < n$ *or* $m = n$, $\det(A) = 0$, *then for any small positive vector* $c = (c_1, c_2, \ldots, c_n)$, $0 < c_i < \varepsilon$, $\mathscr{R}(A, c)$ *contains a nonzero integer point. In other words, the following m inequalities*

$$\left|\sum_{j=1}^{n} a_{ij}x_j\right| < \varepsilon,\ 1 \leq i \leq m.$$

*There exists a nonzero integer solution* $x = (x_1, x_2, \ldots, x_n) \in \mathbb{Z}^n$.

***Proof*** When $\varepsilon > 0$ given, then $\mathrm{Vol}(\mathscr{R}(A, c)) = \infty > 2^n$. By Lemma 7.6, $\mathscr{R}(A, c)$ contains at least one nonzero zero point.

Let $A \in \mathbb{R}^{m\times n}$ be a matrix of order $m \times n$, $c = (c_1, c_2, \ldots, c_n) \in \mathbb{R}^n$ is a positive vector, that is $\forall\, c_i > 0$, write $A = (a_{ij})_{m\times n}$, $\mathscr{R}'(A, c)$ is defined as the set of solutions $x = (x_1, x_2, \ldots, x_n)$ of the following linear inequality:

$$\begin{cases} \left|\sum\limits_{j=1}^{n} a_{1j}x_j\right| \le c_1, \\ \left|\sum\limits_{j=1}^{n} a_{ij}x_j\right| < c_i, \quad i = 2, \ldots, m. \end{cases} \tag{7.8}$$

When $A \in \mathbb{R}^{n\times n}$ is a reversible square matrix, we discuss the nonzero integral point in symmetric convex body $\mathscr{R}'(A, c)$.

**Lemma 7.8** *If $A \in \mathbb{R}^{n\times n}$ is a reversible matrix and $c = (c_1, c_2, \ldots, c_n)$ is a positive vector, when*

$$c_1c_2\cdots c_n \ge |\det(A)|, \tag{7.9}$$

*Then $\mathscr{R}'(A, c)$ contains a nonzero integer point.*

***Proof*** When $c_1c_2\cdots c_n > |\det(A)|$, because of

$$\mathrm{Vol}(\mathscr{R}'(A, c)) = \frac{2^n c_1c_2\cdots c_n}{|\det(A)|} > 2^n,$$

by Lemma 7.6 and 7.7, then the proposition holds, we only discuss the case when the equal sign of formula (7.9) holds.

Let $\varepsilon$ be any positive real number, $0 < \varepsilon < 1$, then by Lemma 7.7, there is a nonzero integral solution $x^{(\varepsilon)} = (x_1^{(\varepsilon)}, x_2^{(\varepsilon)}, \ldots, x_n^{(\varepsilon)}) \in \mathbb{Z}^n$ satisfies

$$\begin{cases} \left|\sum\limits_{j=1}^{n} a_{1j}x_j^{(\varepsilon)}\right| \le c_1 + \varepsilon \le c_1 + 1, \\ \left|\sum\limits_{j=1}^{n} a_{ij}x_j^{(\varepsilon)}\right| < c_i, \quad 2 \le i \le n. \end{cases} \tag{7.10}$$

And there is an upper bound $B$ independent of $\varepsilon$, which satisfies

$$|x_j^{(\varepsilon)}| \le B, \quad 1 \le j \le n.$$

The integral point $x^{(\varepsilon)}$ satisfying the above bounded condition is finite, so there must be a nonzero integral point $x \neq 0$, which holds (7.10) for any $\varepsilon > 0$. Let $\varepsilon \to 0$, then the Lemma holds.

In the following discussion, we make the following restrictions on $\mathscr{R} \subset \mathbb{R}^n$:

$$\mathscr{R} \text{ is a symmetric convex body, } \mathscr{R} \text{ is bounded, and } \mathscr{R} \text{ is a closed subset of } \mathbb{R}^n. \tag{7.11}$$

Obviously, when $A$ is an $n$-order reversible square matrix, for any positive vector $c = (c_1, c_2, \ldots, c_n)$, $\mathscr{R}(A, c)$ satisfies the above restriction (7.11), but $\mathscr{R}'(A, c)$ does not because $\mathscr{R}'(A, c)$ is not closed.

**Definition 7.2** If $\mathscr{R} \subset \mathbb{R}^n$ satisfies the restriction (7.11), then for any $x \in \mathbb{R}^n$, define the distance function $F(x)$ as

$$F(x) = F_{\mathscr{R}}(x) = \inf\{\lambda | \lambda > 0, \lambda^{-1} x \in \mathscr{R}\}. \tag{7.12}$$

By definition, it is obvious that we have the following ordinary conclusions:

(i) $F(x) = 0 \Leftrightarrow x = 0$;
(ii) If $A$ is a reversible $n$-order square matrix, the distance function defined by $\mathscr{R}(A, c)$ is

$$F(x) = \max_{1 \leq i \leq n} c_i^{-1} \left| \sum_{j=1}^{n} a_{ij} x_j \right|. \tag{G1.12'}$$

Property (i) can be derived from the boundedness of $\mathscr{R}$, and property (ii) can be derived directly from the definition of $\mathscr{R}(A, c)$. Later we will see that $0 \leq F(x) < \infty$ holds for all $x \in \mathbb{R}^n$. The main property of distance function $F(x)$ is the following Lemma.

**Lemma 7.9** *If $F(x)$ is a distance function defined by $\mathscr{R}$ satisfying the constraints, then*

*(i) Let $\lambda \geq 0$, then $x \in \lambda\mathscr{R} \Leftrightarrow F(x) \leq \lambda$;*
*(ii) $F(\lambda x) = |\lambda| F(x)$ holds for all $\lambda \in \mathbb{R}$, $x \in \mathbb{R}^n$;*
*(iii) $F(x + y) \leq F(x) + F(y)$, $\forall\, x, y \in \mathbb{R}^n$.*

***Proof*** Since $\mathscr{R}$ is closed, by the definition, $F^{-1}(x)x \in \mathscr{R}$. Thus, if $\lambda \geq F(x)$, by Lemma 7.3, then

$$\lambda^{-1} x = \frac{F(x)}{\lambda} \cdot F^{-1}(x) x, \quad \left| \frac{F(x)}{\lambda} \right| \leq 1.$$

We have $\lambda^{-1} x \in \mathscr{R} \Rightarrow x \in \lambda\mathscr{R}$. Conversely, if $\lambda < F(x) \Rightarrow \lambda^{-1} x \notin \mathscr{R}$. So when $x \in \lambda{-}\mathscr{R}$, there must be $\lambda \geq F(x)$, (i) holds.

(ii) is ordinary. Because $|\lambda|^{-1} F^{-1}(x) \lambda x \in \mathscr{R}$. There is

$$F(\lambda x) \leq |\lambda| F(x).$$

Conversely, let $\delta = F(\lambda x)$, because of $\delta^{-1}\lambda x \in \mathscr{R}$, you might as well let $\lambda > 0$, thus

$$F(x) \leq \frac{\delta}{\lambda} \Longrightarrow \lambda F(x) \leq F(\lambda x).$$

So there is $F(\lambda x) = |\lambda| F(x)$, (ii) holds.

To prove (iii), we let $\mu_1 = F(x), \mu_2 = F(y), \Longrightarrow \mu_1^{-1}x \in \mathscr{R}, \mu_2^{-1}y \in \mathscr{R}$. By Lemma 7.4, we have

$$(\mu_1 + \mu_2)^{-1}(x + y) = \frac{\mu_1}{\mu_1 + \mu_2}(\mu_1^{-1}x) + \frac{\mu_2}{\mu_1 + \mu_2}(\mu_2^{-1}y) \in \mathscr{R}.$$

Thus

$$F(x + y) \leq \mu_1 + \mu_2.$$

The Lemma holds.

Let the volume of $\mathscr{R} \in \mathbb{R}^n$ be $V > 0$, there are $n$ linearly independent vectors $\{\alpha_1, \alpha_2, \ldots, \alpha_n\}$ in $\mathscr{R}$ to form a set of bases of $\mathbb{R}^n$. For any real number $\mu_1, \mu_2, \ldots, \mu_n$, by Lemma 7.9, we have

$$\begin{aligned} F(\mu_1\alpha_1 + \cdots + \mu_n\alpha_n) &\leq |\mu_1|F(\alpha_1) + |\mu_2|F(\alpha_2) + \cdots + |\mu_n|F(\alpha_n) \\ &\leq |\mu_1| + |\mu_2| + \cdots + |\mu_n|. \end{aligned}$$

Because $\alpha_i \in \mathscr{R} \Rightarrow F(\alpha_i) \leq 1$, so the above formula holds. That proves for $\forall\, x \in \mathbb{R}^n \Rightarrow F(x) \leq \infty$.

**Corollary 7.2** *Let* $\mathscr{R} \subset \mathbb{R}^n$ *meet the limiting conditions (7.11), and* $\mathrm{Vol}(\mathscr{R}) > 0$, *then*

*(i)* $\forall\, x \in \mathbb{R}^n$, *there is* $\lambda$ *such that* $x \in \lambda\mathscr{R}$;
*(ii) Let* $\{\alpha_1, \alpha_2, \ldots, \alpha_n\} \subset \mathscr{R}$ *be a set of bases of* $\mathbb{R}^n$, *then*

$$\left\{ \sum_{i=1}^{n} \mu_i\alpha_i \,\middle|\, |\mu_1| + |\mu_2| + \cdots + |\mu_n| \leq 1 \right\} \subset \mathscr{R}.$$

***Proof*** Because $F(x) < \infty$, so by (i) of Lemma 7.9, we can directly deduce the conclusion of (i) and (ii) given directly by Lemma 7.4.

Now let $j$ be a subscript, and we define $\lambda_j$ as

$$\lambda_j = \min\{\lambda \geq 0 | \lambda\mathscr{R} \text{ contains } j \text{ linear independent integral points in } \mathbb{R}^n\}, \quad (7.13)$$

and $\lambda_j$ is called the $j$th continuous minimum of $\mathscr{R}$. By Lemma 7.3, $\lambda\mathscr{R} \subset \lambda'\mathscr{R}$, if $0 \leq \lambda \leq \lambda'$. Therefore, $\lambda$ increases continuously, then $\lambda\mathscr{R}$ can always contain any set of desired vectors. Therefore, the existence of $\lambda_j$ is proof.

By Lemma 7.6, let $V$ be the volume of $\mathscr{R}$, then $\mathrm{Vol}(\lambda\mathscr{R}) = \lambda^n V$, for the first continuous minimum $\lambda_1$, we have the following estimation

$$\lambda_1^n V \leq 2^n. \tag{7.14}$$

For $\lambda_j$ $(j \geq 2)$, there is no explicit upper bound estimation, but we have the following conclusions.

**Lemma 7.10** *Let $\mathscr{R} \subset \mathbb{R}^n$ be a convex body satisfying the limiting condition (7.11), $V = \mathrm{Vol}(\mathscr{R})$, $\lambda_1, \lambda_2, \ldots, \lambda_n$ be n continuous minima of $\mathscr{R}$, then we have*

$$\frac{2^n}{n!} \leq V\lambda_1\lambda_2\cdots\lambda_n \leq 2^n. \tag{7.15}$$

***Proof*** We only prove the left inequality of the above formula, and we continuously select the linear independent whole point $x^{(1)}, x^{(2)}, \ldots, x^{(j)}$ such that $x^{(j)} \in \lambda_j\mathscr{R}$, and $x^{(j)}$ $x^{(1)}, x^{(2)}, \ldots, x^{(j-1)}$ is linearly independent. Let $x^{(j)}=(x_{j1}, x_{j2}, \ldots, x_{jn}) \in \mathbb{Z}^n$. Because matrix $A = (x_{ji})_{n\times n}$ is an integer matrix, and $\det(A) \neq 0$, so

$$|\det(A)| \geq 1.$$

By Lemma 7.9, for any constant $\mu_1, \mu_2, \ldots, \mu_n$, we have

$$\begin{aligned} &F(\mu_1 x^{(1)} + \mu_2 x^{(2)} + \cdots + \mu_n x^{(n)}) \\ &\leq |\mu_1|F(x^{(1)}) + |\mu_2|F(x^{(2)}) + \cdots + |\mu_n|F(x^{(n)}) \\ &\leq |\mu_1|\lambda_1 + |\mu_2|\lambda_2 + \cdots + |\mu_n|\lambda_n. \end{aligned}$$

Thus, if $|\mu_1|\lambda_1 + |\mu_2|\lambda_2 + \cdots + |\mu_n|\lambda_n \leq 1$, then

$$\mu_1 x^{(1)} + \mu_2 x^{(2)} + \cdots + \mu_n x^{(n)} \in \mathscr{R}.$$

So set

$$\mathscr{R}_1 = \{\mu_1 x^{(1)} + \mu_2 x^{(2)} + \cdots + \mu_n x^{(n)} \,||\, \mu_1|\lambda_1 + |\mu_2|\lambda_2 + \cdots + |\mu_n|\lambda_n \leq 1\} \subset \mathscr{R}.$$

The volume of the left set $\mathscr{R}_1$ is

$$\begin{aligned} \mathrm{Vol}(\mathscr{R}_1) &= \int\cdots\int\limits_{|\mu_1|\lambda_1+|\mu_2|\lambda_2+\cdots+|\mu_n|\lambda_n\leq 1} \mathrm{d}\mu_1\mathrm{d}\mu_2\cdots\mathrm{d}\mu_n = \frac{2^n|\det(A)|}{n!\lambda_1\cdots\lambda_n} \\ &\geq \frac{2^n}{n!\lambda_1\cdots\lambda_n}. \end{aligned}$$

So there is

$$\frac{2^n}{n!\lambda_1\cdots\lambda_n} \leq \mathrm{Vol}(\mathscr{R}_1) \leq \mathrm{Vol}(\mathscr{R}) = V.$$

Therefore, the left inequality of (7.15) holds. The proof of the right inequality is quite complex and is omitted here. Interested readers can refer to the classic works (1963, 1971) of J. W. S. Cassels.

An important application of the above geometry of numbers is to solve the problem of rational approximation of real numbers, which is called Diophantine approximation in classical number theory. The main conclusion of this section is the following simultaneous rational approximation theorem of $n$ real numbers.

**Theorem 7.1** *Let $\theta_1, \theta_2, \ldots, \theta_n$ be any $n$ real numbers, $\theta_i \neq 0$, then for any positive number $N > 1$, there are nonzero positive integers $q$ and $p_1, p_2, \ldots, p_n$ to satisfy*

$$\begin{cases} |q\theta_i - p_i| < N^{-\frac{1}{n}}, \quad 1 \leq i \leq n; \\ |q| \leq N. \end{cases} \tag{7.16}$$

***Proof*** The proof of the theorem is a simple application of Minkowski's linear type theorem (see Lemma 7.8). Let $A \in \mathbb{R}^{(n+1)\times(n+1)}$ be an $(n+1)$-order reversible square matrix, defined as

$$A = \begin{pmatrix} -1 & 0 & \cdots\cdots & 0 & \theta_1 \\ 0 & -1 & \cdots\cdots & 0 & \theta_2 \\ \cdots & \cdots & \cdots & \cdots & \cdots \\ 0 & 0 & \cdots\ 0 & -1 & \theta_n \\ 0 & \cdots\cdots & 0 & 0 & -1 \end{pmatrix}.$$

Obviously $|\det(A)| = 1$. Let $(n+1)$-dimensional positive vector $c = (N^{-\frac{1}{n}}, N^{-\frac{1}{n}}, \ldots, N^{-\frac{1}{n}}, N)$, because

$$c_1 c_2 \cdots c_n c_{n+1} = N^{-1} \cdot N = 1 \geq |\det(A)|.$$

So by Lemma 7.8, the symmetric convex body $\mathscr{R}'(A, c)$ defined by $A$ and $c$ has a nonzero integral point $x = (p_1, p_2, \ldots, p_n, q) \neq 0$. We prove $q \neq 0$. Because $x \neq 0$, if $q = 0$, then $p_k \neq 0$ $(1 \leq k \leq n)$, therefore, the $k$-th inequality in Eq. (7.16) will produce the following contradiction,

$$1 \leq |q\theta_k - p_k| < N^{-\frac{1}{n}} < 1.$$

So $q \neq 0$, we complete the proof of Theorem 7.1.

**Corollary 7.3** *Let $\theta_1, \ldots, \theta_n$ be any $n$ real numbers, then for any $\varepsilon > 0$, there is rational number $\frac{p_i}{q}$ $(1 \leq i \leq n)$ satisfies*

$$\left|\theta_i - \frac{p_i}{q}\right| < \frac{\varepsilon}{q}. \tag{7.17}$$

***Proof*** Any $\varepsilon > 0$ given, let $N^{-\frac{1}{n}} < \varepsilon$, Formula (7.17) can be derived directly from Theorem 7.1.

## 7.2 Basic Properties of Lattice

Lattice is one of the most important concepts in modern cryptography. Most of the so-called anti-quantum computing attacks are lattice based cryptosystems. What is a lattice? In short, a lattice is a geometry in $n$-dimensional Euclidean space $\mathbb{R}^n$, for example $L = \mathbb{Z}^n \subset \mathbb{R}^n$, then $\mathbb{Z}^n$ is a lattice in $\mathbb{R}^n$, which is called an integer lattice or a trivial lattice. If $\mathbb{Z}^n$ is rotated once, we get the concept of a general lattice in $\mathbb{R}^n$, which is a geometric description of a lattice, next, we give an algebraic precise definition of a lattice.

**Definition 7.3** Let $L \subset \mathbb{R}^n$ be a nonempty subset, which is called a lattice in $\mathbb{R}^n$, if

(i) $L$ is an additive subgroup of $\mathbb{R}^n$;
(ii) There is a positive constant $\lambda = \lambda(L) > 0$, such that

$$\min\{|x| \,|\, x \in L, x \neq 0\} = \lambda, \tag{7.18}$$

$\lambda = \lambda(L)$ is called the minimal distance of a lattice $L$.

By Definition 7.3, a lattice is simply a discrete additive subgroup in $\mathbb{R}^n$, in which the minimum distance $\lambda = \lambda(L)$ is the most important mathematical quantity of the lattice. Obviously, we have

$$\lambda = \min\{|x - y| \,|\, x \in L, y \in L, x \neq y\}, \tag{7.19}$$

Equation (7.19) shows the reason why $\lambda$ is called the minimal distance of a lattice. If $x \in L$ and $|x| = \lambda$, $x$ is called the shortest vector of $L$.

In order to obtain a more explicit and concise mathematical expression of any lattice, we can regard an additive subgroup as a $\mathbb{Z}$-module. First, we prove that any lattice is a finitely generated $\mathbb{Z}$-module.

**Lemma 7.11** *Let $L \subset \mathbb{R}^n$ be a lattice and $\{\alpha_1, \alpha_2, \ldots, \alpha_m\} \subset L$ be a set of vectors in $L$, then $\{\alpha_1, \alpha_2, \ldots, \alpha_m\}$ is linearly independent in $\mathbb{R}$ if and only if $\{\alpha_1, \alpha_2, \ldots, \alpha_m\}$ is linearly independent in $\mathbb{Z}$.*

***Proof*** If $\{\alpha_1, \alpha_2, \ldots, \alpha_m\}$ is linearly independent in $\mathbb{R}$, it is obviously linearly independent in $\mathbb{Z}$. conversely, if $\{\alpha_1, \alpha_2, \ldots, \alpha_m\}$ is linearly independent in $\mathbb{Z}$, that is, any linear combination

$$a_1\alpha_1 + \cdots + a_m\alpha_m = 0, \quad a_i \in \mathbb{Z},$$

we have $a_1 = a_2 = \cdots = a_m = 0$, then the linear combination in $\mathbb{R}$ is equal to 0, that is

$$\theta_1\alpha_1 + \theta_2\alpha_2 + \cdots + \theta_m\alpha_m = 0, \quad \theta_i \in \mathbb{R}. \tag{7.20}$$

We prove $\theta_1 = \theta_2 = \cdots = \theta_m = 0$. By Lemma 7.1, for sufficiently large $N > 1$, there are positive integers $q \neq 0$ and $p_1, p_2, \ldots, p_m$ such that

$$\begin{cases} |q\theta_i - p_i| < N^{-\frac{1}{m}}, \quad 1 \le i \le m; \\ q \le N. \end{cases}$$

By (7.20), we have

$$\begin{aligned} |p_1\alpha_1 + \cdots + p_m\alpha_m| &= |(q\theta_1 - p_1)\alpha_1 + \cdots + (q\theta_m - p_m)\alpha_m| \\ &\le N^{-\frac{1}{m}}(|\alpha_1| + \cdots + |\alpha_m|) \\ &\le N^{-\frac{1}{m}} \max_{1\le i\le m} |\alpha_i|. \end{aligned}$$

Let $\lambda$ be the minimal distance of $L$ and $\varepsilon > 0$ be a sufficiently small positive number, we choose

$$N > \max\left\{\varepsilon^{-m}, \max_{1\le i\le m} \frac{|\alpha_i|^m}{\lambda^m}\right\},$$

then $N^{-\frac{1}{m}} < \varepsilon$, and

$$N^{-\frac{1}{m}} \max_{1\le i\le m} |\alpha_i| < \lambda.$$

Thus

$$|p_1\alpha_1 + \cdots + p_m\alpha_m| < \lambda.$$

Notice that $p_1\alpha_1 + \cdots + p_m\alpha_m \in L$, so $p_1\alpha_1 + \cdots + p_m\alpha_m = 0$. Since $\{\alpha_1, \alpha_2, \ldots, \alpha_m\}$ is linearly independent on $\mathbb{Z}$, $p_1 = p_2 = \cdots = p_m = 0$ is derived. For any $i$, $1 \le i \le m$, we get $|\theta_i| \le |q\theta_i| < N^{-\frac{1}{m}} < \varepsilon$. Since $\varepsilon$ is any small positive number, there is $\theta_1 = \theta_2 = \cdots = \theta_m = 0$. This proves that $\{\alpha_1, \alpha_2, \ldots, \alpha_m\}$ is also linearly independent in $\mathbb{R}$. Lemma 7.11 holds.

From the above lemma, any lattice $L$ in $\mathbb{R}^n$ is a finitely generated $\mathbb{Z}$-module. Let $\{\beta_1, \beta_2, \ldots, \beta_m\} \subset L$ be a set of $\mathbb{Z}$-bases in $L$, then $L$ as the rank of $\mathbb{Z}$-module satisfies

$$\mathrm{rank}(L) = m \le n, \tag{7.21}$$

and

$$L = \left\{\sum_{i=1}^{m} a_i\beta_i | a_i \in \mathbb{Z}\right\}. \tag{7.22}$$

If $\{\beta_1, \beta_2, \ldots, \beta_m\}$ is a $\mathbb{Z}$-basis of $L$ and each $\beta_i$ is regarded as a column vector, then the matrix

$$B = [\beta_1, \beta_2, \ldots, \beta_m] \in \mathbb{R}^{n\times m}, \quad \text{rank}(B) = m.$$

Equation (7.22) can be written as

$$L = L(B) = \{Bx | x \in \mathbb{Z}^m\} \subset \mathbb{R}^n. \tag{7.23}$$

We take $L$ as the $\mathbb{Z}$-modules, $m$ as the rank of lattice $L$, $B \in \mathbb{R}^{n\times m}$ as the generating matrix of lattice $L$, and $\{\beta_1, \beta_2, \ldots, \beta_m\}$ as a set of generating bases of $L$.

If $\{\alpha_1, \alpha_2, \ldots, \alpha_m\} \subset \mathbb{R}^n$ is any $m$ column vectors in $\mathbb{R}^n$, the Gram matrix of $A = [\alpha_1, \alpha_2, \ldots, \alpha_m] \in \mathbb{R}^{n\times m}$, $\{\alpha_1, \alpha_2, \ldots, \alpha_m\}$ is defined as

$$T = (\langle \alpha_i, \alpha_j \rangle)_{m\times m}.$$

Obviously, we have $T = A'A$, where $A'$ is the transpose matrix of $A$.

**Lemma 7.12** *Let $A \in \mathbb{R}^{n\times m}$, $b \in \mathbb{R}^n$ ($m \le n$ is not required), then*

*(i) Let $x_0 \in \mathbb{R}^m$ be a solution of $A'Ax = A'b$, then*

$$|Ax_0 - b|^2 = \min_{x\in\mathbb{R}^m} |Ax - b|^2.$$

*(ii)* rank$(A'A) = $ rank$(A)$, *and homogeneous linear equations $Ax = 0$ and $A'Ax = 0$ have the same solution.*

*(iii) $A'Ax = A'b$ always has a solution $x \in \mathbb{R}^m$, and when* rank$(A) = m$, *the solution is unique*

$$x = (A'A)^{-1}A'b.$$

***Proof*** First we prove (i). Let $x_0 \in \mathbb{R}^m$ satisfies $A'Ax_0 = A'b$, then for any $x \in \mathbb{R}^m$, we have

$$Ax - b = (Ax_0 - b) + A(x - x_0) = \gamma + \gamma_1 \in \mathbb{R}^n.$$

We prove that $\gamma$ and $\gamma_1$ are two orthogonal vectors in $\mathbb{R}^n$. Because

$$\begin{aligned}(A(x - x_0))'(Ax_0 - b) & \\ &= (x - x_0)'A'(Ax_0 - b) \\ &= (x - x_0)'(A'Ax_0 - A'b) = 0.\end{aligned}$$

So $\gamma \perp \gamma_1$, by Pythagorean theorem, we have

$$|Ax - b|^2 = |Ax_0 - b|^2 + |A(x - x_0)|^2 \ge |Ax_0 - b|^2.$$

So (i) holds.

To prove (ii), let $V_A$ be the solution space of $Ax = 0$ and $V_{A'A}$ the solution space of $A'Ax = 0$, let's prove $V_A = V_{A'A}$. First, there is $V_A \subset V_{A'A}$. Conversely, let $x \in V_{A'A}$, that is $A'Ax = 0$, then

$$x'A'Ax = 0 \Rightarrow (Ax)'Ax = \langle Ax, Ax\rangle = 0.$$

The above formula holds if and only if $Ax = 0$, so $x \in V_A$. There is $V_A = V_{A'A}$. Notice that

$$\begin{cases} \dim V_A = m - \text{rank}(A) \\ \dim V_{A'A} = m - \text{rank}(A'A). \end{cases}$$

So $\text{rank}(A) = \text{rank}(A'A)$, (ii) holds. To prove (iii), $b \in \mathbb{R}^n$ given, then the rank of the augmented matrix of linear equation system $A'Ax = A'b$ is

$$\begin{aligned} \text{rank}[A'A, A'b] &= \text{rank}(A'[A, b]) \\ &\le \text{rank}(A') = \text{rank}(A) = \text{rank}(A'A). \end{aligned}$$

Therefore, the augmented matrix and the coefficient matrix have the same rank, so the linear equations have solutions. When $\text{rank}(A) = m$, then $\text{rank}(A'A) = m$, that is, $A'A$ is a reversible $m$-order square matrix, thus

$$x = (A'A)^{-1} \cdot A'b, \Longrightarrow \text{the solution is unique.}$$

Lemma 7.12 holds!

**Lemma 7.13** *$A \in \mathbb{R}^{n\times m}$, and $\text{rank}(A) = m$, then $A'A$ is a positive definite real symmetric matrix of order $m$, so there is a real orthogonal matrix $P \in \mathbb{R}^{m\times m}$ of order $m$ satisfies*

$$P'A'AP = \begin{vmatrix} \delta_1 & \cdots & & 0 \\ \vdots & \delta_2 & & \\ \vdots & \cdots & \ddots & \\ 0 & & & \delta_m \end{vmatrix}, \tag{7.24}$$

*where $\delta_i > 0$ is the $m$ eigenvalues of $A'A$.*

***Proof*** $\text{rank}(A) = m \Rightarrow m \le n$. Let $T = A'A$, then $T$ is a symmetric matrix of order $m$. Let $x \in \mathbb{R}^m$ be $m$ arguments, quadratic form

$$x'Tx = x'A'Ax = (Ax)'(Ax) = \langle Ax, Ax\rangle \ge 0.$$

Because $\text{rank}(A) = m$, the above formula if and only if when $x = 0$, $x'Tx = 0$. So $T$ is a positive definite matrix. From the knowledge of linear algebra, there is an orthogonal matrix of order $m$, $P \Rightarrow P'TP$ is a diagonal matrix, that is

$$P'TP = \text{diag}\{\delta_1, \delta_2, \ldots, \delta_m\}.$$

Because $P'TP$ and $T$ have the same eigenvalue, $\delta_1, \delta_2, \ldots, \delta_m$ is the eigenvalue of $T$, and $\forall\, \delta_i > 0$. The Lemma holds.

Lemma 7.12 is called the least square method in linear algebra, its significance is to find a vector $x_0$ with the shortest length in the set $\{Ax - b|x \in \mathbb{R}^m\}$ for a given $n \times m$-order matrix $A$ and a given vector $b \in \mathbb{R}^n$. Lemma 7.12 gives an effective algorithm, that is, to solve the linear equations $A'Ax = A'b$, and $x_0$ is the solution of the equations, Lemma 7.13 is called the diagonalization of quadratic form. Now, the main results are as follows:

**Theorem 7.2** *Let $L \subset \mathbb{R}^n$ be a lattice,* $\text{rank}(L) = m$ $(m \leq n)$, *if and only if there is a real matrix $B \in \mathbb{R}^{n\times m}$ of order $n \times m$,* $\text{rank}(B) = m$, *such that*

$$L = \{Bx|x \in \mathbb{Z}^m\} = \left\{\sum_{i=1}^{m} a_i\beta_i|a_i \in \mathbb{Z}\right\}, \tag{7.25}$$

*where $B = [\beta_1, \beta_2, \ldots, \beta_m]$, each $\beta_i \in \mathbb{R}^n$ is a column vector.*

***Proof*** Equation (7.23) proves the necessity of the condition, and we only prove the sufficiency of the condition. If a subset $L$ in $\mathbb{R}^n$ is given by Eq. (7.25), it is obvious that $L$ is an additive subgroup of $\mathbb{R}^n$, because any $\alpha = Bx_1, \beta = Bx_2$, where $x_1, x_2 \in \mathbb{Z}^m$, then $x = x_1 - x_2 \in \mathbb{Z}^n$, and

$$\alpha - \beta = B(x_1 - x_2) = Bx \in L.$$

So we only prove the discreteness of $L$. Let $T = B'B$, then from Lemma 7.13, $T$ is a positive definite real symmetric matrix, let $\delta_1, \delta_2, \ldots, \delta_m$ be the eigenvalue of $T$, then

$$\delta = \min\{\delta_1, \delta_2, \ldots, \delta_m\} > 0.$$

We prove

$$\min_{\substack{x\in\mathbb{Z}^m \\ x\neq 0}} |Bx| \geq \sqrt{\delta} > 0. \tag{7.26}$$

By Lemma 7.13, there is an orthogonal matrix $P$ of order $m$ such that

$$P'TP = \text{diag}\{\delta_1, \delta_2, \ldots, \delta_m\}.$$

For any given $x \in \mathbb{Z}^m$, $x \neq 0$. We have

$$|Bx|^2 = x'Tx = x'P(P'TP)P'x \geq \delta|P'x|^2 = \delta|x|^2.$$

Because $x \neq 0$, then $|x|^2 \geq 1$, so

$$|Bx|^2 \geq \delta, \quad \forall\, x \in \mathbb{Z}^m, \quad x \neq 0.$$

This shows that the distance between any two different points in $L$ is $\geq \delta > 0$. Therefore, in a sphere with 0 as the center and $r$ as the radius, the number of points in $L$ is finite. In these finite vectors, there is a $\alpha \in L, \Rightarrow$

$$|\alpha| = \min_{\substack{x \in L \\ x \neq 0}} |x| = \lambda \geq \delta > 0.$$

According to the definition of lattice, $L$ is a lattice in $\mathbb{R}^n$, the Lemma holds.

It can be directly deduced from the above theorem

**Corollary 7.4** *Let $L = L(B) \subset \mathbb{R}^n$ be a lattice of* rank$(L) = m$*, $\lambda$ be the minimum distance of $L$, $B \in \mathbb{R}^{n\times m}$, $\delta$ be the minimum eigenvalue of $B'B$, then $\lambda \geq \sqrt{\delta}$.*

**Definition 7.4** $L \subset \mathbb{R}^n$ is a lattice, and rank$(L) = n$, call $L$ is a full rank lattice of $\mathbb{R}^n$.

By Theorem 7.2, a sufficient and necessary condition for a full rank lattice with $L$ as $\mathbb{R}^n$ is the existence of a reversible square matrix $B \in \mathbb{R}^{n\times n}$, $\det(B) \neq 0$, such that

$$L = L(B) = \left\{ \sum_{i=1}^{n} a_i \beta_i | a_i \in \mathbb{Z}, 1 \leq i \leq n \right\} = \{Bx | x \in \mathbb{Z}^n\}. \tag{7.27}$$

If $L = L(B)$ is a full rank lattice, define $d = d(L)$ as

$$d = d(L) = |\det(B)|, \tag{7.28}$$

call $d$ is the determinant of $L$. $d = d(L)$ is the second most important mathematical quantity of a lattice. The lattice we discuss below is always assumed to be a full rank lattice.

For a lattice (full rank lattice), the generating matrix is not unique, but $d = d(L)$ is unique. To prove this, first define the so-called unimodular matrix. Define

$$SL_n(\mathbb{Z}) = \{A = (a_{ij})_{n\times n} | a_{ij} \in \mathbb{Z}, \det(A) = \pm 1\}, \tag{7.29}$$

Obviously, $SL_n(\mathbb{Z})$ forms a group under the multiplication of the matrix, because the $n$-order identity matrix $I_n \in SL_n(\mathbb{Z})$, and $A_1 \in SL_n(\mathbb{Z})$, $A_2 \in SL_n(\mathbb{Z})$, then $A = A_1A_2 \in SL_n(\mathbb{Z})$. Specially, if $A \in SL_n(\mathbb{Z})$,$A = (a_{ij})_{n\times n}$, then the inverse matrix of $A$

$$A^{-1} = \pm \begin{vmatrix} a_{11}^* & a_{12}^* & \cdots & a_{1n}^* \\ a_{21}^* & a_{22}^* & \cdots & a_{2n}^* \\ \cdots & \cdots & \cdots & \cdots \\ a_{n1}^* & \cdots & \cdots & a_{nn}^* \end{vmatrix} \in SL_n(\mathbb{Z}),$$

where $a_{ij}^*$ is the algebraic cofactor of $a_{ij}$.

**Lemma 7.14** *$L = L(B) \subset \mathbb{R}^n$ is a lattice (full rank lattice), $B_1 \in \mathbb{R}^{n\times n}$, then $L = L(B) = L(B_1)$ if and only if there is a unimodular matrix $U \in SL_n(\mathbb{Z}) \Rightarrow B = B_1U$.*

***Proof*** If $B = B_1U$, $U \in SL_n(\mathbb{Z})$, we prove $L(B) = L(B_1)$. Let $\alpha = B_1x \in L(B_1)$, where $x \in \mathbb{Z}^n$, then

$$\alpha = B_1x = B_1UU^{-1}x = BU^{-1}x.$$

Because of $U^{-1}x \in \mathbb{Z}^n$, then $\alpha \in L(B)$, that is $L(B_1) \subset L(B)$. Similarly, if $\alpha = Bx$, $x \in \mathbb{Z}^n$, then

$$\alpha = Bx = B_1Ux, \text{ where } Ux \in \mathbb{Z}^n.$$

Thus, $\alpha \in L(B_1)$, that is $L(B) = L(B_1)$.

Conversely, if $L(B) = L(B_1)$, let $B = [\beta_1, \beta_2, \ldots, \beta_n]$, $B_1 = [\alpha_1, \alpha_2, \ldots, \alpha_n]$, transition matrix

$$(\beta_1, \beta_2, \ldots, \beta_n) = (\alpha_1, \alpha_2, \ldots, \alpha_n)U.$$

Obviously by $\beta_i \in L(B_1)(1 \le i \le n)$, $U$ is an integer matrix. and

$$(\alpha_1, \alpha_2, \ldots, \alpha_n) = (\beta_1, \beta_2, \ldots, \beta_n)U_1.$$

Because $\alpha_i \in L(B)(1 \le i \le n)$, $U_1$ also is an integer matrix. Because of

$$(\beta_1, \beta_2, \ldots, \beta_n) = (\alpha_1, \alpha_2, \ldots, \alpha_n)U = (\beta_1, \beta_2, \ldots, \beta_n)U_1U.$$

We have $U_1U = I_n$, thus $\det(U) = \pm 1$, that is $U \in SL_n(\mathbb{Z})$, $B = B_1U$, the Lemma holds.

By Lemma 7.14, $B$, $B_1$ are any two generating matrices of a lattice $L$, then

$$|\det(B)| = |\det(B_1)| = d = d(L).$$

That is, the determinant $d(L)$ of a lattice is an invariant.

For a lattice (full rank lattice) $L \subset \mathbb{R}^n$, the dual lattice of $L$ is defined as

$$L^* = \{\alpha \in \mathbb{R}^n | \langle \alpha, \beta \rangle \in \mathbb{Z}, \forall\, \beta \in L\}. \tag{7.30}$$

**Lemma 7.15** *Let $L = L(B)$ be a lattice, then the dual lattice of $L$ is $L^* = L((B^{-1})')$, that is, if $B$ is the generating matrix of $L$, then $(B^{-1})'$ is the generating matrix of $L^*$.*

***Proof*** Let

$$L((B^{-1})') = \{(B^{-1})'y | y \in \mathbb{Z}^n\}.$$

any $\alpha \in L((B^{-1})')$, $\alpha = (B^{-1})'y$, $y \in \mathbb{Z}^n$, $\beta \in L$, $\beta = Bx$, $x \in \mathbb{Z}^n$, then

$$\langle \alpha, \beta \rangle = \alpha'\beta = y'B^{-1}Bx = y'x \in \mathbb{Z}.$$

That means $L((B^{-1})') \subset L^*$. Conversely, any $\alpha \in L^*$, for all $\beta \in L$, there is $\langle \alpha, \beta \rangle \in \mathbb{Z}$. So let $B = [\beta_1, \beta_2, \ldots, \beta_n]$, then

$$\left\langle \alpha, \sum_{i=1}^{n} x_i \beta_i \right\rangle = \sum_{i=1}^{n} x_i \langle \alpha, \beta_i \rangle \in \mathbb{Z}, \forall\, x_i \in \mathbb{Z},$$

therefore, for each generating vector $\beta_i (1 \leq i \leq n)$, there is $\langle \alpha, \beta_i \rangle \in \mathbb{Z}$. Write $\alpha = (y_1, y_2, \ldots, y_n)$,

$$\langle \alpha, \beta_i \rangle \in \mathbb{Z}, \Longrightarrow B' \begin{pmatrix} y_1 \\ \vdots \\ y_n \end{pmatrix} = \begin{pmatrix} x_1 \\ \vdots \\ x_n \end{pmatrix} \in \mathbb{Z}^n.$$

Thus

$$\begin{pmatrix} y_1 \\ \vdots \\ y_n \end{pmatrix} \in (B')^{-1} \begin{pmatrix} x_1 \\ \vdots \\ x_n \end{pmatrix}.$$

That is $\alpha \in L((B')^{-1})$. Because $B \cdot B^{-1} = I_n, \Longrightarrow (B^{-1})'B' = I_n$, thus $(B^{-1})' = (B')^{-1}$. So $\alpha \in L((B^{-1})')$, that is $L^* \subset L((B^{-1})')$. We have $L^* = L((B^{-1})')$. The Lemma holds.

By Lemma 7.15, we immediately have the following corollary.

**Corollary 7.5** *Let $L = L(B)$ be a full rank lattice, $L^*$ is the dual lattice of $L$, then*

$$d(L^*) = d^{-1}(L).$$

An equivalence relation in $\mathbb{R}^n$ can be defined by using a lattice $L$, for all $\alpha, \beta \in \mathbb{R}^n$, we define

$$\alpha \equiv \beta (\text{mod } L) \Longleftrightarrow \alpha - \beta \in L.$$

Obviously, this is an equivalent relation, called the congruence relation of mod $L$.

**Definition 7.5** Let $F \subset \mathbb{R}^n$ be a subset, and call $F$ the basic region of a lattice (full rank lattice) $L$, if

(i) $\forall\, x \in \mathbb{R}^n$, there is a $\alpha \in F \Rightarrow x \equiv \alpha (\text{mod } L)$,
(ii) Any $\alpha_1, \alpha_2 \in F$, then $\alpha_1 \not\equiv \alpha_2 (\text{mod } L)$.

By definition, the basic neighborhood of a lattice is the representative element set of the additive quotient group $\mathbb{R}^n / L$. Therefore, a basic neighborhood of any $L$ forms an additive group under mod $L$.

**Lemma 7.16** *Let $L = L(B)$ be a full rank lattice, then*

*(i) Any two basic neighborhoods $F_1$ and $F_2$ of $L$ are isomorphic additive groups* (mod $L$).
*(ii) $F = \{Bx | x = (x_1, x_2, \ldots, x_n)'$, and $0 \leq x_i < 1, 1 \leq i \leq n\}$ is a basic neighborhood of $L(B)$.*

*(iii)* $\text{Vol}(F) = d = d(L)$.

***Proof*** (i) is trivial, because

$$F_1 \cong \mathbb{R}^n/L,\ F_2 \cong \mathbb{R}^n/L, \Longrightarrow F_1 \cong F_2.$$

To prove (ii), let $B = [\beta_1, \beta_2, \dots, \beta_n]$, then $\{\beta_1, \beta_2, \dots, \beta_n\}$ is a set of bases of $\mathbb{R}^n$, $\forall\, \alpha \in \mathbb{R}^n$, $\alpha$ can be expressed as a linear combination of $\beta_1, \beta_2, \dots, \beta_n$, that is

$$\alpha = \sum_{i=1}^{n} a_i \beta_i, \forall\, a_i \in \mathbb{R}.$$

Let $[\alpha]_B = \sum_{i=1}^{n} [a_i]\beta_i$, $\{\alpha\}_B = \alpha - [\alpha]_B$, then $\{\alpha\}_B$ can be expressed as

$$\{\alpha\}_B = B \begin{pmatrix} x_1 \\ x_2 \\ \vdots \\ x_n \end{pmatrix}, \text{ where } 0 \leq x_i < 1, \quad 1 \leq i \leq n.$$

That is $\{\alpha\}_B \in F$. Because $\alpha - \{\alpha\}_B = [\alpha]_B \in L$, so for any $\alpha \in \mathbb{R}^n$, there is a $\{\alpha\}_B \in F$, such that

$$\alpha \equiv \{\alpha\}_B (\text{mod } L).$$

And two points $\alpha = Bx$ and $\beta = By$ in $F$, then

$$\alpha - \beta = B(x - y) = Bz.$$

where $z = (z_1, z_2, \dots, z_n)$, $|z_i| < 1$. So $\alpha \not\equiv \beta (\text{mod } L)$, if $\alpha \neq \beta$. So $F$ is a basic neighborhood of $L$.

Let's prove (iii). Because all basic neighborhoods of $L$ are isomorphic, they have the same volume, (iii) gives a specific basic region $F$ of $L$, so we can only prove $\text{Vol}(F) = d = d(L)$. Obviously,

$$\text{Vol}(F) = \int \cdots \int_{y=(y_1, y_2, \dots, y_n) \in F} \mathrm{d}y_1 \mathrm{d}y_2 \cdots \mathrm{d}y_n$$

make variable substitution $Bx = y$ and calculate the Jacobi of the vector value

$$\mathrm{d}y_1 \mathrm{d}y_2 \cdots \mathrm{d}y_n = d(\lambda) \mathrm{d}x_1 \cdots \mathrm{d}x_n.$$

Thus

$$\mathrm{Vol}(F) = \int_0^1 \cdots \int_0^1 d(\lambda)\mathrm{d}x_1 \cdots \mathrm{d}x_n = d(L).$$

We have completed the proof of Lemma 7.16.

Next, we discuss the gram Schmidt orthogonalization algorithm. If $B = [\beta_1, \beta_2, \ldots, \beta_n]$ is the generation matrix of $L$, $\{\beta_1, \beta_2, \ldots, \beta_n\}$ can be transformed into a set of orthogonal bases $\{\beta_1^*, \beta_2^*, \ldots, \beta_n^*\}$, where $\beta_1^* = \beta_1$, and

$$\beta_i^* = \beta_i - \sum_{j=1}^{i-1} \frac{\langle \beta_i, \beta_j^* \rangle}{\langle \beta_j^*, \beta_j^* \rangle} \beta_j^*, \tag{7.31}$$

$\{\beta_1^*, \beta_2^*, \ldots, \beta_n^*\}$ is called the orthogonal basis corresponding to $\{\beta_1, \beta_2, \ldots, \beta_n\}$. $B^* = [\beta_1^*, \ldots, \beta_n^*]$ is the orthogonal matrix corresponding to $B$. For any $1 \leq i \leq n$, denote

$$\begin{cases} u_{ii} = 1, u_{ij} = 0, \text{ when } j > i. \\ u_{ij} = \dfrac{\langle \beta_i, \beta_j^* \rangle}{|\beta_j^*|^2}, \text{ when } 1 \leq j \leq i \leq n. \\ U = (u_{ij})_{n \times n}. \end{cases} \tag{7.32}$$

Then $U$ is a lower triangular matrix, and

$$\begin{pmatrix} \beta_1 \\ \beta_2 \\ \vdots \\ \beta_n \end{pmatrix} = U \begin{pmatrix} \beta_1^* \\ \beta_2^* \\ \vdots \\ \beta_n^* \end{pmatrix}. \tag{7.33}$$

If both sides are transposed at the same time, there is

$$(\beta_1, \beta_2, \ldots, \beta_n) = (\beta_1^*, \beta_2^*, \ldots, \beta_n^*)U'. \tag{7.34}$$

Therefore, $U'$ is the transition matrix between two groups of bases.

**Lemma 7.17** *Let $L = L(B) \subset \mathbb{R}^n$ be a lattice, $B = [\beta_1, \beta_2, \ldots, \beta_n]$ is the generating matrix, $B^* = [\beta_1^*, \beta_2^*, \ldots, \beta_n^*]$ is the corresponding orthogonal matrix, $d = d(L)$ is the determinant of $L$, then we have*

$$d = \prod_{i=1}^n |\beta_i^*| \leq \prod_{i=1}^n |\beta_i|. \tag{7.35}$$

***Proof*** By (7.24), we have $B = B^*U$, because $\det(U) = 1$, so

$$\det(B) = \det(B^*).$$

By the definition,

$$\begin{aligned} d^2 &= \det(B'B) = \det(U(B^*)'B^*U') \\ &= \det((B^*)'B^*) \\ &= \det(\text{diag}\{|\beta_1^*|^2, |\beta_2^*|^2, \ldots, |\beta_n^*|^2\}). \end{aligned}$$

So there is

$$d = \prod_{i=1}^{n} |\beta_i^*|.$$

In order to prove the inequality on the right of Eq. (7.35), we only prove

$$|\beta_i^*| \leq |\beta_i|, \quad 1 \leq i \leq n. \tag{7.36}$$

Because $\beta_i = \sum_{j=1}^{i} u_{ij}\beta_j^*$, then

$$\begin{aligned} |\beta_i|^2 = \langle \beta_i, \beta_i \rangle &= \left\langle \sum_{j=1}^{i} u_{ij}\beta_j^*, \sum_{j=1}^{i} u_{ij}\beta_j^* \right\rangle \\ &= \sum_{j=1}^{i} u_{ij}^2 \langle \beta_j^*, \beta_j^* \rangle \\ &= \langle \beta_i^*, \beta_i^* \rangle + \sum_{j=1}^{i-1} u_{ij}^2 \langle \beta_j^*, \beta_j^* \rangle. \end{aligned}$$

Therefore, the inequality on the right of (7.35) holds, the Lemma is proved.

Equation (7.35) is usually called Hadamard inequality, and we give another proof here.

In order to define the concept of continuous minima on a lattice $L$, we record the minimum distance on $L$ with $\lambda_1$. That is $\lambda_1 = \lambda(L)$. Another definition of $\lambda_1$ is the minimum positive real number $r$, so that the linear space formed by $L \cap \text{Ball}(0, r)$ is a one-dimensional space, where

$$\text{Ball}(0, r) = \{x \in \mathbb{R}^n \mid |x| \leq r\}$$

is a closed sphere with 0 as the center and $r$ as the radius. The concept of $n$ continuous minima $\lambda_1, \lambda_2, \ldots, \lambda_n$ in $L$ can be given.

**Definition 7.6** Let $L = L(B) \subset \mathbb{R}^n$ be a full rank lattice, the $i$-th continuous minimum $\lambda_i$ is defined as

$$\lambda_i = \lambda_i(L) = \inf\{r \mid \dim(\text{span}(L \cap \text{Ball}(0, r))) \geq i\}.$$

The following lemma is a useful lower bound estimate of the minimum distance $\lambda_1$.

**Lemma 7.18** *$L = L(B) \subset \mathbb{R}^n$ is a lattice (full rank lattice), $B^* = [\beta_1^*, \beta_2^*, \ldots, \beta_n^*]$ is the corresponding orthogonal basis, then*

$$\lambda_1 = \lambda(L) \geq \min_{1\leq i\leq n} |\beta_i^*|. \tag{7.37}$$

***Proof*** For $\forall\, x \in \mathbb{Z}^n$, $x \neq 0$, we prove

$$|Bx| \geq \min_{1\leq i\leq n} |\beta_i^*|,\ x \in \mathbb{Z}^n,\ x \neq 0.$$

Let $x = (x_1, x_2, \ldots, x_n) \neq 0$, $j$ be the largest subscript $\Rightarrow x_j \neq 0$, then

$$|\langle Bx, \beta_j^*\rangle| = \left|\left\langle \sum_{i=1}^{j} x_i\beta_j, \beta_j^* \right\rangle\right| = |x_j||\beta_j^*|^2.$$

Because when $i < j$,

$$\langle \beta_i, \beta_j^*\rangle = 0,\ \text{and}\ \langle \beta_j, \beta_j^*\rangle = \langle \beta_j^*, \beta_j^*\rangle.$$

On the other hand,

$$|\langle Bx, \beta_j^*\rangle| \leq |Bx||\beta_j^*|.$$

So

$$|Bx| \geq |x_j||\beta_j^*| \geq \min_{1\leq i\leq n} |\beta_i^*|.$$

Lemma 7.18 holds!

**Corollary 7.6** *The continuous minimum $\lambda_1, \lambda_2, \ldots, \lambda_n$ of a lattice L is reachable, that is, it exists $\alpha_i \in L \Rightarrow |\alpha_i| = \lambda_i$, $1 \leq i \leq n$.*

***Proof*** The lattice points contained in ball Ball$(0, \delta)$ with center 0 and radius $\delta$ $(\delta > \lambda_i)$ are finite, because in a bounded region (finite volume), if there are infinite lattice points, there must be a convergent subsequence, but the distance between any different two points in $L$ is greater than or equal to $\lambda_1$, which indicates that

$$|L \cap \text{Ball}(0, \delta)|\langle\infty,\ \delta\rangle\lambda_i$$

has finite lattice points, it's not hard for us to find $\alpha_1 \in L \Rightarrow |\alpha_1| = \lambda_1$, $\alpha_2 \in L \Rightarrow |\alpha_2| = \lambda_2$,…,$|\alpha_n| = \lambda_n$. The Corollary holds.

In Sect. 7.1, the geometry of numbers is relative to the integer lattice $\mathbb{Z}^n$; next, we extend the main results to the general full rank lattice.

**Lemma 7.19** (Compare with Lemma 7.5) *$L = L(B) \subset \mathbb{R}^n$ is a lattice (full rank lattice), $\mathscr{R} \subset \mathbb{R}^n$, if $\mathrm{Vol}(\mathscr{R}) > d(L)$, then there are two different points in $\mathscr{R}$, $\alpha \in \mathscr{R}$, $\beta \in \mathscr{R} \Rightarrow \alpha - \beta \in L$.*

***Proof*** Let $F$ be a basic region of $L$, that is

$$F = \{Bx | x = (x_1, \ldots, x_n), 0 \leq |x_i| < 1, 1 \leq i \leq n\}.$$

Obviously, $\mathbb{R}^n$ can be divided into the following disjoint subsets,

$$\begin{aligned} \mathbb{R}^n &= \cup_{\alpha \in L}\{\alpha + y | y \in F\} \\ &= \cup_{\alpha \in L}\{\alpha + F\}. \end{aligned}$$

For a given lattice point $\alpha \in L$, define

$$\mathscr{R}_\alpha = \mathscr{R} \cap \{\alpha + F\} = \alpha + D_\alpha, \; D_\alpha \subset F.$$

Therefore, $\mathscr{R}$ can be divided into the following disjoint subsets,

$$\mathscr{R} = \cup_{\alpha \in L}\mathscr{R}_\alpha, \Rightarrow \mathrm{Vol}(\mathscr{R}) = \sum_{\alpha \in L}\mathrm{Vol}(\mathscr{R}_\alpha) = \sum_{\alpha \in L}\mathrm{Vol}(D_\alpha).$$

If for any $\alpha, \beta \in L, \alpha \neq \beta, D_\alpha \cap D_\beta = \varnothing$, then

$$\mathrm{Vol}(\mathscr{R}) = \mathrm{Vol}(\cup_{\alpha \in L} D_\alpha) \leq \mathrm{Vol}(F) = d(L),$$

contradicts assumptions. So it must exist $\alpha, \beta \in L, \alpha \neq \beta, \Longrightarrow D_\alpha \cap D_\beta \neq \varnothing$. Let $x \in D_\alpha \cap D_\beta$, then $\alpha + x \in \mathscr{R}, \beta + x \in \mathscr{R}$. And

$$(\alpha + x) - (\beta + x) = \alpha - \beta \in L.$$

The Lemma holds.

**Lemma 7.20** (Compare with 7.6) *Let L be a full rank lattice, $\mathscr{R} \subset \mathbb{R}^n$ is a symmetric convex body. And $\mathrm{Vol}(\mathscr{R}) > 2^n d(L)$, then $\mathscr{R}$ contains a nonzero lattice point, that is $\exists\, \alpha \in L$, $\alpha \neq 0$, such that $\alpha \in \mathscr{R}$.*

***Proof*** Let

$$\frac{1}{2}\mathscr{R} = \{x | 2x \in \mathscr{R}\}.$$

Then

$$\mathrm{Vol}\left(\frac{1}{2}\mathscr{R}\right) = 2^{-n}\mathrm{Vol}(\mathscr{R}) > d(L).$$

By 7.19, there is $x \in \frac{1}{2}\mathscr{R}$, $y \in \frac{1}{2}\mathscr{R}, \Longrightarrow x - y \in L$. And because $2x \in \mathscr{R}$, $2y \in \mathscr{R}$, $\mathscr{R}$ is a symmetric convex body, by Lemma 7.4,

$$\frac{1}{2}(2x - 2y) = x - y \in \mathscr{R}.$$

The Lemma holds.

**Corollary 7.7** *Let L be a full rank lattice, $\lambda(L) = \lambda_1$ is the minimum distance of L. Then*

$$\lambda_1 = \lambda(L) \leq \sqrt{n}(d(L))^{\frac{1}{n}}. \tag{7.38}$$

***Proof*** First we prove

$$\text{Vol}(\text{Ball}(0, r)) \geq \left(\frac{2r}{\sqrt{n}}\right)^n. \tag{7.39}$$

This is because Ball$(0, r)$ contains the following cubes

$$\left\{x \in \mathbb{R}^n | x = (x_1, \ldots, x_n), \forall\ |x_i| < \frac{r}{\sqrt{n}}\right\} \subset \text{Ball}(0, r).$$

By the definition, there are no nonzero lattice points in open ball Ball$(0, \lambda_1)$, by Lemma 7.20, because Ball$(0, \lambda_1)$ is a symmetrical convex body, there is

$$\text{Vol}(\text{Ball}(0, \lambda_1)) \leq 2^n d(L).$$

Thus

$$\left(\frac{2\lambda_1}{\sqrt{n}}\right)^n \leq 2^n d(L).$$

That is

$$\lambda_1 \leq \sqrt{n}(d(L))^{\frac{1}{n}}.$$

The Corollary holds.

Combined with Eq. (7.37), we obtain the estimation of the upper and lower bounds of the minimum distance of a lattice,

$$\min_{1 \leq i \leq n} |\beta_i^*| \leq \lambda(L) \leq \sqrt{n}(d(L))^{\frac{1}{n}}. \tag{7.40}$$

**Lemma 7.21** *Let $L \subset \mathbb{R}^n$ be a lattice (full rank lattice), $\lambda_1, \lambda_2, \ldots, \lambda_n$ is the continuous minimum of L, $d = d(L)$ is the determinant of L, then*

$$\lambda_1 \lambda_2 \ldots \lambda_n \leq n^{\frac{n}{2}} d(L). \tag{7.41}$$

***Proof*** Let $\{\alpha_1, \alpha_2, \ldots, \alpha_n\} \subset L$, and $|\alpha_i| = \lambda_i$ is a set of bases of $\mathbb{R}^n$. Let

$$T = \left\{y \in \mathbb{R}^n | \sum_{i=1}^{n} \left(\frac{\langle y, \alpha_i^* \rangle}{\lambda_i |\alpha_i^*|}\right)^2 < 1\right\}, \tag{7.42}$$

where $\{\alpha_1^*, \alpha_2^*, \ldots, \alpha_n^*\}$ is the orthogonal basis corresponding to $\{\alpha_1, \alpha_2, \ldots, \alpha_n\}$. Let's prove that $T$ does not contain any nonzero lattice points. Let $y \in L$, $y \neq 0$, let $k$ be the largest subscript so that $|y| \geq \lambda_k$, then

$$y \in \text{Span}(\alpha_1^*, \alpha_2^*, \ldots, \alpha_k^*) = \text{Span}(\alpha_1, \alpha_2, \ldots, \alpha_k).$$

Because if $y$ is linearly independent of $\alpha_1, \alpha_2, \ldots, \alpha_k$, then

$$k + 1 \leq \dim(\text{Span}(\alpha_1, \alpha_2, \ldots, \alpha_k, y) \cap \text{Ball}(0, |y|)).$$

$\lambda_{k+1} \leq |y|$ is obtained from the definition of $\lambda_{k+1}$, which contradicts the definition of $k$. By $y \in \text{Span}(\alpha_1, \alpha_2, \ldots, \alpha_k)$,

$$\begin{aligned}\sum_{i=1}^{n} \left( \frac{\langle y, \alpha_i^* \rangle}{\lambda_i |\alpha_i^*|} \right)^2 &= \sum_{i=1}^{k} \left( \frac{\langle y, \alpha_i^* \rangle}{\lambda_i |\alpha_i^*|} \right)^2 \\ &\geq \frac{1}{\lambda_k^2} \sum_{i=1}^{k} \frac{\langle y, \alpha_i^* \rangle^2}{|\alpha_i^*|^2} = \frac{1}{\lambda_k^2} |y|^2 \geq 1.\end{aligned}$$

Therefore $y \notin T$, by Lemma 7.20, because $T$ is a symmetric convex body, thus

$$\text{Vol}(T) \leq 2^n d.$$

On the other hand,

$$\begin{aligned}\text{Vol}(T) &= \left( \prod_{i=1}^{n} \lambda_i \right) \cdot \text{Vol}(\text{Ball}(0, 1)) \\ &\geq \prod_{i=1}^{n} \lambda_i \left( \frac{2}{\sqrt{n}} \right)^n.\end{aligned}$$

So

$$\prod_{i=1}^{n} \lambda_i \leq n^{\frac{n}{2}} d.$$

Lemma 7.21 holds.

The above lemma shows that the upper bound (7.38) of $\lambda_1$ is valid for $\lambda_i$ in the sense of geometric average.

Finally, we discuss the computational difficulties on the lattice. These problems are the main scientific basis and technical support in the design of trap gate function, and they are also the cornerstone of the security of lattice cryptography.

1. Shortest vector problem SVP

Lattice $L$ is a discrete geometry in $\mathbb{R}^n$, we know that its minimum distance $\lambda_1 = \lambda(L)$ is the length of the shortest vector in $L$. How to find its shortest vector $u_0 \in L$

for any full rank lattice $L$, $\Longrightarrow$

$$|u_0| = \min_{x\in L, x\neq 0} |x| = \lambda_1.$$

It is the so-called shortest vector calculation problem. At present, there are insurmountable difficulties in theory and calculation, because we only know the existence of $u_0$, but we can't calculate $u_0$. Second, the current main research focuses on the approximation of the shortest vector. The so-called shortest vector approximation is to find a nonzero vector $u \in L$ on $L$, $\Longrightarrow$

$$|u| \leq r(n)\lambda_1, \ \ u \in L, u \neq 0,$$

where $r(n) \geq 1$ is called the approximation coefficient, which only depends on the dimension of lattice $L$.

In 1982, H. W. Lenstra, A. K. Lenstra and L. Lovasz creatively developed a set of algorithms in (1982) to effectively solve the approximation problem of the shortest vector, which is the famous LLL algorithm in lattice theory. The computational complexity of LLL algorithm is polynomial for the whole lattice, and the approximation coefficient $r(n) = 2^{\frac{n-1}{2}}$. How to improve the approximation coefficient in LLL algorithm to the polynomial coefficient of $n$ is the main research topic at present. For example, Schnorr's work in 1987 and Gama and Nguyen's work (2008a, 2008b) are very representative, but they are still far from the polynomial function, so the academic circles generally speculate:

Conjecture 1: there is no polynomial algorithm that can approximate the shortest vector so that the approximation coefficient $r(n)$ is a polynomial function of $n$.

2. Closest vector problem CVP

Let $L \subset \mathbb{R}^n$ be a lattice, $t \in \mathbb{R}^n$ is an arbitrary given vector, and it is easy to prove that there is a lattice point $u_t \in L$, $\Longrightarrow$

$$|u_t - t| = \min_{x\in L} |x - t|,$$

$u_t$ is called the nearest lattice point (vector) of $t$. When $t = 0$ is a zero vector, $u_0$ is the shortest vector of $L$, so the adjacent vector problem is a general form of the shortest vector problem. Similarly, we only know the existence of the adjacent vector $u_t$, and there is no definite algorithm to find $u_t$ instead of the approximation problem of the adjacent vector, $x \in L$, if

$$|x - t| \leq r_1(n)|u_t - t|,$$

then $x$ is called the approximation coefficient, which is the approximation adjacent vector of $r_1(n)$, in 1986, Babai proposed an effective algorithm to approximate the adjacent vector in Babai (1986), and its approximation coefficient $r_1(n)$ is generally of the same order as the approximation coefficient $r(n)$ of the shortest vector.

There are many other difficult computational problems on lattice, such as the Successive Shortest vector problem, which is essentially to find a deterministic algorithm to approximate each $\alpha_i \in L$, where $|\alpha_i| = \lambda_i$ is the continuous minimum of $L$. However, SVP and CVP are commonly used in lattice cryptosystem design and analysis, and most of the research is based on the integer lattice.

## 7.3 Integer Lattice and $q$-Ary Lattice

**Definition 7.7** A full rank lattice $L$ is called an integer lattice, if $L \subset \mathbb{Z}^n$, an integer lattice $L$ is called a $q$-ary lattice, if $q\mathbb{Z}^n \subset L \subset \mathbb{Z}^n$, where $q \geq 1$ is a positive integer.

It is easy to see from the definition that a lattice $L = L(B)$ is an integer lattice $\Leftrightarrow B \in \mathbb{Z}^{n\times n}$ is an integer square matrix, so the determinant $d = d(L)$ of an entire lattice $L$ is a positive integer.

**Lemma 7.22** *Let $L = L(B) \subset \mathbb{Z}^n$ be an integer lattice, $d = d(L)$ is the determinant of $L$, then $d\mathbb{Z}^n \subset L \subset \mathbb{Z}^n$, therefore, an integer lattice is always a $d$-ary lattice $(d = q)$.*

***Proof*** Let $\alpha \in d\mathbb{Z}^n$, let's prove that $\alpha \in L$, that is, $\alpha = Bx$ always has the solution of the entire vector $x \in \mathbb{Z}^n$. Let $B^{-1}$ be the inverse matrix of $B$, then

$$B^{-1} = \frac{1}{\det(B)}B^* = \frac{1}{\det(B)}\begin{bmatrix} b_{11}^* & b_{12}^* & \cdots & b_{1n}^* \\ b_{21}^* & b_{22}^* & \cdots & b_{2n}^* \\ \cdots & \cdots & \cdots & \cdots \\ b_{n1}^* & b_{n2}^* & \cdots & b_{nn}^* \end{bmatrix},$$

where $B = (b_{ij})_{n\times n}$, $b_{ij}^*$ is the algebraic cofactor of $b_{ij}$. Because $B \in \mathbb{Z}^{n\times n}$, so $B^* \in \mathbb{Z}^{n\times n}$, thus $dB^{-1} = \pm B^* \in \mathbb{Z}^{n\times n}$, write $\alpha = d\beta$, then $\beta \in \mathbb{Z}^n$, and

$$x = B^{-1}\alpha = dB^{-1}\beta = \pm B^*\beta \in \mathbb{Z}^n.$$

Thus $\alpha \in L$. That is $d\mathbb{Z}^n \subset L$, the Lemma holds.

The following lemma is a simple conclusion in algebra. For completeness, we prove the following.

**Lemma 7.23** *Let $L$ be a $q$-ary lattice, $\mathbb{Z}_q$ is the residual class rings* mod $q$*, then*

*(i) $\mathbb{Z}^n/q\mathbb{Z}^n \cong \mathbb{Z}_q^n$ (additive group isomorphism).*
*(ii) $\mathbb{Z}^n/L \cong \mathbb{Z}_q^n/_{L/q\mathbb{Z}^n}$ (additive group isomorphism). Therefore, $L/q\mathbb{Z}^n$ is a linear code on $\mathbb{Z}_q^n$.*

***Proof*** $\alpha = (a_1, a_2, \ldots, a_n) \in \mathbb{Z}^n$, $\beta = (b_1, b_2, \ldots, b_n) \in \mathbb{Z}^n$, if $\forall\ a_i \equiv b_i (\bmod q)$, we write $\alpha \equiv \beta (\bmod q)$. For any $\alpha \in \mathbb{Z}^n$, define

$$\bar{\alpha} = (\bar{a_1}, \bar{a_2}, \ldots, \bar{a_n}) \in \mathbb{Z}_q^n,$$

where $\bar{a_i}$ is the minimum nonnegative residue of $a_i \bmod q$, and thus, we have $\alpha \equiv \bar{\alpha} (\bmod q)$. Define mapping $\sigma : \mathbb{Z}^n \xrightarrow{\sigma} \mathbb{Z}_q^n$ as $\sigma(\alpha) = \bar{\alpha}$, this is a surjection, and

$$\sigma(\alpha + \beta) = \bar{\alpha} + \bar{\beta} = \sigma(\alpha) + \sigma(\beta).$$

Therefore, $\sigma$ is a full group homomorphism. Obviously Ker$\sigma = q\mathbb{Z}^n$, therefore, by the isomorphism theorem of groups, we have

$$\mathbb{Z}^n / q\mathbb{Z}^n \cong \mathbb{Z}_q^n.$$

Because of $q\mathbb{Z}^n \subset L \subset \mathbb{Z}^n$, then by the isomorphism theorem of groups,

$$\mathbb{Z}^n / L \cong \mathbb{Z}^n / q\mathbb{Z}^n / L / q\mathbb{Z}^n \cong \mathbb{Z}_q^n / L / q\mathbb{Z}^n.$$

The Lemma holds.

Next, we will prove that $\mathbb{Z}^n / L$ is a finite group. Therefore, we first discuss the elementary transformation of matrix. The so-called elementary transformation of matrix refers to elementary row transformation and elementary column transformation, specifically refers to the following three kinds of elementary transformations:

(1) Transform two rows or two columns of matrix $A$:

$$\begin{cases} \sigma_{ij}(A)\text{-Transform rows } i \text{ and } j \text{ of } A \\ \tau_{ij}(A)\text{-Transform columns } i \text{ and } j \text{ of } A \end{cases}$$

(2) A row or column multiplied by $-1$ by $A$:

$$\begin{cases} \sigma_{-i}(A)\text{-Multiply row } i \text{ of } A \text{ by } -1 \\ \tau_{-i}(A)\text{-Multiply column } i \text{ of } A \text{ by } -1 \end{cases}$$

(3) Add the $k$ times of a row (column) to another row (column), $k \in \mathbb{R}$, in many cases, we require $k \in \mathbb{Z}$ to be an integer:

$$\begin{cases} \sigma_{ki+j}(A)\text{-Add } k \text{ times of row } i \text{ of } A \text{ to row } j \\ \tau_{ki+j}(A)\text{-Add } k \text{ times of column } i \text{ of } A \text{ to column } j \end{cases}$$

The $n$-order identity matrix is represented by $I_n$, the matrix obtained by the above elementary transformation of $I_n$ is called elementary matrix. We note that all elementary matrices are unimodular matrices (see (7.29)), and

$$\begin{cases} \sigma_{ij}(A) = \sigma_{ij}(I_n)A, \ \tau_{ij}(A) = A\tau_{ij}(I_n) \\ \sigma_{-i}(A) = \sigma_{-i}(I_n)A, \ \tau_{-i}(A) = A\tau_{-i}(I_n) \\ \sigma_{ki+j}(A) = \sigma_{ki+j}(I_n)A, \ \tau_{ki+j}(A) = A\tau_{ki+j}(I_n) \end{cases} \tag{7.43}$$

That is, elementary row transformation for $A$ is equal to multiplying the corresponding elementary matrix from the left, and elementary column transformation for $A$ is equal to multiplying the corresponding elementary matrix from the right.

**Lemma 7.24** *Let $L = L(B) \subset \mathbb{Z}^n$ be an integer lattice, then $\mathbb{Z}^n/L$ is a finite group, and*

$$|\mathbb{Z}^n/L| = d(L).$$

***Proof*** According to the knowledge of linear algebra, an integer square matrix $B \in \mathbb{Z}^n$ can always be transformed into a lower triangular matrix by elementary row transformation; that is, there is a unimodular matrix $U \in SLn(\mathbb{Z})$, so that

$$UB = \begin{bmatrix} * & 0 & \cdots & 0 \\ * & * & \cdots & 0 \\ \cdots & \cdots & \cdots & \cdots \\ * & * & \cdots & * \end{bmatrix}.$$

Then the elementary column transformation of $UB$ can always be transformed into an upper triangular matrix, so it is a diagonal matrix; that is, there is a unimodular matrix $U_1 \in SLn(\mathbb{Z})$, $\Rightarrow$

$$UBU_1 = \text{diag}\{\delta_1, \delta_2, \ldots, \delta_n\}.$$

where $\delta_i \neq 0$, $\delta_i \in \mathbb{Z}$, and

$$d(L) = |\det(UBU_1)| = \prod_{i=1}^{n} |\delta_i|.$$

Let $L(UBU_1)$ be an integral lattice generated by $UBU_1$, we have quotient group isomorphism

$$\mathbb{Z}^n/L(UBU_1) \cong \oplus_{i=1}^{n}\mathbb{Z}/|\delta_i|\mathbb{Z} = \oplus_{i=1}^{n}\mathbb{Z}_{|\delta_i|}.$$

Thus

$$|\mathbb{Z}^n/L(UBU_1)| = \prod_{i=1}^{n} |\delta_i| = d(L).$$

Because of $L(B) = L(BU_1)$ and $L(B) \cong L(UB)$, Thus $L(B) \cong L(UBU_1)$, so

$$|\mathbb{Z}^n/L(B)| = |\mathbb{Z}^n/L(UBU_1)| = d(L).$$

Lemma 7.24 holds.

An integer square matrix $B = (bij)_{n\times n} \in \mathbb{Z}^{n\times n}$ is called Hermite normal form matrix, if $B$ is an upper triangular matrix, that is $b_{ij} = 0, 1 \le j < i \le n$, and

$$b_{ii} \ge 1, 0 \le b_{ij} < b_{ii}, 1 \le i < j \le n. \tag{7.44}$$

A Hermite normal form matrix, referred to as HNF matrix.

**Definition 7.8** $L = L(B) \subset \mathbb{Z}^n$ is an integer lattice, and $B$ is the HNF matrix, which is called the HNF basis of $L$, denote as $B = \text{HNF}(L)$.

The following lemma proves that a whole lattice has a unique HNF basis, so it is reasonable to use HNF($L$) to represent HNF basis.

**Lemma 7.25** *Let $L \subset \mathbb{Z}^n$ be an integer lattice, then there is a unique HNF matrix $B \Rightarrow L = L(B)$.*

***Proof*** Let $L = L(A)$, $A$ is the generating matrix of $L$, by using the elementary column transformation, $A$ can be transformed into an upper triangular matrix, that is

$$AU_1 = \begin{bmatrix} c_{11} & c_{12} & \cdots & c_{1n} \\ 0 & c_{22} & \cdots & c_{2n} \\ \cdots & \cdots & \cdots & \cdots \\ 0 & 0 & \cdots & c_{nn} \end{bmatrix}, \quad U_1 \in SLn(\mathbb{Z}).$$

where $C_{ii} > 0, 1 \le i \le n$, if $AU_1$ is transformed continuously, there is a unimodular matrix $U_2, \Rightarrow AU_1U_2 = B$ is the HNF matrix, because $L(B) = L(AU_1U_2)$, know that $L$ has HNF base $B$.

Let's prove the uniqueness of HNF base $B$ if there are two HNF matrices $B_1$,$B_2 \Rightarrow L(B_1) = L(B_2)$, then from Lemma 7.14, there is a unimodular matrix $U \in SLn(\mathbb{Z})$ such that $B_1 = B_2U$; that is, the elementary column transformation defined by formula (7.43) can be continuously implemented on $B_2$ to obtain $B_1$, but for $B_2$, any column transformation $\tau_{ij}$ ,$\tau_{-i}$ and $\tau_{ki+j}$ is not a HNF matrix, so $U = I_n$ is a unit matrix, that is $B_1 = B_2$. The Lemma holds.

**Lemma 7.26** *Let $L = L(B)$ be an integer lattice, $B = (b_{ij})_{n\times n}$ is a HNF matrix, $B^* = [\beta_1^*, \beta_2^*, \ldots, \beta_n^*]$ is the orthogonal basis corresponding to $B = [\beta_1, \beta_2, \ldots, \beta_n]$, then*

$$B^* = [\beta_1^*, \beta_2^*, \ldots, \beta_n^*] = \text{diag}\{b_{11}, b_{22}, \ldots, b_{nn}\}$$

*is a diagonal matrix.*

***Proof*** We prove $\beta_i^* = (0, 0, \ldots, b_{ii}, 0, \ldots, 0)'$, induction of $i$, when $i = 1$, $\beta_1^* = \beta_1 = (b_{11}, 0, \ldots, 0)'$. The proposition holds, if for $j \le i$, there is $\beta_j^* = (0, 0, \ldots, b_{jj}, 0, \ldots, 0)'$ holds, then when $i + 1$, by (7.31), there is

$$
\begin{aligned}
\beta_{i+1}^* &= \beta_{i+1} - \sum_{j=1}^{i} \frac{\langle \beta_{i+1}, \beta_j^* \rangle}{|\beta_j^*|^2} \beta_j^* \\
&= \beta_{i+1} - \sum_{j=1}^{i} \frac{b_{j(i+1)}}{b_{jj}} \beta_j^* \\
&= \begin{pmatrix} b_{1(i+1)} \\ b_{2(i+1)} \\ \vdots \\ b_{i(i+1)} \\ b_{(i+1)(i+1)} \\ \vdots \\ 0 \end{pmatrix} - \begin{pmatrix} b_{1(i+1)} \\ b_{2(i+1)} \\ \vdots \\ b_{i(i+1)} \\ 0 \\ \vdots \\ 0 \end{pmatrix} = \begin{pmatrix} 0 \\ 0 \\ \vdots \\ 0 \\ b_{(i+1)I(i+1)} \\ \vdots \\ 0 \end{pmatrix}.
\end{aligned}
$$

Thus the proposition holds.

Next, we discuss $q$-ary lattices, where $q \geq 1$ is a positive integer, the following two $q$-ary lattices are often used in lattice cryptosystems.

**Definition 7.9** Let $\mathbb{Z}_q$ be a residue class ring mod $q$, $A \in \mathbb{Z}_q^{n\times m}$, the following two $q$-ary lattices are defined as

$$\Lambda_q(A) = \{y \in \mathbb{Z}^m | \text{there is } x \in \mathbb{Z}^n \Rightarrow y \equiv A^{'} x (\bmod q)\}, \tag{7.45}$$

and

$$\Lambda_q^{\perp}(A) = \{y \in \mathbb{Z}^m | Ay \equiv 0 (\bmod q)\}. \tag{7.46}$$

By the definition: $\Lambda_q(A) \subset \mathbb{Z}^m$ and $\Lambda_q^{\perp}(A) \subset \mathbb{Z}^m$ is an $m$-dimensional integer lattice. And any $\alpha \in q\mathbb{Z}^m$, then there is $x = 0 \in \mathbb{Z}^m$, $\Rightarrow \alpha \equiv A^{'} x (\bmod q)$, and $A\alpha \equiv 0 (\bmod q)$, there is

$$\begin{cases} q\mathbb{Z}^m \subset \Lambda_q(A) \subset \mathbb{Z}^m \\ q\mathbb{Z}^m \subset \Lambda_q^{\perp}(A) \subset \mathbb{Z}^m. \end{cases}$$

That is, $\Lambda_q(A)$ and $\Lambda_q^{\perp}(A)$ are $q$-element lattices of dimension $m$.

**Lemma 7.27** *We have*

$$\begin{cases} \Lambda_q^{\perp}(A) = q\Lambda_q(A)^* \\ \Lambda_q(A) = q\Lambda_q^{\perp}(A)^* \end{cases}$$

***Proof*** Any $\alpha \in \Lambda_q(A)^*$, by the definition, then

$$\langle y, \alpha \rangle \in \mathbb{Z}, \forall\, y \in \Lambda_q(A).$$

And

$$\langle y, \alpha \rangle = y^{'}\alpha \in \mathbb{Z} \Rightarrow y^{'}\alpha \equiv 0 (\text{mod } 1).$$

There is

$$y^{'}q\alpha \equiv 0 (\text{mod } q), \forall\, y \in \Lambda_q(A).$$

Because $y \in \Lambda_q(A)$, thus there is $x \in \mathbb{Z}^n \Rightarrow y \equiv A^{'}x (\text{mod } q)$, from the above formula,

$$x^{'}Aq\alpha \equiv 0 (\text{mod } q), \forall\, x \in \mathbb{Z}^n.$$

Thus

$$Aq\alpha \equiv 0 (\text{mod } q), \Rightarrow q\alpha \in \Lambda_q^{\perp}(A).$$

We prove

$$q\Lambda_q(A)^* \subset \Lambda_q^{\perp}(A).$$

Conversely, if $y \in \Lambda_q^{\perp}(A)$, we have

$$Ay \equiv 0 (\text{mod } q) \Rightarrow A\left(\frac{1}{q}y\right) \equiv 0 (\text{mod } 1).$$

Any $\alpha \in \Lambda_q(A)$, let $x \in \mathbb{Z}^n$, $\alpha \equiv A^{'}x (\text{mod } q)$, then

$$\left\langle \alpha, \frac{1}{q}y \right\rangle = x^{'}A\left(\frac{1}{q}y\right) \equiv 0 (\text{mod } 1), \forall\, x \in \mathbb{Z}^n.$$

We have

$$\frac{1}{q}y \in \Lambda_q(A)^* \Rightarrow y \in q\Lambda_q(A)^*.$$

That is

$$\Lambda_q^{\perp}(A) \subset q\Lambda_q(A)^*.$$

Thus, $\Lambda_q^{\perp}(A) = q\Lambda_q(A)^*$. Similarly, the second equation can be proved.

**Lemma 7.28** *Let $q$ be a prime, $A \in \mathbb{Z}_q^{n\times m}$, $m \geq n$, and* $\text{rank}(A) = n$*, then*

$$|\det(\Lambda_q^{\perp}(A))| = q^n, \tag{7.47}$$

*and*

$$|\det(\Lambda_q(A))| = q^{m-n}. \tag{7.48}$$

***Proof*** In finite field $\mathbb{Z}_q$, $\text{rank}(A) = n$, then the linear equation system $Ay = 0$ has exactly $q^{m-n}$ solutions, from which we can get

$$|\Lambda_q^{\perp}(A)/q\mathbb{Z}^m| = q^{m-n}.$$

By Lemma 7.23,

$$|\mathbb{Z}^m/\Lambda_q^{\perp}(A)| = |\mathbb{Z}_q^m/\Lambda_q^{\perp}(A)/q\mathbb{Z}^m| = q^n.$$

By Lemma 7.24,

$$|\det(\Lambda_q^{\perp}(A))| = |\mathbb{Z}^m/\Lambda_q^{\perp}(A)| = q^n.$$

So (7.47) holds. By Corollary 7.5 of the previous section, we have

$$|\det(\Lambda_q^{\perp}(A)^*)| = q^{-n}.$$

By Lemma 7.27,

$$|\det(\Lambda_q(A))| = q^m |\det(\Lambda_q^{\perp}(A)^*)| = q^{m-n}.$$

The Lemma holds.

## 7.4 Reduced Basis

In lattice theory, Reduced basis and corresponding LLL algorithm are the most important contents, which have an important impact on computational algebra, computational number theory and other neighborhoods, and are recognized as one of the most important computational methods in recent 100 years. In order to introduce Reduced basis and LLL algorithm, we recall the gram Schmidt orthogonalization process summarized by Eqs. (7.31)–(7.34). Let $\{\beta_1, \beta_2, \dots, \beta_n\} \subset \mathbb{R}^n$ be a set of bases corresponding to $\mathbb{R}^n$ , $\{\beta_1^*, \beta_2^*, \dots, \beta_n^*\}$ is the corresponding Gram–Schmidt orthogonal basis, where

$$\beta_1^* = \beta_1,\ \beta_i^* = \beta_i - \sum_{j=1}^{i-1} \frac{\langle \beta_i, \beta_j^* \rangle}{\langle \beta_j^*, \beta_j^* \rangle} \beta_j^*,\ 1 < i \le n. \tag{7.49}$$

The above formula can be written as

$$\beta_i = \sum_{j=1}^{i} \frac{\langle \beta_i, \beta_j^* \rangle}{\langle \beta_j^*, \beta_j^* \rangle} \beta_j^*,\ 1 \le i \le n. \tag{7.50}$$

There is

**Lemma 7.29** *Let $\{\beta_1, \beta_2, \dots, \beta_n\}$ be a set of bases of $\mathbb{R}^n$, $\{\beta_1^*, \beta_2^*, \dots, \beta_n^*\}$ is the corresponding Gram–Schmidt orthogonal basis, $L(\beta_1, \beta_2, \dots, \beta_k) = \mathrm{Span}\{\beta_1, \beta_2, \dots, \beta_k\}$ is a linear subspace extended by $\beta_1, \beta_2, \dots, \beta_k$, then*

*(i)*

$$L(\beta_1, \beta_2, \ldots, \beta_k) = L(\beta_1^*, \beta_2^*, \ldots, \beta_k^*),\ 1 \le k \le n. \tag{7.51}$$

*(ii) For* $1 \le i \le n$, *there is*

$$\begin{cases} \langle \beta_i, \beta_k^* \rangle = 0, & \text{when } k > i; \\ \langle \beta_i, \beta_k \rangle = \langle \beta_k^*, \beta_k^* \rangle, & \text{when } k = i. \end{cases} \tag{7.52}$$

*(iii)* $\forall\, x \in \mathbb{R}^n$, $x = \sum_{i=1}^n x_i \beta_i^*$, *then*

$$x_i = \frac{\langle x, \beta_i^* \rangle}{\langle \beta_i^*, \beta_i^* \rangle},\ 1 \le i \le n. \tag{7.53}$$

***Proof*** The above three properties can be derived directly from Eq. (7.49) or (7.50).

Let $U = (U_{ij})_{n \times n}$, where

$$U_{ij} = \frac{\langle \beta_i, \beta_j^* \rangle}{\langle \beta_j^*, \beta_j^* \rangle}, \Rightarrow U_{ij} = 0,\ \text{when } j > i.\ U_{ii} = 1. \tag{7.54}$$

Therefore, $U$ is the lower triangular matrix with element 1 on the diagonal, and

$$\begin{bmatrix} \beta_1 \\ \beta_2 \\ \vdots \\ \beta_n \end{bmatrix} = U \begin{bmatrix} \beta_1^* \\ \beta_2^* \\ \vdots \\ \beta_n^* \end{bmatrix}. \tag{7.55}$$

$U$ is called the coefficient matrix when $\{\beta_1, \beta_2, \ldots, \beta_n\}$ is orthogonalized.

Let's introduce the concept of orthogonal projection: suppose $V \subset \mathbb{R}^k \subset \mathbb{R}^n (1 \le k \le n)$, the orthogonal complement space $V^\perp$ of $V$ in $\mathbb{R}^k$ is

$$V^\perp = \{x \in \mathbb{R}^k | \langle x, \alpha \rangle = 0, \forall\, \alpha \in V\}. \tag{7.56}$$

Because $\mathbb{R}^k = V \oplus V^\perp$, so $\forall\, x \in \mathbb{R}^k$, the only can be expressed as

$$x = \alpha + \beta, \text{ where } \alpha \in V, \beta \in V^\perp.$$

$\alpha$ is called the orthogonal projection of $x$ on subspace $V$, obviously $|x|^2 = |\alpha|^2 + |\beta|^2$.

**Lemma 7.30** *Let* $\{\beta_1, \beta_2, \ldots, \beta_n\}$ *be a set of bases of* $\mathbb{R}^n$ *and* $\{\beta_1^*, \beta_2^*, \ldots, \beta_n^*\}$ *be the corresponding orthogonal basis,* $1 \le k \le n$, *then* $\beta_k^*$ *is the orthogonal projection of* $\beta_k$ *on the orthogonal complement space* $V$ *of the subspace* $L(\beta_1, \beta_2, \ldots, \beta_{k-1})$ *of* $L(\beta_1, \beta_2, \ldots, \beta_k)$.

***Proof*** When $k = 1$, the proposition is trivial, if $k > 1$, then by Lemma 7.29,

$$L(\beta_1, \beta_2, \ldots, \beta_{k-1}) = L(\beta_1^*, \beta_2^*, \ldots, \beta_{k-1}^*).$$

Therefore, the orthogonal complement space $V = L(\beta_k^*)$ of $L(\beta_1, \beta_2, \ldots, \beta_{k-1})$ in $L(\beta_1, \beta_2, \ldots, \beta_{k-1}, \beta_k)$ is a one-dimensional space, because of

$$\beta_k = \beta_k^* + \sum_{j=1}^{k-1} u_{kj}\beta_j^*,$$

and

$$\left\langle \beta_k^*, \sum_{j=1}^{k-1} u_{kj}\beta_j^* \right\rangle = 0.$$

So $\beta_k^*$ is the orthogonal projection of $\beta_k$ on $V$. The Lemma holds.

Next, we discuss the transformation law of the corresponding orthogonal basis when making the elementary column transformation of the base matrix $[\beta_1, \beta_2, \ldots, \beta_n]$.

**Lemma 7.31** *Let* $\{\beta_1, \beta_2, \ldots, \beta_n\} \subset \mathbb{R}^n$ *is a set of bases,* $\{\beta_1^*, \beta_2^*, \ldots, \beta_n^*\}$ *is the corresponding orthogonal basis,* $A = (u_{ij})_{n\times n}$ *is the coefficient matrix. Exchange* $\beta_{k-1}$ *with* $\beta_k$ *to get a set of bases* $\{\alpha_1, \alpha_2, \ldots, \alpha_n\}$ *of* $\mathbb{R}^n$*, where*

$$\alpha_{k-1} = \beta_k, \alpha_k = \beta_{k-1}, \alpha_i = \beta_i, \text{ when } i \neq k-1, k.$$

*Let* $\{\alpha_1^*, \alpha_2^*, \ldots, \alpha_n^*\}$ *be the corresponding orthogonal basis and* $A_1 = (v_{ij})_{n\times n}$ *be the corresponding coefficient matrix, then we have*

*(i)* $\alpha_i^* = \beta_i^*$, *if* $i \neq k-1, k$.

*(ii)*

$$\begin{cases} \alpha_{k-1}^* = \beta_k^* + u_{kk-1}\beta_{k-1}^* \\ \alpha_k^* = \beta_{k-1}^* - v_{kk-1}\beta_{k-1}^*. \end{cases}$$

*(iii)* $v_{ij} = u_{ij}$, *if* $1 \leq j < i \leq n$, *and* $\{i, j\} \bigcap \{k, k-1\} = \varnothing$.

*(iv)*

$$\begin{cases} v_{ik-1} = u_{ik-1}v_{kk-1} + u_{ik}\frac{|\beta_k^*|^2}{|\alpha_{k-1}^*|^2}, & i > k. \\ v_{ik} = u_{ik-1} - u_{ik}u_{kk-1}, & i > k. \end{cases}$$

*(v)* $v_{k-1j} = u_{kj}, v_{kj} = u_{k-1j}$, $1 \leq j < k-1$.

***Proof*** If $1 \leq i < k-1$, or $k < i \leq n$, then the orthogonal complement space in $L(\alpha_1, \alpha_2, \ldots, \alpha_i) = L(\beta_1, \beta_2, \cdots, \beta_i)$,

$$V = L^{\perp}(\alpha_1, \alpha_2, \ldots, \alpha_{i-1}) = L^{\perp}(\beta_1, \beta_2, \ldots, \beta_{i-1}).$$

Therefore, the orthogonal projection of $\alpha_i^*$ as $\alpha_i = \beta_i$ on $V$ is the same as that of $\beta_i^*$ as $\beta_i$ on $V$, that is $\alpha_i^* = \beta_i^* (i \neq k-1, k)$, (i) holds.

To prove (ii), because $\alpha_{k-1}^*$ is the orthogonal projection of $\beta_k (= \alpha_{k-1})$ on the orthogonal complement space $L(\beta_{k-1}^*)$ of $L(\beta_1, \beta_2, \ldots, \beta_{k-2})$, because of

$$\begin{aligned}\beta_k^* &= \beta_k - \sum_{j=1}^{k-1} u_{kj}\beta_j^* \\ &= \beta_k - u_{kk-1}\beta_{k-1}^* - \sum_{j=1}^{k-2} u_{kj}\beta_j^*,\end{aligned}$$

and $L(\beta_1, \beta_2, \ldots, \beta_{k-2}) = L(\beta_1^*, \beta_2^*, \ldots, \beta_{k-2})^*$, there is

$$\alpha_{k-1}^* = \beta_k^* + u_{kk-1}\beta_{k-1}^*.$$

Similarly, $\alpha_k^*$ is the orthogonal projection of $\beta_{k-1}^*$ on $L(\alpha_{k-1}^*)$, thus

$$\alpha_k^* = \beta_{k-1}^* - v_{kk-1}\alpha_{k-1}^*.$$

where

$$\begin{aligned}v_{kk-1} &= \frac{\langle \beta_{k-1}^*, \alpha_{k-1}^* \rangle}{|\alpha_{k-1}^*|^2} \\ &= \frac{\langle \beta_{k-1}^*, u_{kk-1}\beta_{k-1}^* \rangle}{|\alpha_{k-1}^*|^2} \\ &= u_{kk-1}\frac{|\beta_{k-1}^*|^2}{|\alpha_{k-1}^*|^2},\end{aligned}$$

thus (ii) holds. Similarly, other properties can be proved. Lemma 7.31 holds.

**Lemma 7.32** *Let $\{\beta_1, \beta_2, \ldots, \beta_n\}$ be a set of bases of $\mathbb{R}^n$, $\{\beta_1^*, \beta_2^*, \ldots, \beta_n^*\}$ be the corresponding orthogonal basis, and $A = (u_{ij})_{n\times n}$ be the coefficient matrix. For any $k \geq 2$, if we replace $\beta_k$ with $\beta_k - r\beta_{k-1}$ and keep the other $\beta_i$ unchanged $(i \neq k)$, we get a new set of bases.*

$$\{\alpha_1, \alpha_2, \ldots, \alpha_n\} = \{\beta_1, \beta_2, \ldots, \beta_{k-1}, \beta_k - r\beta_{k-1}, \beta_{k+1}, \ldots, \beta_n\}.$$

*Let $\{\alpha_1^*, \alpha_2^*, \ldots, \alpha_n^*\}$ be the corresponding orthogonal basis and $A_1 = (v_{ij})_{n\times n}$ the corresponding coefficient matrix, then we have*

*(i)* $\alpha_i^* = \beta_i^*$, $\forall\ 1 \leq i \leq n$, *that is, $\beta_i^*$ remains unchanged.*
*(ii)* $v_{ij} = u_{ij}$, *if* $1 \leq j < i \leq n$, $i \neq k$.
*(iii)*

$$\begin{cases} v_{kj} = u_{kj} - r u_{k-1,j}, & \text{if } j < k-1 \\ v_{kk-1} = u_{kk-1} - r, & \text{if } j = k-1. \end{cases}$$

***Proof*** When $i < k$, or $i > k$, $\alpha_i^* = \beta_i^*$ is trivial, to prove (i), only prove when $i = k$. Because $\alpha_k^*$ is the orthogonal projection of $\alpha_k = \beta_k - r\beta_{k-1}$ in the orthogonal complement space $L(\alpha_k^*) = L(\beta_k^*)$ of $L(\beta_1, \beta_2, \ldots, \beta_{k-1}) = L(\alpha_1, \alpha_2, \ldots, \alpha_{k-1})$,

$$\begin{aligned}\beta_k^* &= \beta_k - \sum_{j=1}^{k-1} u_{kj}\beta_j^* \\ &= \beta_k - r\beta_{k-1} - \left(\sum_{j=1}^{k-2} u_{kj}\beta_j^* + (u_{kk-1} - r)\beta_{k-1}^*\right) \\ &= \alpha_k - \left(\sum_{j=1}^{k-2} u_{kj}\beta_j^* + (u_{kk-1} - r)\beta_{k-1}^*\right).\end{aligned}$$

This proves that $\alpha_k^* = \beta_k^*$. Thus (i) holds. To prove (ii), when $i \neq k$, we have

$$v_{ij} = \frac{\langle \alpha_i, \alpha_j^* \rangle}{|\alpha_j^*|^2} = \frac{\langle \beta_i, \beta_j^* \rangle}{|\beta_j^*|^2} = u_{ij},$$

that is (ii) holds. When $i = k$,

$$\begin{aligned}v_{kj} &= \frac{\langle \alpha_k, \alpha_j^* \rangle}{|\alpha_j^*|^2} \\ &= \frac{\langle \beta_k - r\beta_{k-1}, \beta_j^* \rangle}{|\alpha_j^*|^2} (1 \leq j < k \leq n) \\ &= \frac{\langle \beta_k, \beta_j^* \rangle}{|\beta_j^*|^2} - r\frac{\langle \beta_{k-1}, \beta_j^* \rangle}{|\beta_j^*|^2} \\ &= u_{kj} - ru_{k-1j}.\end{aligned}$$

The above formula holds for all $1 \leq j \leq k-1$, thus (iii) holds, the Lemma holds.

Next, we introduce the concept of a set of Reduced bases of $\mathbb{R}^n$.

**Definition 7.10** Let $\{\beta_1, \beta_2, \ldots, \beta_n\} \subset \mathbb{R}$ be a set of bases, $\{\beta_1^*, \beta_2^*, \ldots, \beta_n^*\}$ be the corresponding orthogonal basis, $A = (u_{ij})_{n\times n}$ be the coefficient matrix, and $\{\beta_1, \beta_2, \ldots, \beta_{n-1}\}$ be a set of Reduced bases of $\mathbb{R}^n$, if

$$\begin{cases} \text{(i) } |u_{ij}| \leq \frac{1}{2}, \forall\, 1 \leq j < i \leq n. \\ \text{(ii) } |\beta_i^* - u_{kk-1}\beta_{i-1}^*|^2 \geq \frac{3}{4}|\beta_{i-1}^*|^2, \forall\, 1 < i \leq n. \end{cases} \tag{7.57}$$

A set of Reduced bases of $\mathbb{R}^n$ is sometimes called Lovisz Reduced bases, which is of great significance in lattice theory. The important result of this section is that any lattice $L$ in $\mathbb{R}^n$ has Reduced bases, and the method to calculate the Reduced bases is the famous LLL algorithm.

**Theorem 7.3** *Let $L \subset \mathbb{R}^n$ be a lattice(full rank lattice), then there is a generating matrix $B = [\beta_1, \beta_2, \ldots, \beta_n]$ of $L$, where $\{\beta_1, \beta_2, \ldots, \beta_n\}$ is a Reduced basis of $\mathbb{R}^n$ and will also be a Reduced basis of lattice $L = L(B)$.*

***Proof*** Let $B = [\beta_1, \beta_2, \ldots, \beta_n]$, $L = L(B)$, first we prove

$$|u_{kk-1}| \leq \frac{1}{2}, \forall\, 1 \leq k. \tag{7.58}$$

If there is a $k > 1$, then the above formula does not hold, let $r$ be the nearest integer of $u_{kk-1}$, obviously,

$$|u_{kk-1} - r| \leq \frac{1}{2}.$$

In $\{\beta_1, \beta_2, \ldots, \beta_n\}$, replace $\beta_k$ with $\beta_k - r\beta_{k-1}$, thus by Lemma 7.32,

$$u_{kj} \to u_{kj} - ru_{k-1j},\ \ 1 \leq j \leq k.$$

Specially, when $j = k - 1$,

$$u_{kk-1} \to u_{kk-1} - r,$$

under the new basis, all $\beta_i$ and $u_{ij}(1 \leq j < i \neq k)$ remain unchanged, so Eq. (7.58) holds under the new basis.

In the second step of LLL algorithm, we prove that

$$|\beta_k^* - u_{kk-1}\beta_{k-1}^*|^2 \geq \frac{3}{4}|\beta_{k-1}^*|^2, \forall\, 1 < k \leq n. \tag{7.59}$$

By (7.4),

$$|\beta_k^* + u_{kk-1}\beta_{k-1}^*|^2 = |\beta_k^* - u_{kk-1}\beta_{k-1}^*|^2.$$

Therefore, the sign in the absolute value on the right of Eq. (7.59) can be changed arbitrarily. If there is a $k$, $1 < k \leq n$ such that (7.59) does not hold, that is

$$|\beta_k^* + u_{kk-1}\beta_{k-1}^*|^2 < \frac{3}{4}|\beta_{k-1}^*|^2. \tag{7.60}$$

In this case, if $\beta_k$ and $\beta_{k-1}$ are exchanged and the other $\beta_i$ remains unchanged, there is a new set of bases $\{\alpha_1, \alpha_2, \ldots, \alpha_n\}$, the corresponding orthogonal basis $\{\alpha_1^*, \alpha_2^*, \ldots, \alpha_n^*\}$ and the coefficient matrix $A_1 = (v_{ij})_{n \times n}$, where

$$\alpha_i = \beta_i (i \neq k - 1, k),\ \alpha_{k-1} = \beta_k,\ \alpha_k = \beta_{k-1}.$$

Let's prove that under the new base $\{\alpha_1, \alpha_2, \ldots, \alpha_n\}$, there is

$$|\alpha_k^* + v_{kk-1}\alpha_{k-1}^*|^2 \geq \frac{3}{4}|\alpha_{k-1}^*|^2, \tag{7.61}$$

by Lemma 7.31,

$$\begin{cases} \alpha_{k-1}^* = \beta_k^* + u_{kk-1}\beta_{k-1}^* \\ \alpha_k^* = \beta_{k-1}^* - v_{kk-1}\beta_{k-1}^*. \end{cases}$$

By (7.60), we have

$$|\alpha_{k-1}^*|^2 < \frac{3}{4}|\alpha_k^* + v_{kk-1}\alpha_{k-1}^*|^2.$$

That is

$$|\alpha_k^* + v_{kk-1}\alpha_{k-1}^*|^2 > \frac{4}{3}|\alpha_{k-1}^*|^2 > \frac{3}{4}|\alpha_{k-1}^*|^2.$$

Thus (7.61) holds. Using the above method continuously, it can be proved that formula (7.59) is valid for $\forall\, k > 1$, however, when $k$ is replaced by $k-1$, the new $\beta_{k-1}^*$ is replaced by

$$\beta_{k-1}^* \to \beta_{k-1}^* + u_{k-1k-2}\beta_{k-2}^* = \overline{\beta_{k-1}^*}.$$

We have to prove (7.59), it remains unchanged when $k-1$ is used instead of $k$. In fact,

$$\begin{aligned} |\beta_k^* + u_{kk-1}\overline{\beta_{k-1}^*}|^2 &= |\beta_k^* + u_{kk-1}(\beta_{k-1}^* + u_{k-1k-2}\beta_{k-2}^*)|^2 \\ &= |\beta_k^* + u_{kk-1}\beta_{k-1}^*|^2 + |u_{kk-1}u_{k-1k-2}\beta_{k-2}^*|^2 \\ &\geq \frac{3}{4}(|\beta_{k-1}^*|^2 + u_{kk-1}^2|u_{k-1k-2}\beta_{k-2}^*|^2) \\ &\geq \frac{3}{4}(|\beta_{k-1}^*|^2 + |u_{k-1k-2}\beta_{k-2}^*|^2) \\ &= \frac{3}{4}|\beta_{k-1}^* + u_{k-1k-2}\beta_{k-2}^*|^2 \\ &= \frac{3}{4}|\overline{\beta_{k-1}^*}|^2. \end{aligned}$$

Therefore, Eq. (7.59) does not change when the transformation of commutative vector is carried out continuously; that is, Eq. (7.59) holds for all $k$, $1 < k \leq n$.

The third step of the LLL algorithm, let's prove that

$$|u_{kj}| \leq \frac{1}{2}, \forall\, 1 \leq j < k \leq n. \tag{7.62}$$

When $j = k-1$, (7.58) is the (7.62). For given $k$, $1 < k \leq n$, if (7.62) does not hold, let $l$ be the largest subscript $\Rightarrow |u_{kl}| > \frac{1}{2}$. Let $r$ be the nearest integer to $u_{kl}$, then $|u_{kl} - r| \leq \frac{1}{2}$. Replace $\beta_k$ with $\beta_k - r\beta_l$, from Lemma 7.32, all $\beta_i^*$ remain unchanged and the coefficient matrix is changed to:

$$\begin{cases} u_{kj} = u_{kj} - ru_{lj}, 1 \le j < l \\ u_{kl} = u_{kl} - r. \end{cases}$$

While the other $u_{ij}$ remains unchanged, at this time,

$$|u_{kl} - r| = |v_{kl}| \le \frac{1}{2}.$$

So we have Eq. (7.62) for all $1 \le j < k \le n$.

The above matrix transformation is equivalent to multiplying a unimodular matrix from the right, so the Reduced basis $B \Rightarrow L = L(B)$ of lattice $L$ is finally obtained. We complete the proof of Theorem 7.3.

**Lemma 7.33** *Let $L = L(B)$ be a lattice, $B$ is a Reduced basis of $L$, and $B^* = [\beta_1^*, \beta_2^*, \ldots, \beta_n^*]$ is the corresponding orthogonal basis, then for any $1 \le j < i \le n$, we have*

$$|\beta_j^*|^2 \le 2^{i-j}|\beta_i^*|^2.$$

***Proof*** Because $B = [\beta_1, \beta_2, \ldots, \beta_n]$ is a Reduced basis, then

$$|\beta_k^* + u_{kk-1}\beta_{k-1}^*|^2 \ge \frac{3}{4}|\beta_{k-1}^*|^2.$$

Thus

$$|\beta_k^* + u_{kk-1}\beta_{k-1}^*|^2 = |\beta_k^*|^2 + u_{kk-1}^2|\beta_{k-1}^*|^2 \ge \frac{3}{4}|\beta_{k-1}^*|^2.$$

There is

$$\begin{aligned} |\beta_k^*|^2 &= \frac{3}{4}|\beta_{k-1}^*|^2 - u_{kk-1}^2|\beta_{k-1}^*|^2 \\ &\ge \frac{3}{4}|\beta_{k-1}^*|^2 - \frac{1}{4}|\beta_{k-1}^*|^2 \\ &= \frac{1}{2}|\beta_{k-1}^*|^2. \end{aligned}$$

So when $1 \le j < i \le n$ given, we have

$$\begin{aligned} |\beta_i^*|^2 &\ge \frac{1}{2}|\beta_{i-1}^*|^2 \\ &\ge \frac{1}{4}|\beta_{i-2}^*|^2 \\ &\ge \cdots \\ &\ge 2^{-(i-j)}|\beta_j^*|^2, \end{aligned}$$

thus

$$|\beta_j^*|^2 \le 2^{i-j}|\beta_i^*|^2.$$

***Remark 7.3*** In the definition of Reduced base, the coefficient $\frac{3}{4}$ on the left of the second inequality of (7.57) can be replaced by any $\delta$, where $\frac{1}{4} < \delta < 1$. Specially, Babai pointed out in (1986) that the second inequality of Eq. (7.57) can be replaced by the following weaker inequality,

$$|\beta_i^*| \leq \frac{1}{2}|\beta_{i-1}^*|. \tag{7.63}$$

Let's discuss the computational complexity of the LLL algorithm. Let $B = \{\beta_1, \beta_2, \ldots, \beta_n\}$ be any set of bases, for any $0 \leq k \leq n$, we define

$$d_0 = 1, d_k = \det(\langle\beta_i, \beta_j\rangle_{k\times k}). \tag{7.64}$$

If $\{\beta_1^*, \beta_2^*, \ldots, \beta_n^*\}$ is the orthogonal basis corresponding to $\{\beta_1, \beta_2, \ldots, \beta_n\}$, there is obviously

$$d_k = \prod_{i=1}^{k} |\beta_i^*|^2, 0 < k \leq n. \tag{7.65}$$

Thus, $d_i$ is a positive number, and $d_n = d(L)^2$. Let

$$D = \prod_{k=1}^{n-1} d_k, \tag{7.66}$$

We first prove that $d_k (0 < k \leq n)$ and $D$ have lower bounds.

**Lemma 7.34** *Let*

$$m(L) = \lambda(L)^2 = \min\{|x|^2 : x \in L, x \neq 0\}.$$

*Then*

$$d_k \geq \left(\frac{3}{4}\right)^{\frac{k(k-1)}{2}} m(L)^k, 1 \leq k \leq n.$$

***Proof*** The determinant of $k$-dimensional lattice $L_k = L(\beta_1, \beta_2, \ldots, \beta_k) \subset \mathbb{R}^k (1 \leq k \leq n)$ has

$$d^2(L_k) = d_k.$$

By the conclusion of Cassels (1971), there is a nonzero lattice point $x$ in $L_k$, which satisfies $x \in L_k, x \neq 0$, and

$$|x|^2 \leq \left(\frac{4}{3}\right)^{\frac{k-1}{2}} d_k^{\frac{1}{k}}. \tag{7.67}$$

Then

$$d_k \geq \left(\frac{3}{4}\right)^{\frac{k(k-1)}{2}} m(L_k)^k$$
$$\geq \left(\frac{3}{4}\right)^{\frac{k(k-1)}{2}} (m(L))^k.$$

The Lemma holds.

Another important conclusion of this section is that for the integer lattice $L$ estimation, the computational complexity of the Reduced basis of the integer lattice is obtained by using the LLL algorithm. We prove that the LLL algorithm on the integer lattice is polynomial.

**Theorem 7.4** *Let $L = L(B) \subset \mathbb{Z}^n$ be an integer lattice, $B = [\beta_1, \beta_2, \ldots, \beta_n]$ is the generating matrix, suppose $N$ satisfies*

$$\max_{1\leq i\leq n} |\beta_i|^2 \leq N.$$

*Then the computational complexity of the Reduced basis of L obtained by B using the LLL algorithm is*

$$\text{Time(LLL algorithm)} = O(n^4 \log N).$$

*The binary digits of all integers in the LLL algorithm are $O(n \log N)$, so the computational complexity of the LLL algorithm on the integer lattice is polynomial.*

***Proof*** By (7.36), we have

$$|\beta_i^*| \leq |\beta_i|,\ 1 \leq i \leq n.$$

where $\{\beta_1^*, \beta_2^*, \ldots, \beta_n^*\}$ is the orthogonal basis corresponding to $\{\beta_1, \beta_2, \ldots, \beta_n\}$, then by (7.65) and (7.66), we have

$$d_k = \prod_{i=1}^{k} |\beta_i^*|^2 \leq \prod_{i=1}^{k} |\beta_i|^2 \leq N^k,\ 1 \leq k \leq n.$$

And

$$1 \leq D \leq N^{\frac{n(n-1)}{2}}. \tag{7.68}$$

The inequality on the left of the above formula is because of $d_k \in \mathbb{Z}$, and $d_k \geq 1$, by (7.66), then $D \geq 1$. Therefore, $O(n)$ arithmetic operations are required in the first step of the LLL algorithm, $O(n^3)$ arithmetic operations are required in the second and third steps, and the number of bit operations per algorithm operation is $\leq$ Time (calculate $D$), thus

$$\text{Time(LLL algorithm)} \leq O(n^3)\text{Time(calculate } D) = O(n^4 \log N).$$

Therefore, the first conclusion of Theorem 7.4 is proved. The second conclusion is more complex, we will omit it. Interested readers can refer to the original (1982) of A. K. Lenstra, H. W. Lenstra and L. Lovasz.

## 7.5 Approximation of SVP and CVP

The most important application of lattice Reduced basis and LLL algorithm is to provide approximation algorithms for the shortest vector problem and the shortest adjacent vector problem, and obtain some approximate results. Firstly, we prove the following Lemma.

**Lemma 7.35** *Let $\{\beta_1, \beta_2, \ldots, \beta_n\}$ be a Reduced basis of a lattice $L$, $\{\beta_1^*, \beta_2^*, \ldots, \beta_n^*\}$ be the corresponding orthogonal basis, and $d(L)$ be the determinant of $L$, then we have*

*(i)*

$$d(L) \leq \prod_{i=1}^{n} |\beta_i| \leq 2^{\frac{n(n-1)}{4}} d(L). \tag{7.69}$$

*(ii)*

$$|\beta_1| \leq 2^{\frac{n-1}{4}} d(L)^{\frac{1}{n}}. \tag{7.70}$$

***Proof*** The inequality on the left of (i), called Hadamard inequality, has been given by Lemma 7.17. The inequality on the right of (i) gives an upper bound of $\prod_{i=1}^{n} |\beta_i|$, by Lemma 7.33,

$$|\beta_j^*| \leq 2^{\frac{i-j}{2}} |\beta_i^*|, \ 1 \leq j < i \leq n. \tag{7.71}$$

Thus

$$\beta_i = \beta_i^* + \sum_{j=1}^{i-1} u_{ij} \beta_j^*.$$

We get

$$\begin{aligned} |\beta_i|^2 &= |\beta_i^*|^2 + \sum_{j=1}^{i-1} u_{ij}^2 |\beta_j^*|^2 \\ &\leq |\beta_i^*|^2 + \frac{1}{4} \sum_{j=1}^{i-1} |\beta_j^*|^2 \\ &\leq \left(1 + \frac{1}{4} \sum_{j=1}^{i-1} 2^{i-j}\right) |\beta_i^*|^2 \end{aligned} \tag{7.72}$$

$$= \left(1 + \frac{1}{4}(2^i - 2)\right) |\beta_i^*|^2$$
$$\leq 2^{i-1} |\beta_i^*|^2.$$

There is

$$\prod_{i=1}^{n} |\beta_i|^2 \leq \prod_{i=1}^{n} 2^{i-1} |\beta_i^*|^2$$
$$= 2^{\sum_{i=0}^{n-1} i} \prod_{i=1}^{n} |\beta_i^*|^2$$
$$= 2^{\frac{n}{2}(n-1)} \prod_{i=1}^{n} |\beta_i^*|^2$$
$$= 2^{\frac{n}{2}(n-1)} (d(L))^2.$$

So

$$\prod_{i=1}^{n} |\beta_i| \leq 2^{\frac{n}{4}(n-1)} d(L).$$

We have (7.69) holds. To prove (iii), by (7.72) and (7.71), then

$$|\beta_j|^2 \leq 2^{j-1} |\beta_j^*|^2 \leq 2^{j-1} 2^{i-j} |\beta_i^*|^2 = 2^{i-1} |\beta_i^*|^2. \tag{7.73}$$

For all $1 \leq j \leq i \leq n$, especially,

$$|\beta_1^*| \leq 2^{i-1} |\beta_i^*|^2, \ 1 \leq i \leq n.$$

Thus

$$|\beta_1|^{2n} \leq 2^{\sum_{i=0}^{n} (i-1)} \prod_{i=1}^{n} |\beta_i^*|^2$$
$$= 2^{\frac{n}{2}(n-1)} (d(L))^2.$$

So

$$|\beta_1| \leq 2^{\frac{n-1}{4}} d(L)^{\frac{1}{n}}.$$

Lemma 7.35 holds!

The following theorem shows that if $\{\beta_1, \beta_2, \ldots, \beta_n\}$ is a set of Reduced bases of a lattice $L$, then $\beta_1$ is the approximation vector of the shortest vector $u_0$ of lattice $L$, and the approximation coefficient $r_n = 2^{n-1}$.

**Theorem 7.5** *Let $L = L(B) \subset \mathbb{R}^n$ be a lattice (full rank lattice), $B = [\beta_1, \beta_2, \ldots, \beta_n]$ is a set of Reduced bases of $L$, $\lambda_1 = \lambda(L)$ is the minimal distance of $L$, then*

$$|\beta_1| \le 2^{\frac{n-1}{2}}\lambda_1 = 2^{\frac{n-1}{2}}\lambda(L). \tag{7.74}$$

***Proof*** We only prove that for $\forall\, x \in L, x \neq 0$, there is

$$|\beta_1|^2 \le 2^{n-1}|x|^2, \forall\, x \in L, x \neq 0. \tag{7.75}$$

When $x \in L, x \neq 0$ given, let

$$x = \sum_{i=1}^{n} r_i\beta_i = \sum_{i=1}^{n} r_i^{'}\beta_i^*, r_i \in \mathbb{Z}, r_i^{'} \in \mathbb{R}, 1 \le i \le n.$$

Let $k$ be the largest subscript $\Rightarrow r_k \neq 0$, thus $r_k = r_k^{'}$. So

$$|x|^2 \ge r_k^2|\beta_k^*|^2 \ge |\beta_k^*|^2 \ge 2^{1-k}|\beta_1|^2. \tag{7.76}$$

Thus

$$|\beta_1|^2 \le 2^{k-1}|x|^2 \le 2^{n-1}|x|^2, x \in L, x \neq 0.$$

That is (7.75) holds, thus Theorem 7.5 holds.

The following results show that not only the shortest vector, the whole Reduced basis vector is the approximation vector of the Successive Shortest vector of the lattice.

**Lemma 7.36** *Let $L \subset \mathbb{R}^n$ be a lattice, $\{\beta_1, \beta_2, \ldots, \beta_n\}$ is a Reduced base of $L$, let $\{x_1, x_2, \ldots, x_t\} \subset L$ be $t$ linearly independent lattice points, then*

$$|\beta_j|^2 \le 2^{n-1}\max\{|x_1|^2, |x_2|^2, \ldots, |x_t|^2\}. \tag{7.77}$$

*For all $1 \le j \le t$ holds.*

***Proof*** Write

$$x_j = \sum_{i=1}^{n} r_{ij}\beta_i, r_{ij} \in \mathbb{Z}, 1 \le i \le n, 1 \le j \le t.$$

For fixed $j$, let $i(j)$ the largest positive integer $i \Rightarrow r_{ij} \neq 0$, by (7.76), we have

$$|x_j|^2 \ge |\beta_{i(j)}^*|^2,\ 1 \le j \le t.$$

Change the order of $x_j$ to ensure $i(1) \le i(2) \le \cdots \le i(t)$, then $j \le i(j)$, for $\forall\, 1 \le j \le t$ holds. Otherwise, the assumption that

$$\{x_1, x_2, \ldots, x_n\} \subset L(\beta_1, \beta_2, \ldots, \beta_{j-1})$$

is linearly independent of $x_1, x_2, \ldots, x_j$ is contradictory. Thus $j \le i(j)$. By (7.73) of Lemma 7.35, then

$$\begin{aligned}|\beta_j|^2 &\le 2^{i(j)-1}|\beta^*_{i(j)}|^2 \\ &\le 2^{n-1}|\beta^*_{i(j)}|^2 \\ &\le 2^{n-1}|x_j|^2, \forall\, 1 \le j \le t.\end{aligned}$$

Thus (7.77) holds, the Lemma holds.

***Remark 7.4*** We give a proof of $r_k = r_k^{'}$ in Theorem 7.5, because $k$ is the largest subscript $\Rightarrow r_k \ne 0$, so

$$x = \sum_{i=1}^{k} r_i \beta_i = \sum_{i=1}^{k} r_i^{'} \beta_i^*.$$

By (7.52) and (7.53),

$$r_k^{'} = \frac{\langle x, \beta_k^* \rangle}{|\beta_k^*|^2}, r_k = \frac{\langle x, \beta_k^* \rangle}{\langle \beta_k, \beta_k^* \rangle}.$$

Because $\langle \beta_k, \beta_k^* \rangle = \langle \beta_k^*, \beta_k^* \rangle$, so

$$r_k^{'} = \frac{\langle x, \beta_k^* \rangle}{\langle \beta_k, \beta_k^* \rangle} = r_k.$$

In order to discuss the approximation of the Successive Shortest vector of a lattice, let's look at the definitions of the continuous minimum $\lambda_1, \lambda_2, \ldots, \lambda_n$ and the Successive Shortest vector of a lattice, by Definition 7.6 and Corollary 7.6 in Sect. 7.2, the continuous minimum $\lambda_1, \lambda_2, \ldots, \lambda_n$ of a full rank lattice is reachable, for all $1 \le i \le n$, there is

$$|\alpha_i| = \lambda_i,\ \alpha_i \in L, 1 \le i \le n.$$

For a Successive Shortest vector called $\alpha_1, \alpha_2, \ldots, \alpha_n$, $|\alpha_i|$ is the shortest under the condition that $\alpha_i$ is linearly independent of $\{\alpha_1, \alpha_2, \ldots, \alpha_{i-1}\}$.

**Theorem 7.6** *Let $\{\beta_1, \beta_2, \ldots, \beta_n\}$ be a Reduced basis of lattice $L$, and $\lambda_1, \lambda_2, \ldots, \lambda_n$ be the continuous minimum of $L$, then we have*

$$|\beta_i|^2 \le 2^{n-1}\lambda_i, 1 \le i \le n. \tag{7.78}$$

***Proof*** We make an induction of $i$. Because $\{\beta_1, \beta_2, \ldots, \beta_i\}$ is an Reduced basis of lattice $L_i$ in $\mathbb{R}^i$, the proposition is obviously true when $i = 1$ (see Theorem 7.5 ). If the proposition holds for $i - 1$, then by Lemma 7.36,

$$|\beta_i^*|^2 \le 2^{n-1}\max\{\lambda_1, \lambda_2, \ldots, \lambda_i\} = 2^{n-1}\lambda_i.$$

Therefore, (7.78) holds for all $i$. The Theorem holds.

Next, we choose the Reduced basis to solve the shortest adjacent vector problem (CVP). For any given $t \in \mathbb{R}^n$, because there are only finite lattice points in one lattice $L$ in the Ball$(t, r)$ with $t$ as the center and $r$ as the radius, there is a lattice point $u_t$ closest to $t$, that is

$$|u_t - t| = \min_{x \in L, x \neq t} |x - t|. \tag{7.79}$$

We use the Reduced basis to find a lattice point $\omega \in L \Rightarrow$

$$|\omega - t| \leq r_1(n)|u_t - t|, \tag{7.80}$$

$\omega$ is called an approximation of the nearest lattice point $u_t$, and $r_1(n)$ is called an approximation coefficient. According to Babai (1986), to solve the approximation of the nearest lattice point $u_t$, we adopt the following two technical means:

(A) rounding off: $\forall\ x \in \mathbb{R}^n$, $[\beta_1, \beta_2, \ldots, \beta_n] = B$ is a Reduced base of lattice $L$. The discard vector $[x]_B$ of $x$ is defined as follows, let

$$x = \sum_{i=1}^{n} x_i \beta_i,\ x_i \in \mathbb{R},$$

Let $\delta_i$ be the nearest integer to $x_i$, then define

$$[x]_B = \sum_{i=1}^{n} \delta_i \beta_i, \tag{7.81}$$

$[x]_B$ is called the discard vector of $x$ under base $B$, write $x = [x]_B + \{x\}_B$, then

$$\{x\}_B \in \left\{ \sum_{i=1}^{n} a_i \beta_i | -\frac{1}{2} < a_i \leq \frac{1}{2}, 1 \leq i \leq u \right\}.$$

(B) Adjacent plane

Let $U = \sum_{i=1}^{n-1} \mathbb{R}\beta_i = L(\beta_1, \beta_2, \ldots, \beta_{n-1}) \subset \mathbb{R}^n$ be an $n-1$-dimensional subspace, $L^{'} = \sum_{i=1}^{n} \mathbb{Z}\beta_i \subset L$ be a sublattice of $L$, and $v \in L$, call $U + v$ is an affine plane of $\mathbb{R}^n$. When $x \in \mathbb{R}^n$ given, if the distance between $x$ and $U + v$ is the smallest, $U + v$ is called the nearest affine plane of $x$.

Let $x^{'}$ be the orthogonal projection of $x$ in the nearest affine plane $U + v$, let $y \in L^{'}$ be the vector closest to $x - v$ in $L^{'}$, and let $w = y + v$ be the approximation of the vector closest to $x$ in $L$.

Let $L(\beta_1, \beta_2, \ldots, \beta_n) \subset \mathbb{R}^n$ be a lattice, $\{\beta_1^*, \beta_2^*, \ldots, \beta_n^*\}$ is the corresponding orthogonal basis. $\forall\ x \in \mathbb{R}^n$, write $x = \sum_{i=1}^{n} x_i \beta_i^*$, $x_i \in \mathbb{R}^n$, $\delta_i$ represents the nearest integer of $x_i$, according to the nearest plane method, we take (see Lemma 7.43 below).

$$\begin{cases} U = L(\beta_1^*, \beta_2^*, \ldots, \beta_{n-1}^*) = L(\beta_1, \beta_2, \ldots, \beta_{n-1}) \\ r = \delta_n \beta_n \in L \\ x^{'} = \sum\limits_{i=1}^{n-1} x_i \beta_i^* + \delta_n \beta_n^* \\ y \text{ is a sublattice The grid point closest to } x - v \text{ in } L^{'} = \sum\limits_{i=1}^{n-1} \mathbb{Z}\beta_i \\ \omega = y + v \end{cases} \tag{7.82}$$

We prove that

**Theorem 7.7** *Let $L = L(B) \subset \mathbb{R}^n$ be a lattice, $B = [\beta_1, \beta_2, \ldots, \beta_n]$ is a Reduced base of $L$, for $\forall\, x \in \mathbb{R}^n$ given, the adjacent plane method produces a lattice point $\omega = y + v$ adjacent to $x$ in $L$ (by (7.82)), satisfies*

$$|w - x| \le 2^{\frac{n}{2}} |u_x - x|, \tag{7.83}$$

*where $u_x$ is given by Eq. (7.79) and further*

$$|x - \omega| \le 2^{\frac{n}{2}-1} |\beta_n^*|. \tag{7.84}$$

***Proof*** If $n = 1$, then $B = \theta \in \mathbb{R}$, $\theta \ne 0$. Let $x \in \mathbb{R}$, $x = x_1\theta$, $L = n\theta$, then when $n \in \mathbb{Z}$,

$$|x - n\theta| = |x_1\theta - n\theta| = |x_1 - n||\theta| \ge |x_1 - \delta||\theta|,$$

where $\delta$ is the nearest integer to $x_1$, let $\omega = \delta\theta$, then

$$|x - \omega| = |x_1 - \delta||\theta| \le |x - n\theta|, \ \forall\, n \in \mathbb{Z}.$$

So $\omega = \delta\theta$ is the lattice point closest to $x$ in $L$, so $\omega = u_x \in L$, that is

$$|x - \omega| = |u_x - x|.$$

Thus (7.83) holds.

Let $n \ge 2$, we observe (see (7.82)), $v = \delta_n \beta_n$, $x^{'} = \sum_{i=1}^{n-1} x_i \beta_i^* + \delta_n \beta_n^*$, then

$$|x - x^{'}| = |x_n - \delta_n||\beta_n^*| \le \frac{1}{2}|\beta_n^*|, \tag{7.85}$$

since the distance between affine planes $\{u + z | z \in L\}$ is at least $|\beta_n^*|$, and $|x - x^{'}|$ is the distance between $x$ and the nearest affine plane, there is

$$|x - x^{'}| \le |u_x - x|. \tag{7.86}$$

Let $\omega = y + v = y + \delta_n \beta_n \in L$, we prove that

$$|x-\omega|^2 = |x-x^{'}|^2 + |x^{'}-\omega|^2. \tag{7.87}$$

Because $x - x^{'} = (x_n - \delta_n)\beta_n^*$, $x^{'} - \omega = x^{'} - v - y \in u$, so $(x - x^{'}) \perp (x^{'} - \omega)$. Therefore, by the Pythagorean theorem, (7.87) holds. By induction, we have (see (7.79))

$$|x-\omega|^2 \leq \frac{1}{4}(|\beta_1^*|^2 + |\beta_2^*|^2 + \cdots + |\beta_n^*|^2).$$

By (7.71),

$$|\beta_i^*|^2 \leq 2^{n-i}|\beta_n^*|^2.$$

Thus

$$\begin{aligned}|x-\omega|^2 &\leq \frac{1}{4}|\beta_n^*|^2(1+2+2^2+\cdots+2^{n-1})\\ &= \frac{1}{4}(2^n-1)|\beta_n^*|^2\\ &\leq 2^{n-2}|\beta_n^*|^2.\end{aligned}$$

There is

$$|x-\omega| \leq 2^{\frac{n}{2}-1}|\beta_n^*|, \tag{7.88}$$

that is (7.84) holds. To prove (7.83), we have two situations:

Case 1: if $u_x \in U + x$,

In this case, $u_x - v \in U \Rightarrow u_x - v \in L^{'}$ is the lattice point closest to $x^{'} - v$ in $L$, so there is

$$|x^{'}-\omega| = |x^{'}-v-y| \leq C_{n-1}|x^{'}-u_x| \leq C_{n-1}|x-u_x|,$$

where $C_n = 2^{\frac{n}{2}}$. By (7.87), we have

$$|x-\omega|^2 \leq (1+(C_{n-1})^2)^{\frac{1}{2}}|x-u| < C_n|x-u|.$$

The proposition holds.

Case 2: If $u_x \notin U + x$, then

$$|x-u_x| \geq \frac{1}{2}|\beta_n^*|.$$

By (7.88), we get

$$|x-\omega| < 2^{\frac{n}{2}}|x-u_x|.$$

Thus, Theorem 7.7 holds.

Comparing Theorems 7.6 and 7.7, when $x = 0$, the approximation coefficient of Theorem 7.6 is $2^{\frac{n-1}{2}}$, for general $x \in \mathbb{R}^n$, there is an additional factor $\sqrt{2}$ in the approximation coefficient. Using the rounding off technique, we can give an approximation to adjacent vectors, another main result in this section is

**Theorem 7.8** *Let $B = [\beta_1, \beta_2, \ldots, \beta_n]$ be a Reduced basis of $L$, $x \in \mathbb{R}^n$ given arbitrarily, $u_x \in L$ is the lattice point closest to $x$, and $[x]_B$ is given by Eq. (7.82), then $\omega = [x]_B \in L$, and*

$$|x - [x]_B| \leq \left(1 + 2n\left(\frac{9}{2}\right)^{\frac{n}{2}}\right)|x - u_x|. \tag{7.89}$$

By Theorem 7.8, $[x]_B \in L$ is an approximation of the nearest lattice point $u_x$, and the approximation coefficient is $\gamma_1(n) = 1 + 2n\left(\frac{9}{2}\right)^{\frac{n}{2}}$, it is a little worse than the approximation coefficients generated by adjacent planes, but the approximation vector is relatively simple. In lattice cryptosystem, $[x]_B$ as input information has higher efficiency. To prove Theorem 7.8, we need the following Lemma.

**Lemma 7.37** *Let $B = [\beta_1, \beta_2, \ldots, \beta_n]$ is a Reduced base of $\mathbb{R}^n$, $\theta_k$ represents the angle between vector $\beta_k$ and subspace $U_k$, where*

$$U_k = \sum_{i \neq k} \mathbb{R}\beta_i. \tag{7.90}$$

*Then for each $k$, $1 \leq k \leq n$, we have*

$$\sin\theta_k \geq \left(\frac{\sqrt{2}}{3}\right)^n. \tag{7.91}$$

***Proof*** $1 \leq k \leq n$ given, $\forall\, m \in U_k$, we prove

$$|\beta_k| \leq \left(\frac{9}{2}\right)^{\frac{n}{2}}|m - \beta_k|, m \in U_k. \tag{7.92}$$

Because

$$\sin\theta_k = \min_{m \in U_k}\frac{|m - \beta_k|}{|\beta_k|},$$

so by (7.92), $\Rightarrow$ (7.91), the Lemma holds. To prove (7.92), let $\{\beta_1^*, \beta_2^*, \ldots, \beta_n^*\}$ be the orthogonal basis corresponding to the Reduced basis Reduced $\{\beta_1, \beta_2, \ldots, \beta_n\}$, then $m \in U_k$ can express as

$$m = \sum_{i \neq k} a_i\beta_i = \sum_{j=1}^{n} b_j\beta_j^*, a_i, b_j \in \mathbb{R}.$$

Write

$$m = (a_1, \ldots, a_n) \begin{bmatrix} \beta_1 \\ \beta_2 \\ \vdots \\ \beta_n \end{bmatrix} = (a_1, \ldots, a_n) U \begin{bmatrix} \beta_1^* \\ \beta_2^* \\ \vdots \\ \beta_n^* \end{bmatrix}.$$

where $a_k = 0$, $U$ is the transition matrix of Gram–Schmidt orthogonalization (see (7.87)). Then for any $1 \leq j \leq n$, $1 \leq k \leq n$, there is

$$b_j = \sum_{i \neq k} a_i u_{ij}, \ \beta_k = \sum_{i=1}^{n} u_{ki} \beta_i^*.$$

So

$$m - \beta_k = \sum_{j=1}^{n} \gamma_j \beta_j^*, \text{ where } \gamma_j = b_j - u_{kj}.$$

Let $a_k = -1$, then

$$\gamma_j = \sum_{i=1}^{n} a_i u_{ij} = a_j + \sum_{i=j+1}^{n} a_i u_{ij}. \tag{7.93}$$

Therefore, Eq. (7.92) can be rewritten as

$$|\beta_k|^2 = \sum_{j=1}^{k} u_{kj}^2 |\beta_j^*|^2 \leq \left(\frac{9}{2}\right)^{\frac{n}{2}} \sum_{j=1}^{n} \gamma_j^2 |\beta_j^*|^2. \tag{7.94}$$

Let us first prove the following assertion:

$$\sum_{j=k}^{n} \gamma_j^2 \geq \left(\frac{2}{3}\right)^{2(n-k)}. \tag{7.95}$$

If the above formula does not hold, i.e.,

$$\sum_{j=k}^{n} \gamma_j^2 < \left(\frac{2}{3}\right)^{2(n-k)}.$$

Then for all $j$, $k \leq j \leq n$, there is

$$\gamma_j^2 < \left(\frac{2}{3}\right)^{2(n-k)} \Rightarrow |\gamma_j| < \left(\frac{2}{3}\right)^{(n-k)}. \tag{7.96}$$

By (7.93),

$$\begin{cases} \gamma_n = a_n \\ \gamma_{n-1} = a_{n-1} + a_n u_{nn-1} \\ \gamma_{n-2} = a_{n-2} + a_{n-1} u_{n-1n-2} + a_n u_{nn-2} \\ \cdots \\ \gamma_k = a_k + a_{k+1} u_{k+1k} + \cdots + a_n u_{nk} \end{cases}$$

We can prove

$$|a_j| < \left(\frac{3}{2}\right)^{n-j} \cdot \left(\frac{2}{3}\right)^{n-k}. \tag{7.97}$$

Because when $j = n$, $a_n = \gamma_n$, (7.96) ensures that (7.97) holds. Reverse induction of $j (k \le j \le n)$, by (7.93),

$$\begin{aligned} |a_j| &= |\gamma_j - \sum_{i=j+1}^{n} a_i u_{ij}| \le |\gamma_j| + \sum_{i=j+1}^{n} \frac{|a_i|}{2} \\ &< \left(\frac{2}{3}\right)^{n-k} + \frac{1}{2} \sum_{i=j+1}^{n} \left(\frac{3}{2}\right)^{n-i} \left(\frac{2}{3}\right)^{n-k} \\ &= \left(\frac{2}{3}\right)^{n-k} + \frac{1}{2} \left(\frac{2}{3}\right)^{n-k} \sum_{i=0}^{n-j-1} \left(\frac{3}{2}\right)^{i} \\ &= \left(\frac{2}{3}\right)^{n-k} + \left(\frac{2}{3}\right)^{n-k} \left(\left(\frac{3}{2}\right)^{n-j} - 1\right) \\ &= \left(\frac{3}{2}\right)^{n-j} \left(\frac{2}{3}\right)^{n-k}. \end{aligned}$$

Therefore, under the assumption of (7.96), we have (7.97). Take $j = k$ in (7.97), then $|a_k| < 1$, but $a_k = -1$, this contradiction shows that Formula (7.96) does not hold, thus (7.95) holds.

We now prove Formula (7.94) to complete the proof of Lemma. By Lemma 7.33,

$$|\beta_k^*|^2 \ge 2^{j-k} |\beta_j^*|^2, \ 1 \le j \le k \le n.$$

And

$$|\beta_k^*|^2 \le 2^{j-k} |\beta_j^*|^2, \ 1 \le k \le j \le n.$$

Therefore, there is an estimate on the left of Eq. (7.94)

$$\begin{aligned}\sum_{j=1}^{k} u_{kj}^2 |\beta_j^*|^2 &\leq |\beta_k^*|^2 \sum_{j=1}^{k} u_{kj}^2 2^{k-j} \\ &\leq \frac{1}{4} |\beta_k^*|^2 \sum_{j=1}^{k} 2^{k-j} \\ &= \frac{1}{4} |\beta_k^*|^2 (2^k - 1) \\ &< 2^k |\beta_k^*|^2.\end{aligned}$$

On the other hand, there is an estimate on the right of (7.94),

$$\begin{aligned}\sum_{j=1}^{n} \gamma_j^2 |\beta_j^*|^2 &\geq \sum_{j=k}^{n} \gamma_j^2 |\beta_j^*|^2 \\ &\geq \sum_{j=k}^{n} \gamma_j^2 2^{k-j} |\beta_i^*|^2 \\ &\geq 2^{k-n} |\beta_k^*|^2 \sum_{j=k}^{n} \gamma_j^2 \\ &\geq 2^{k-n} \left(\frac{2}{3}\right)^{2(n-k)} |\beta_k^*|^2 \\ &\geq \left(\frac{2}{9}\right)^{\frac{n}{2}} |\beta_k^*|^2.\end{aligned}$$

Thus (7.94) holds, we complete the proof of Lemma 7.37.

Now we give the proof of 7.8:

***Proof*** (*The proof of Theorem* 7.8) Let $B = \{\beta_1, \beta_2, \ldots, \beta_n\}$ be a Reduced basis of lattice $L = L(B)$, $1 \leq k \leq n$ given, $U_k$ is a linear subspace generated by $B - \{\beta_k\}$, by Lemma 7.37, we have

$$|\beta_1| \leq \left(\frac{9}{2}\right)^{\frac{n}{2}} |m - \beta_k|, \forall\, m \in U_k. \tag{7.98}$$

Let $x \in \mathbb{R}^n$, $\omega = [x]_B \in L$, then

$$x - \omega = x - [x]_B = \sum_{i=1}^{n} c_i \beta_i, \ |c_i| \leq \frac{1}{2} (1 \leq i \leq n).$$

Let $u_x$ be the nearest grid point to $x$ in $L$, and let

$$u_x - \omega = \sum_{i=1}^{n} a_i \beta_i, \, a_i \in \mathbb{Z}.$$

We prove

$$|u_x - \omega| \leq 2n \left(\frac{9}{2}\right)^{\frac{n}{2}} |u_x - x|. \tag{7.99}$$

Might as well make $u_x \neq \omega$, and suppose

$$|a_k \beta_k| = \max_{1 \leq j \leq n} |a_j \beta_j| > 0.$$

Obviously,

$$|u_x - \omega| \leq n |a_k \beta_k|. \tag{7.100}$$

On the other hand,

$$u_x - x = (u_x - \omega) + (\omega - x) = \sum_{i=1}^{n} (a_i + c_i) \beta_i = (a_k + c_k)(\beta_k - m).$$

where

$$m = -\frac{1}{a_k + c_k} \sum_{j \neq k} (a_j + c_j) \beta_j \in U_k.$$

By (7.99),

$$|u_x - x| = |a_k + c_k||\beta_k - m| \geq \frac{1}{2} \left(\frac{2}{9}\right)^{\frac{n}{2}} |\beta_k||a_k|.$$

There is

$$|a_k \beta_k| \leq 2 \left(\frac{9}{2}\right)^{\frac{n}{2}} |u_x - x|.$$

So

$$|u_x - \omega| \leq 2n \left(\frac{9}{2}\right)^{\frac{n}{2}} |u_x - x|.$$

That is (7.99) holds, finally,

$$|x - \omega| \leq |x - u_x| + |u_x - \omega| \leq \left(1 + 2n \left(\frac{9}{2}\right)^{\frac{n}{2}}\right) |x - u_x|.$$

We complete the proof of Theorem 7.8.

## 7.6 GGH/HNF Cryptosystem

Lattice-based cryptosystem is the main research object of postquantum cryptography. Since it was first proposed in 1996, it has only a history of more than 20 years. Among them, the representative technologies are Ajtai-Dwork cryptosystem, GGH cryptosystem, McEliece-Niederreiter cryptosystem and NTRU cryptosystem based on algebraic code theory. We will introduce them, respectively, below.

GGH cryptosystem is a cryptosystem based on lattice theory proposed by Goldreich, Goldwasser and Halevi in 1997. It is generally considered that it is a new public key cryptosystem to replace RSA in the postquantum cryptosystem era.

Let $L \subset \mathbb{Z}^n$ be an integer lattice, $B$ and $R$ are two generating matrices of $L$, that is

$$L = L(B) = L(R).$$

Because there is a unique HNF base in $L$ (see Lemma 3.4). Let $B = \text{HNF}(L)$ be HNF matrix, $B$ as public key and $R$ as private key. Let $v \in \mathbb{Z}^n$ be an integer point, $e \in \mathbb{R}^n$ is an error vector. Let $\sigma$ be a parameter vector. Take $e = \sigma$ or $e = -\sigma$, they each chose with a probability of $\frac{1}{2}$.

Encryption: for the plaintext $v \in \mathbb{Z}^n$ encoded and input and the error vector randomly selected according to the parameter vector $\sigma$, the public key $B$ is used for encryption. The encryption function $f_{B,\sigma}$ is defined as

$$f_{B,\sigma}(v, e) = Bv + e = c \in \mathbb{R}^n. \tag{7.101}$$

Decryption: decrypt cryptosystem text $c$ with private key $R$, because $c \in \mathbb{R}^n$ , $R = [\alpha_1, \alpha_2, \dots, \alpha_n]$, then $c$ can be expressed in $\{\alpha_1, \alpha_2, \dots, \alpha_n\}$ linearity,

$$c = \sum_{i=1}^{n} x_i \alpha_i, \; x_i \in \mathbb{R}.$$

Let $\delta_i$ be the nearest integer to $x_i$, define (see (7.81))

$$[c]_R = \sum_{i=1}^{n} \delta_i \alpha_i \in L. \tag{7.102}$$

Define the decryption function as

$$\begin{cases} f_{B,\sigma}^{-1}(c) = B^{-1}[c]_R = v, \\ e = c - Bv. \end{cases} \tag{7.103}$$

In order to verify the correctness of decryption function $f_{B,\sigma}^{-1}$, we first prove the following simple Lemma. For any $x \in \mathbb{R}^n$, and $R = [\alpha_1, \alpha_2, \dots, \alpha_n] \in \mathbb{R}^{n\times n}$ is any set of bases of $\mathbb{R}^n$, if $x = (a_1, a_2, \dots, a_n) \in \mathbb{R}^n$, $\gamma_i$ represents the integer closest to

$a_i$, then define (see (7.7))

$$[x] = (\gamma_1, \gamma_2, \ldots, \gamma_n) \in \mathbb{Z}^n. \tag{7.104}$$

Write $x = \sum_{i=1}^{n} x_i\alpha_i$, $\delta_i$ is the nearest integer to $x_i$, then define (see (7.102))

$$[x]_R = \sum_{i=1}^{n} \delta_i\alpha_i \in L(R). \tag{7.105}$$

**Lemma 7.38** *For* $\forall\, x \in \mathbb{R}^n$, $R \in \mathbb{R}^{n\times n}$ *is a set of bases of* $\mathbb{R}^n$, *we have*

$$[x]_R = R[R^{-1}x].$$

***Proof*** Write

$$x = \begin{bmatrix} a_1 \\ a_2 \\ \vdots \\ a_n \end{bmatrix} \in \mathbb{R}^n \Rightarrow [x] = \begin{bmatrix} \delta_1 \\ \delta_2 \\ \vdots \\ \delta_n \end{bmatrix} \in \mathbb{Z}^n,\ |a_i - \delta_i| \le \frac{1}{2}.$$

If $x = \sum_{i=1}^{n} x_i\alpha_i$ , $R = [\alpha_1, \alpha_2, \ldots, \alpha_n]$, then

$$x = R\begin{bmatrix} x_1 \\ x_2 \\ \vdots \\ x_n \end{bmatrix}, \text{ and } [x]_R = R\begin{bmatrix} \delta_1 \\ \delta_2 \\ \vdots \\ \delta_n \end{bmatrix}, \ \delta_i \text{ is the nearest integer to } x_i.$$

Thus

$$R^{-1}[x]_R = \begin{bmatrix} \delta_1 \\ \delta_2 \\ \vdots \\ \delta_n \end{bmatrix} = [R^{-1}x].$$

Lemma 7.38 holds.

**Theorem 7.9** *Let* $L = L(R) = L(B) \subset \mathbb{Z}^n$ *be an integer lattice,* $B$ *is the public key,* $R$ *is the private key,* $v \in \mathbb{Z}^n$ *is plaintext,* $e$ *is the error vector. If and only if* $[R^{-1}e] \neq 0$,

$$f_{B,\sigma}^{-1}(c) \neq v.$$

***Proof*** By the definition, cryptosystem text $c = Bv + e = f_{B,\sigma}(v, e)$, and

$$f_{B,\sigma}^{-1}(c) \equiv B^{-1}[c]_R = B^{-1}R[R^{-1}c] = T[R^{-1}c]. \tag{7.106}$$

where $T = B^{-1}R \in \mathbb{R}^{n\times n}$ is a unimodular matrix. Because $L(B) = L(R)$, $\Rightarrow$

$$B = RU, U \in SLn(\mathbb{Z}).$$

So

$$B^{-1}R = UR^{-1}R = U = T,$$

that is $T$ is a unimodular matrix. By (7.106),

$$\begin{aligned} T[R^{-1}c] &= T[R^{-1}(Bv + e)] \\ &= T[R^{-1}Bv + R^{-1}e] \\ &= T[T^{-1}v + R^{-1}e]. \end{aligned}$$

Because $T$ is a unimodular matrix, $v \in \mathbb{Z}^n$, so

$$[T^{-1}v + R^{-1}e] = T^{-1}v + [R^{-1}e]. \tag{7.107}$$

Thus

$$T[R^{-1}c] = v + T[R^{-1}e].$$

That is

$$f_{B,\sigma}^{-1}(c) = v + T[R^{-1}e].$$

Because $T$ is a unimodular matrix, $T[R^{-1}e] = 0 \Leftrightarrow [R^{-1}e] = 0$, so the Theorem holds.

By Theorem 7.9, whether the GGH cryptographic mechanism is correct or not depends entirely on whether $[R^{-1}e]$ is a 0 vector, that is

$$f_{B,\sigma}^{-1}(c) = v \Leftrightarrow [R^{-1}e] = 0. \tag{7.108}$$

Therefore, when the private key $R$ is given, the selection of error vector $e$ and parameter vector $\sigma$ becomes the key to the correctness of GGH password. Notice that (7.106), if we decrypt with public key $B$, then

$$[B^{-1}c] = [B^{-1}(Bv + e)] = [v + B^{-1}e] = v + [B^{-1}e].$$

Therefore, the basic condition for the security and accuracy of GGH password is

$$\begin{cases} [R^{-1}e] = 0 \\ [B^{-1}e] \neq 0. \end{cases} \tag{7.109}$$

Because the public key $B$ we choose is HNF matrix, $[B^{-1}e] \neq 0$ is easy to satisfy. Let $B = (b_{ij})_{n\times n} \Rightarrow B^{-1} = (c_{ij})_{n\times n}$. Where $c_{ii} = b_{ii}^{-1}$. Let $e = (e_1, e_2, \ldots, e_n)$, each $e_i$ has the same absolute value, that is $|e_i| = \sigma$ , $\sigma$ is the parameter. Thus, $2|e_n| > b_{nn} \Rightarrow [B^{-1}e] \neq 0$. Let's focus on $[R^{-1}e] = 0$.

$\forall\, x = (x_1, x_2, \ldots, x_n) \in \mathbb{R}^n$, define the $L_1$ norm $|x|_1$ and $L_\infty$ norm $|x|_\infty$ of $x$ as

$$|x|_\infty = \max_{1\le i\le n} |x_i|,\ |x|_1 = \sum_{i=1}^{n} |x_i|. \tag{7.110}$$

**Lemma 7.39** *Let $R \in \mathbb{R}^{n\times n}$ be a reversible square matrix, $R^{-1} = \begin{bmatrix}\alpha_1\\ \alpha_2\\ \vdots\\ \alpha_n\end{bmatrix}$, where $\alpha_i$ is the row vector of $R^{-1}$. $e = (e_1, e_2, \ldots, e_n) \in \mathbb{R}^n$, $|e_i| = \sigma$, $\forall\ 1 \le i \le n$, let*

$$\rho = \max_{1\le i\le n} |\alpha_i|(|\alpha_i|_1) \tag{7.111}$$

*be the maximum of the $L_1$ norm of $n$ row vectors of $R^{-1}$, then when $\sigma < \frac{1}{2\rho}$, we have $[R^{-1}e] = 0$.*

***Proof*** Suppose $\alpha_i = (c_{i1}, c_{i2}, \ldots, c_{in})$, the $i$-th component of $R^{-1}e$ can be written as

$$\left|\sum_{i=1}^{n} c_{ij}e_j\right| \le \sigma \sum_{j=1}^{n} |c_{ij}| = \sigma|\alpha_i|_\infty \le \sigma\rho.$$

If $\sigma < \frac{1}{2\rho}$, then each component of $R^{-1}e$ is $< \frac{1}{2}$, there is $[R^{-1}e] = 0$.

**Lemma 7.40** *$R \in \mathbb{R}^{n\times n}$ , $R^{-1} = \begin{bmatrix}\alpha_1\\ \alpha_2\\ \vdots\\ \alpha_n\end{bmatrix}$, let $\max_{1\le i\le n} |\alpha_i|_\infty = \frac{\gamma}{\sqrt{n}}$, then the probability of $[R^{-1}e] \neq 0$ is*

$$P\{[R^{-1}e] \neq 0\} \le 2n \exp\left(-\frac{1}{8\sigma^2\gamma^2}\right). \tag{7.112}$$

*where $\sigma$ is the parameter, error vector $e = (e_1, \ldots, e_n)$, $|e_i| = \sigma$.*

***Proof*** Let $R^{-1} = (c_{ij})_{n\times n}$, $R^{-1}e = \begin{bmatrix}a_1\\ a_2\\ \vdots\\ a_n\end{bmatrix}$, where $a_i = \sum_{j=1}^{n} c_{ij}e_j$.

Because $|c_{ij}| \le \frac{\gamma}{\sqrt{n}}$, $|e_j| = \sigma$, then $c_{ij}e_j$ is in interval $[-\frac{\gamma\sigma}{\sqrt{n}}, \frac{\gamma\sigma}{\sqrt{n}}]$; therefore, by Hoeffding inequality, we have

$$P\left\{|a_i| > \frac{1}{2}\right\} = P\left\{\left|\sum_{j=1}^{n} c_{ij} e_j\right| > \frac{1}{2}\right\} < 2\exp\left(-\frac{1}{8\sigma^2\gamma^2}\right).$$

To satisfy $[R^{-1}e] \neq 0$, then only one of the above conditions $\{|a_i| > \frac{1}{2}\}$ is true. Thus

$$\begin{aligned} P\{[R^{-1}e] \neq 0\} &= P\left\{\bigcup_{i=1}^{n}\left\{|a_i| > \frac{1}{2}\right\}\right\} \\ &\leq \sum_{i=1}^{n} P\left\{|a_i| > \frac{1}{2}\right\} \\ &< 2n\exp\left(-\frac{1}{8\sigma^2\gamma^2}\right). \end{aligned}$$

The Lemma holds.

**Corollary 7.8** *For any given $\varepsilon > 0$, when parameter $\sigma$ satisfies*

$$\sigma \leq \left(\gamma\sqrt{8\log\frac{2n}{\varepsilon}}\right)^{-1} \Rightarrow P\{[R^{-1}e] \neq 0\} < \varepsilon. \tag{7.113}$$

In order to have a direct impression of Eq. (7.113), let's give an example. Let $n = 120$, $\varepsilon = 10^{-5}$, when the elements of matrix $R^{-1} = (c_{ij})_{n\times n}$ change in the interval $[-4, 4]$, that is $-4 \leq c_{ij} \leq 4$, then it can be verified that the maximum $L_\infty$ norm of the row vector of $R^{-1}$ is approximately equal to $\frac{1}{30\times\sqrt{120}}$, thus $\gamma = \frac{1}{30}$, by Corollary, when $\sigma \leq (\frac{1}{30}\sqrt{8\log 240 \times 10^5})^{-1} \approx \frac{30}{11.6} \approx 2.6$, we have

$$P\{[R^{-1}e] \neq 0\} < 10^{-5}.$$

It can be seen from the above analysis that GGH cryptosystem does not effectively solve the selection of private key $R$, public key $B$, especially parameter $\sigma$ and error vector. In 2001, Professor Micciancio of the University of California, San Diego further improved GGH cryptosystem by using HNF basis and adjacent plane method. In order to introduce GGH/HNF cryptosystem, we review several important results in the previous sections.

**Lemma 7.41** *Let $L = L(B) \subset \mathbb{R}^n$ be a lattice, $B = [\beta_1, \beta_2, \ldots, \beta_n]$ is the generating base, $B^* = [\beta_1^*, \beta_2^*, \ldots, \beta_n^*]$ is the corresponding orthogonal basis, $\lambda_1 = \lambda(L)$ is the minimum distance of $L$, then*

*(i)*

$$\lambda_1 = \lambda(L) \geq \min_{1\leq i\leq n} |\beta_i^*|. \tag{7.114}$$

*For $L = L(B)$, take parameter $\rho = \rho(B)$ as*

$$\rho = \frac{1}{2} \min_{1 \leq i \leq n} |\beta_i^*|. \tag{7.115}$$

*Then for any* $x \in \mathbb{R}^n$, *there is at most one grid point*

$$\alpha \in L \Rightarrow |x - \alpha| < \rho. \tag{7.116}$$

(ii) *Suppose* $L \subset \mathbb{Z}^n$ *is an integer lattice, then* $L$ *has a unique HNF base* $B$, *that is* $L = L(B)$, $B = (b_{ij})_{n \times n}$, *satisfies*

$$0 \leq b_{ij} < b_{ii}, \text{ when } 1 \leq i < j \leq n, b_{ij} = 0, \text{ when } 1 \leq j < i \leq n.$$

*That is,* $B$ *is an upper triangular matrix, and the corresponding orthogonal basis* $B^*$ *of* $B$ *is a diagonal matrix, that is*

$$B^* = \text{diag}\{b_{11}, b_{22}, \ldots, b_{nn}\}.$$

***Proof*** Equation (7.114) is given by Lemma 7.18 and the property (ii) is given by Lemma 7.26. We only prove that if there is lattice point $\alpha \in L \Rightarrow |x - \alpha| < \rho$, then $\alpha$ is the only one. Let $\alpha_1 \in L$ , $\alpha_2 \in L$, and

$$|\alpha_1 - x| < \rho, |\alpha_2 - x| < \rho \Rightarrow |\alpha_1 - \alpha_2| < 2\rho = \min_{1 \leq i \leq n} |\beta_i^*| \leq \lambda_1.$$

Because $\alpha_1 - \alpha_2 \in L$, this contradicts the definition of $\lambda_1$. There is $\alpha_1 = \alpha_2$.

In the previous section, we introduced Babai's adjacent plane method (see (7.82)). The distance between two subsets $A_1$ and $A_2$ in $\mathbb{R}^n$ is defined as

$$|A_1 - A_2| = \min\{|x - y||x \in A_1, y \in A_2\}.$$

$x \in \mathbb{R}^n$ is a vector, $A \subset \mathbb{R}^n$ is a subset, the distance between $x$ and $A$ is defined as

$$|x - A| = \min\{|x - y||y \in A\}.$$

Suppose $L \in \mathbb{R}^n$ , $B = [\beta_1, \beta_2, \ldots, \beta_n]$ is a generating base, $B^* = [\beta_1^*, \beta_2^*, \ldots, \beta_n^*]$ is the corresponding orthogonal basis. Define subspace

$$\begin{cases} U = L(\beta_1, \beta_2, \ldots, \beta_{n-1}) = \mathbb{R}^{n-1}, & L' = \sum_{i=1}^{n-1} \mathbb{Z}\beta_i \text{ is a sub-lattice.} \\ A_v = U + v, & v \in L. \end{cases}$$

$A_v$ is called an affine plane with $v$ as the representative element. Any $x \in \mathbb{R}^n$, let $A_v$ be the affine plane closest to $x$, that is

$$|x - A_v| = \min\{|x - A_\alpha||\alpha \in L\}.$$

Let $x'$ be the orthogonal projection of $x$ on $A_v$. Because $x' - v \in U = \mathbb{R}^{n-1}$. Recursively let $y \in L'$ be the nearest lattice point to $x' - v$. Then we define the adjacent plane operator $\tau_B$ of $x$ under base $B$ as

$$\tau_B(x) = w = y + v \in L. \tag{7.117}$$

**Lemma 7.42** *Under the above definition, if* $v_1, v_2 \in L$, *and* $A_{v_1} \neq A_{v_2}$, *then*

$$|A_{v_1} - A_{v_2}| \geq |\beta_n^*|. \tag{7.118}$$

***Proof*** $v_1, v_2 \in L$, then it can be given by the linear combination of $\{\beta_1^*, \beta_2^*, \ldots, \beta_n^*\}$, that is

$$\begin{cases} v_1 = \sum_{i=1}^n a_i \beta_i^*, & \text{where } a_i \in \mathbb{R}, a_n \in \mathbb{Z}. \\ v_2 = \sum_{i=1}^n b_i \beta_i^*, & \text{where } b_i \in \mathbb{R}, b_n \in \mathbb{Z}. \end{cases}$$

In order to prove the $n$-th component, $a_n$ and $b_n$ are integers, let

$$v_1 = \sum_{i=1}^n a_i^* \beta_i, \; v_2 = \sum_{i=1}^n b_i^* \beta_i, \; a_i^*, b_i^* \in \mathbb{Z}.$$

Therefore,

$$a_n = \frac{\langle v_1, \beta_n^* \rangle}{|\beta_n^*|^2} = \frac{\langle a_n^* \beta_n, \beta_n^* \rangle}{|\beta_n^*|^2} = a_n^* \in \mathbb{Z}.$$

The above equation uses Eq. (7.52), which can prove $b_n \in \mathbb{Z}$ in the same way. By condition $v_1 - v_2 \notin U$, then $a_n = b_n$, therefore

$$|A_{v_1} - A_{v_2}| = |a_n - b_n||\beta_n^*| \geq |\beta_n^*|.$$

We have completed the proof of Lemma.

**Lemma 7.43** *Under the above definitions and symbols, suppose* $x \in \mathbb{R}^n$, $x = \sum_{i=1}^n \gamma_i \beta_i^*$, $\delta$ *is the nearest integer to* $\gamma_n$, *then*

*(i)*

$$v = \delta \beta_n, \; x' = \sum_{i=1}^{n-1} \gamma_i \beta_i^* + \delta \beta_n^*. \tag{7.119}$$

*That is, the affine plane closest to* $x$ *is* $A_{\beta_n}$, *the orthogonal projection of* $x$ *on* $A_v$ *is* $x'$.

*(ii) Let* $u_x \in L$ *be the lattice point closest to* $x$, *then*

$$|x - x'| \leq |x - u_x|. \tag{7.120}$$

***Proof*** Take $v = \delta\beta_n$, then $v \in L$, we want to prove that the distance between $x$ and $A_v$ is the smallest. Because $x = \sum_{i=1}^{n} \gamma_i \beta_i^*$, so (see (7.119))

$$x - v = \sum_{i=1}^{n-1} \gamma_i' \beta_i^* + (\gamma_n - \delta)\beta_n^*,$$

$$\Longrightarrow |x - A_v| = |x - v - U| \leq |\gamma_n - \delta||\beta_n^*| \leq \frac{1}{2}|\beta_n^*|.$$

Let $v_1 \in L$, $v - v_1 \notin U$, by trigonometric inequality,

$$|x - A_{v_1}| \geq |A_{v_1} - A_v| - |x - A_v| \geq |\beta_n^*| - \frac{1}{2}|\beta_n^*| = \frac{1}{2}|\beta_n^*| \geq |x - A_v|.$$

So it is correct to take $v = \delta\beta_n$. Secondly, we prove that the orthogonal projection $x'$ of $x$ and affine plane $A_v$ is

$$x' = \sum_{i=1}^{n-1} \gamma_i \beta_i^* + \delta\beta_n^*.$$

Let's first prove $x' \in A_v$. Because $v = \delta\beta_n$, and

$$\beta_n = \sum_{i=1}^{n-1} c_i \beta_i^* + \beta_n^* \Rightarrow \delta\beta_n = \sum_{i=1}^{n-1} \delta c_i \beta_i^* + \delta\beta_n^* = v. \tag{7.121}$$

Thus

$$x' - v = \sum_{i=1}^{n-1} (\gamma_i - \delta c_i)\beta_i^* \in U.$$

That is $x' \in U + v = A_v$. And $x - x' = \delta\beta_n^* \Rightarrow (x - x') \perp U$. Because

$$U \bigcap A_v = \varnothing.$$

Then $A_v$ and $U$ are two parallel planes, thus $(x - x') \perp A_v$. This proves that the orthogonal projection of $x$ on $A_v$ is $x'$, and thus (i) holds.

The proof of (ii) is direct. By the definition of $x$ and any affine plane $A_\alpha$, the distance of $\alpha \in L$ satisfies

$$|x - \alpha| \geq |x - A_\alpha|.$$

When $\alpha = v$, because $(x - x') \perp A_v$, thus

$$|x - x'| = |x - A_v| \leq |x - A_\alpha|, \forall\, \alpha \in L.$$

Let $u_x \in L$ be the lattice point closest to $x$, then take $\alpha = u_x$, there is

$$|x - x'| \le |x - A_{u_x}| \le |x - u_x|.$$

The Lemma holds.

**Lemma 7.44** *Let* $L = L(B) \subset \mathbb{R}^n$ *be a lattice,* $x \in \mathbb{R}^n, \alpha \in L$. *If* $|x - \alpha| < \rho$, *where* $\rho = \frac{1}{2}\min\{|\beta_i^*||1 \le i \le n\}$, *then the nearest plane operator* $\tau_B$ *has*

$$\tau_B(x) = \alpha. \tag{7.122}$$

***Proof*** Because of

$$|x - A_\alpha| \le |x - \alpha| < \rho.$$

By Lemma 7.42, $A_\alpha$ is the plane $A_v$ closest to $x$, that is $A_\alpha = A_v$. And $\tau_B(x) = w = y + v$, then we have

$$|x - w| \le |x - \alpha| < \rho. \tag{7.123}$$

By Lemma 7.41, we have $\alpha = w = \tau_B(x)$. The Lemma holds!

Now let's introduce the workflow of GGH/HNF password:

1. $L = L(B) = L(R) \subset \mathbb{Z}^n$ is an integer lattice, $R = [r_1, r_2, \ldots, r_n]$ is the private key, $B = [\beta_1, \beta_2, \ldots, \beta_n]$ is the public key, and is the HNF basis of $L$, where

$$B^* = \text{diag}\{b_{11}, b_{22}, \ldots, b_{nn}\}.$$

We choose the private key $R$ as a particularly good base, that is $\rho = \frac{1}{2}\min\{|r_i^*||1 \le i \le n\}$. Specially, public key $B$ satisfies

$$\frac{1}{2}b_{ii} < \rho, \forall\, 1 \le i \le n.$$

2. Let $v \in \mathbb{Z}^n$ be an integer, $e \in \mathbb{R}^n$ is the error vector, satisfies $|e| < \rho$.
3. Encryption: after any plaintext information $v \in \mathbb{Z}^n$ and error vector $e$ are selected, with $\rho$ as the parameter, the encryption function $f_{B,\rho}$ is defined as

$$f_{B,\rho}(v, e) = Bv + e = c.$$

4. Decryption: We decrypt cryptosystem text $c$ with private key $R$. Decryption is transformed into

$$f_{B,\rho}^{-1}(v, e) = B^{-1}\tau_R(c).$$

where $\tau_R$ is the nearest plane operator defined by $R$.

By Lemma 7.44, when $|e| < \rho, \Rightarrow |c - Bv| = |e| < \rho$, thus

$$B^{-1}\tau_R(c) = B^{-1}\tau_R(Bv + e) = B^{-1}Bv = v. \tag{7.124}$$

This ensures the correctness of decryption.

Comparing GGH with GGH/HNF, they choose the same encryption function, but the decryption transformation is very different. GGH adopts Babai's rounding off method, while GGH/HNF adopts Babai's nearest plane method. There is a certain difference between the two at the selection point of error vector $e$. The error vector $e$ of GGH depends on each component of parameter $\sigma$ and $e$, and $\pm\sigma$. The error vector $e$ of GGH/HNF depends on the parameter $\rho$ as long as the length is less than $\rho$. Therefore, GGH/HNF has greater flexibility in the selection of error vector $e$.

Next, we explain the reason why public key $B$ chooses HNF basis. For any entire lattice $L = L(B) \subset \mathbb{Z}^n$, $B^* = [\beta_1^*, \beta_2^*, \ldots, \beta_n^*]$ is the corresponding orthogonal basis. Using the congruence relation $\bmod L$, we define an equivalent relation in $\mathbb{R}^n$, which is also the equivalent relation between integral points in $\mathbb{Z}^n$. By Lemma 7.24, quotient group $\mathbb{Z}^n/L$ is a finite group, and $|\mathbb{Z}^n/L| = d(L)$. We further give a set of representative elements of $\mathbb{Z}^n/L$. Let

$$F(B^*) = \left\{ \sum_{i=1}^{n} x_i \beta_i^* | 0 \leq x_i < 1 \right\} \tag{7.125}$$

be a parallelogram, it can be compared with the base area $F = F(B)$ of $\mathbb{R}^n/L$ (see Lemma 7.16).

$$F = F(B) = \left\{ \sum_{i=1}^{n} x_i \beta_i | 0 \leq x_i < 1 \right\}.$$

$F$ is just a quadrilateral.

**Lemma 7.45** *For any integer point $\alpha \in \mathbb{Z}^n$, there is a unique $w \in F(B^*)$ such that*

$$\alpha \equiv w (\bmod L).$$

***Proof*** $\alpha \in \mathbb{Z}^n$ is a integer point, then $\alpha$ can be expressed as a linear combination of $B^*$, write

$$\alpha = \sum_{i=1}^{n} a_i \beta_i^*, a_i \in \mathbb{R}.$$

$[a_i]$ represents the largest integer not greater than $a_i$, Suppose

$$w = \sum_{i=1}^{n} a_i \beta_i^* - \sum_{i=1}^{n} [a_i] \beta_i. \tag{7.126}$$

Then

$$\alpha - w = \sum_{i=1}^{n} [a_i] \beta_i \in L \Rightarrow \alpha \equiv w (\bmod L).$$

We prove that $w \in F(B^*)$, linearly express $w$ with the basis vector of $B^*$,

$$w = \sum_{i=1}^{n} b_i \beta_i^*.$$

We can only prove that $0 \le b_i < 1$. By (7.52), it is not difficult to have

$$b_n = \frac{\langle w, \beta_n^* \rangle}{|\beta_n^*|^2} = \frac{(a_n - [a_n])|\beta_n^*|^2}{|\beta_n^*|^2} = a_n - [a_n].$$

Thus $0 \le b_n < 1$, It is not difficult to verify that $\forall\ 1 \le i \le n$, we have $0 \le b_i < 1$ by induction, that is $w \in F(B^*)$. To prove uniqueness. Let

$$w = \sum_{i=1}^{n} a_i \beta_i^*, \text{ where } |a_i| < 1.$$

We prove that if

$$w = 0 (\text{mod } L) \Leftrightarrow w = 0. \tag{7.127}$$

Write $w = \sum_{i=1}^{n} b_i \beta_i$, then by (7.52), there is

$$a_n = \frac{\langle w, \beta_n^* \rangle}{\langle \beta_n^*, \beta_n^* \rangle} = \frac{b_n |\beta_n^*|^2}{|\beta_n^*|^2} = b_n.$$

Because of $w \in L$ and $|b_n| < 1 \Rightarrow b_n = 0$. It is not difficult to have $b_1 = b_2 = \cdots = b_n = 0$ by induction. That is $w = 0$, (7.127) holds.

$\alpha \in \mathbb{Z}^n$, if $w_1 \in F(B^*)$, $w_2 \in F(B^*)$, $\alpha \equiv w_1 (\text{mod } L)$, $\alpha \equiv w_2 (\text{mod } L)$, then

$$w_1 - w_2 \equiv 0 (\text{mod } L).$$

By (7.127), there is $w_1 = w_2$. As can be seen from the above, $w_1 \in F(B^*)$, $w_2 \in F(B^*)$, then when $w_1 \ne w_2$, there is $w_1 \not\equiv w_2 (\text{mod } L)$, that is, the points in $F(B^*)$ are not congruent under mod $L$, the Lemma holds.

From the above lemma, any two points in parallelogram $F(B^*)$ are not congruent mod $L$, therefore, for not congruent lattice points $\alpha_1, \alpha_2 \in L$, then

$$\{F(B^*) + \alpha_1\} \cap \{F(B^*) + \alpha_2\} = \varnothing.$$

Thus, $\mathbb{R}^n$ can be split into

$$\mathbb{R}^n = \cup_{\alpha \in L} F(B^*) + \alpha. \tag{7.128}$$

By Lemma 7.45, any $\alpha \in \mathbb{Z}^n$, there exists a unique $w \in F(B^*) \Rightarrow \alpha \equiv w (\text{mod } L)$, define

$$w = \alpha \bmod L.$$

Then $\alpha \to \alpha \bmod L$ gives a surjection of $\mathbb{Z}^n \to \mathbb{Z}^n \cap F(B^*)$, this mapping is a 1-1 correspondence of $\mathbb{Z}^n/L \to \mathbb{Z}^n \cap F(B^*)$. Because if $\alpha, \beta \in \mathbb{Z}^n$, then

$$\begin{cases} \alpha \equiv \beta (\text{mod } L) \Rightarrow \alpha \bmod L = \beta \bmod L \in \mathbb{Z}^n \cap F(B^*) \\ \alpha \not\equiv \beta (\text{mod } L) \Rightarrow \alpha \bmod L \neq \beta \bmod L. \end{cases}$$

By Lemma 7.24, we obviously have the following Corollary.

**Corollary 7.9** *If $L = L(B) \subset \mathbb{Z}^n$ is an integer lattice, then $F(B^*) \cap \mathbb{Z}^n$ is a representative element set of $\mathbb{Z}^n/L$, and*

$$|F(B^*) \cap \mathbb{Z}^n| = d(L). \tag{7.129}$$

If $B$ is the HNF basis of the whole lattice $L$, then $B^* = \text{diag}\{b_{11}, b_{22}, \ldots, b_{nn}\}$, thus, parallelogram $F(B^*)$ takes the simplest form:

$$F(B^*) = \{(x_1, x_2, \ldots, x_n) | 0 \leq x_i < b_{ii}\}. \tag{7.130}$$

This is a cube with a volume of $d(L)$. Thus

$$\mathbb{Z}^n/L = F(B^*) \cap \mathbb{Z}^n = \{(x_1, x_2, \ldots, x_n) | 0 \leq x_i < b_{ii}, x_i \in \mathbb{Z}\}. \tag{7.131}$$

This is another proof of Lemma 7.24.

$\alpha \bmod L$ is called the reduction vector of $\alpha$ under module $L$, for any $\alpha \in \mathbb{Z}^n$, express that the number of bits of the reduction vector $\alpha \bmod L$ is

$$\sum_{i=1}^{n} \log b_{ii} = \log \prod (b_{ii}) = \log d(L). \tag{7.132}$$

To sum up, the parallelogram of the HNF basis of $L$ has a particularly simple geometry, which is actually a cube, which is very helpful for calculating the reduction vector $x \bmod L$ of an entire point $x \in \mathbb{Z}^n$, the reduction vector is of great significance in the further improvement and analysis of GGH/HNF cryptosystem. For detailed work, please refer to D. Micciancio's paper (Micciancio, 2001) in 2001.

## 7.7 NTRU Cryptosystem

NTRU cryptosystem is a new public key cryptosystem proposed in 1996 by the number theory research unit (NTRU) composed of three digit theorists J. Hoffstein, J. Piper and J. Silverman of Brown University in the USA. Its main feature is that the key generation is very simple, and the encryption and decryption algorithms are much faster than the commonly used RSA and elliptic curve cryptography, NTRU, in particular, can resist quantum computing attacks and is considered to be a potential public key cryptography that can replace RSA in the postquantum cryptography era.

The essence of NTRU cryptographic design is the generalization of RSA on polynomials, so it is called the cryptosystem based on polynomial rings. However, NTRU can give a completely equivalent form by using the concept of $q$-ary lattice, so NTRU is also a lattice based cryptosystem. For simplicity, we start with polynomial rings.

Let $\mathbb{Z}[x]$ be a polynomial ring with integral coefficients and $N \geq 1$ be a positive integer. We define the polynomial quotient ring $R$ as

$$R = \mathbb{Z}[x]/\langle x^N - 1\rangle = \{a_0 + a_1 x + \cdots + a_{N-1}x^{N-1} | a_i \in \mathbb{Z}\}.$$

Any $F(x) \in R$, $F(x)$ can be written as an entire vector,

$$F(x) = \sum_{i=0}^{N-1} F_i x^i = (F_0, F_1, \ldots, F_{N-1}) \in \mathbb{Z}^N. \tag{7.133}$$

In $R$, we define a new operation $\otimes$ called the convolution of two polynomials. Let

$$F(x) = \sum_{i=0}^{N-1} F_i x^i, G(x) = \sum_{i=0}^{N-1} G_i x^i.$$

Define

$$F \otimes G = H(x) = \sum_{i=0}^{N-1} H_i x^i = (H_0, H_1, \ldots, H_{N-1}).$$

For any $k$, $0 \leq k \leq N-1$,

$$\begin{aligned} H_k &= \sum_{i=0}^{k} F_i G_{k-i} + \sum_{i=k+1}^{N-1} F_i G_{N+k-i} \\ &= \sum_{\substack{0 \leq i < N \\ 0 \leq j < N \\ i+j \equiv k (\mathrm{mod}\ N)}} F_i G_j. \end{aligned} \tag{7.134}$$

**Lemma 7.46** *Under the new multiplication, R is a commutative ring with unit elements.*

***Proof*** By (7.134),

$$F \otimes G = G \otimes F, F \otimes (G + H) = F \otimes G + F \otimes H.$$

So $R$ forms a commutative ring under $\otimes$.

If $a \in \mathbb{Z}, 0 \le a \le N-1$, is a constant polynomial in $R$, then

$$a \otimes F = aF = (aF_0, aF_1, \ldots, aF_{N-1}).$$

Therefore, $R$ has the unit element $a = 1$. The Lemma holds..

Let $F(x) = (F_0, F_1, \ldots, F_{N-1}) \in R$. Define

$$\tilde{F} = \frac{1}{N}\sum_{i=0}^{N-1} F_i, \text{ is arithmetic mean of the coefficients of } F. \tag{7.135}$$

The $L^2$ norm (European norm) and $L^\infty$ norm of $F$ are defined as

$$\begin{cases} |F|_2 = (\sum_{i=0}^{N-1}(F_i - \tilde{F})^2)^{\frac{1}{2}} \\ |F|_\infty = \max_{0\le i\le N-1} F_i - \min_{0\le i\le N-1} F_i. \end{cases} \tag{7.136}$$

**Definition 7.11** Let $d_1, d_2$ be two positive integers, and $d_1 + d_2 \le N$, define polynomial set $A(d_1, d_2)$ as

$$A(d_1, d_2) = \{F \in R | F \text{ has } d_1 \text{ coefficients of } 1, d_2 \text{ coefficients of } -1, \text{ other coefficients are } 0\}. \tag{7.137}$$

**Lemma 7.47** *Let* $1 \le d < [\frac{N}{2}]$,

*(i) Suppose* $F \in A(d, d-1)$, *then*

$$|F|_2 = \sqrt{2d - 1 - \frac{1}{N}}.$$

*(ii) If* $F \in A(d, d)$, *then*

$$|F|_2 = \sqrt{2d}.$$

***Proof*** If $F \in A(d, d-1)$, by (7.135), then $\tilde{F} = \frac{1}{N}$, thus

$$\begin{aligned}(|F|_2)^2 &= \sum_{i=0}^{N-1}\left(F_i - \frac{1}{N}\right)^2 \\ &= \sum_{i=0}^{N-1}\left(F_i^2 - \frac{2}{N}F_i + \frac{1}{N^2}\right) \\ &= 2d - 1 - \frac{2}{N} + \frac{1}{N} = 2d - 1 - \frac{1}{N},\end{aligned}$$

so (i) holds. If $F \in A(d, d)$, then $\tilde{F} = 0$, thus

$$(|F|_2)^2 = 2d, \Rightarrow |F|_2 = \sqrt{2d}.$$

The Lemma holds.

The parameters of NTRU cryptosystem are three positive integers, $N, q, p$, where $1 \le p < q$, and $(p, q) = 1$, that is

$$\text{parameter system} = \{(N, q, p) | 1 \le p < q, \text{and } (p, q) = 1\}.$$

When the parameter $(N, q, p)$ is selected, we will discuss the key generation of NTRU.

Key generation. Each NTRU user selects two polynomials $f \in R, g \in R$, $\deg f = \deg g = N - 1$, as private key. Take $f = (f_0, f_1, \dots, f_{N-1}), g = (g_0, g_1, \dots, g_{N-1})$ as the row vector, then $(f, g) \in \mathbb{Z}^{2N} \subset R^{2N}$. Where $f \bmod q$ is reversible as a polynomial on $\mathbb{Z}_q$ and $f \bmod p$ is reversible as a polynomial on $\mathbb{Z}_p$, that is $\exists\ F_q \in \mathbb{Z}_q[x], F_p \in \mathbb{Z}_p[x]$ such that

$$F_q \otimes f \equiv 1 (\bmod q), \text{ and } F_p \otimes f \equiv 1 (\bmod p). \tag{7.138}$$

When the private key $(f, g)$ is selected, the public key $h$ is given by the following formula:

$$h \equiv F_q \otimes g (\bmod q). \tag{7.139}$$

$h$ can be regarded as a polynomial on $\mathbb{Z}_q$. Quotient rings $\mathbb{Z}_q$ and $\mathbb{Z}_p$ are

$$\mathbb{Z}_q = \mathbb{Z}/q\mathbb{Z} = \left\{a \in \mathbb{Z} | -\frac{q}{2} \le a < \frac{q}{2}\right\}.$$

$$\mathbb{Z}_p = \mathbb{Z}/p\mathbb{Z} = \left\{a \in \mathbb{Z} | -\frac{p}{2} \le a < \frac{p}{2}\right\}.$$

Encryption transformation. User $B$ wants to use NTRU to send encrypted information $m$ to user $A$. First, the plaintext $m$ is encoded as $m \in R$, that is $m \in \mathbb{Z}^N$, then take the value under mod $p$, that is

$$m \in \mathbb{Z}_p^N.$$

Then select a polynomial $\phi \in R$, $\deg \phi = N - 1$ at random, then use the public key $h$ of user $A$ for encryption. The encryption function $\sigma$ is

$$\sigma(m) = c \equiv p\phi \otimes h + m (\bmod q). \tag{7.140}$$

$c$ is the cryptosystem text received by user $A$, $c$ is a polynomial on $\mathbb{Z}_q$ and a vector in $\mathbb{Z}_q^N$.

Decryption transformation. After receiving cryptosystem text $c$, user $A$ decrypts it with its own private keys $f$ and $F_p$ and first calculates

$$a \equiv f \otimes c (\bmod q). \tag{7.141}$$

$a$ as a polynomial on $\mathbb{Z}_q$, that is, $a \in \mathbb{Z}_q^N$ is unique. Finally, the decryption transform $\sigma^{-1}$ is

$$\sigma^{-1}(c) \equiv a \otimes F_p (\bmod p). \tag{7.142}$$

Why is the decryption transformation correct? If the parameter selection meets

$$p\phi \otimes h + m \in \mathbb{Z}_q^N. \tag{7.143}$$

Then

$$c = p\phi \otimes h + m. \tag{7.144}$$

Similarly, if $c \otimes f \in \mathbb{Z}_q^N$, then $a = f \otimes c$. By (7.142),

$$a \otimes F_p = F_p \otimes f \otimes c \equiv c = p\phi \otimes h + m (\bmod p).$$

Thus

$$a \otimes F_p \equiv m (\bmod p).$$

Because $m \in \mathbb{Z}_q^N$, so

$$\sigma^{-1}(c) \equiv a \otimes F_p \equiv m (\bmod p), \Rightarrow \sigma^{-1}(c) = m.$$

Therefore, the decryption transformation is correct under the conditions of (7.143) and $c \otimes f \in \mathbb{Z}_q^N$.

NTRU's encryption and decryption transformation cannot guarantee the correct decryption of 100%. Because $a$ is taken out as a polynomial under mod $q$ for decryption operation (see (7.142)). To satisfy (7.144), and $c \otimes f \in \mathbb{Z}_q^N$, then the following formula is necessary,

$$|f \otimes c|_\infty = |f \otimes (p\phi \otimes h + m)|_\infty < q. \tag{7.145}$$

Therefore, as a necessary condition, when the following formula holds, (7.145) holds.

$$|f \otimes m|_\infty \le \frac{q}{4}, \text{ and } |p\phi \otimes g|_\infty \le \frac{q}{4}. \tag{7.146}$$

**Lemma 7.48** *For any $\varepsilon > 0$, there are constants $r_1$ and $r_2 > 0$, depending only on $\varepsilon$ and $N$, for randomly selected polynomial $F, G \in R$, then the probability of satisfying the following formula is $\ge 1 - \varepsilon$, that is*

$$P\{r_1|F|_2|G|_2 \le |F \otimes G|_\infty \le r_2|F|_2|G|_2\} \ge 1 - \varepsilon.$$

***Proof*** See reference Hoffstein et al. (1998) in this chapter.

By Lemma, to satisfy (7.146), we choose three parameters $d_f$, $d_g$ and $d$, where

$$f \in A(d_f, d_f - 1), g \in A(d_g, d_g), \phi \in A(d, d). \tag{7.147}$$

By Lemma 7.47, $|f|_2$, $|g|_2$ and $|\phi|_2$ are known, we choose

$$|f|_2 \cdot |m|_2 \approx \frac{q}{4r_2}, |\phi|_2 \cdot |g|_2 \approx \frac{q}{4pr_2}. \tag{7.148}$$

Then, Eq. (7.146) can be guaranteed to be true (in the sense of probability), so that the success rate of the decryption algorithm will be greater than $1 - \varepsilon$. Thus, (7.148) becomes the main parameter selection index of NTRU.

Next, we use the concept of $q$-element lattice to make an equivalent description of the above NTRU. We first discuss it from the cyclic matrix. Let $T$ and $T_1$ be the following two $N$-order square matrices.

$$T = \begin{pmatrix} 0 & \cdots & 0 & 1 \\ & & & 0 \\ & I_{n-1} & & \vdots \\ & & & 0 \end{pmatrix}, \quad T_1 = \begin{pmatrix} 0 & & & \\ 0 & & I_{n-1} & \\ \vdots & & & \\ 1 & 0 & 0 & 0 \end{pmatrix}.$$

Then $T^N = T_1^N = I_N$, $T_1 = T'$, and $T_1 = T^{-1}$, because $T$ is an orthogonal matrix $\Rightarrow T_1 = T^{N-1}$, where $I_N$ is the $N$-th order identity matrix, let $a = (a_1, a_2, \ldots, a_N) \in \mathbb{R}^N$, it is easy to verify

$$T \cdot \begin{bmatrix} a_1 \\ a_2 \\ a_3 \\ \vdots \\ a_N \end{bmatrix} = \begin{bmatrix} a_N \\ a_1 \\ a_2 \\ \vdots \\ a_{N-1} \end{bmatrix}, \quad (a_1, a_2, a_3, \ldots, a_N)T_1 = (a_N, a_1, a_2, \ldots, a_{N-1}). \tag{7.149}$$

The following general assumptions $a = \begin{bmatrix} a_1 \\ a_2 \\ \vdots \\ a_N \end{bmatrix} \in \mathbb{R}^N$ are the column vector. The $N$-order cyclic matrix $T^*(a)$ generated by $a$ is defined as

$$T^*(a) = [a, Ta, T^2a, \ldots, T^{N-1}a]. \tag{7.150}$$

If $b = (b_1, b_2, \ldots, b_N) \in \mathbb{R}^N$ is a row vector, we define an $N$-order matrix

$$T_1^*(b) = \begin{bmatrix} b \\ bT_1 \\ \vdots \\ bT_1^{N-1} \end{bmatrix}. \tag{7.151}$$

In order to distinguish in the mathematical formula, $T^*(a)$ and $T_1^*(a)$ are sometimes written as $T^*a$ and $T_1^*a$ or $[T^*a]$ and $[T_1^*a]$. Obviously, the transpose of $T^*(a)$ is

$$(T^*(a))' = \begin{bmatrix} a' \\ a'T_1 \\ \vdots \\ a'T_1^{N-1} \end{bmatrix} = T_1^*(a'). \tag{7.152}$$

Equation (7.150) is column vector blocking of cyclic matrix, in order to obtain row vector blocking of cyclic matrix. For any $x \in (x_1, \ldots, x_N) \in \mathbb{R}^N$, we let

$$\overline{x} = (x_N, x_{N-1}, \ldots, x_2, x_1) \Rightarrow \overline{\overline{x}} = x.$$

Similarly, define column vectors $\overline{x}$. So for any column vector $a \in \mathbb{R}^N$, we have

$$T^*(a) = [a, Ta, T^2a, \ldots, T^{N-1}a] = \begin{bmatrix} \overline{a}'T_1 \\ \overline{a}'T_1^2 \\ \vdots \\ \overline{a}'T_1^N \end{bmatrix}. \tag{7.153}$$

On the right side of (7.153) is a cyclic matrix, which is partitioned by rows. We first prove that the transpose of the cyclic matrix is still a cyclic matrix.

**Lemma 7.49** $\forall\, a = \begin{bmatrix} \alpha_1 \\ \alpha_2 \\ \vdots \\ \alpha_N \end{bmatrix} \in \mathbb{R}^N$, *then* $(T^*(a))' = T^*(\overline{T^{-1}a})$.

***Proof*** Let $\alpha = \begin{bmatrix} \alpha_1 \\ \alpha_2 \\ \vdots \\ \alpha_N \end{bmatrix} \in \mathbb{R}^N$, by (7.152), $(T^*(a))' = T_1^*(a')$, where $\alpha' = (\alpha_1, \ldots, \alpha_N)$ is the transpose of $\alpha$, let

$$\beta = (\alpha_1, \alpha_N, \alpha_{N-1}, \ldots, \alpha_2) = \overline{\alpha}'T_1.$$

Easy to verify

$$T_1^*(\beta) = \begin{bmatrix} \beta \\ \beta T_1 \\ \vdots \\ \beta T_1^{N-1} \end{bmatrix} = T^*(\alpha).$$

There is

$$T_1^*(\beta) = (T^*(\beta'))' = T^*(\alpha).$$

Because $\overline{\alpha'} = (\alpha_N, \alpha_{N-1}, \cdots, \alpha_2, \alpha_1)$, and $\beta = \overline{\alpha'} T_1$, so

$$\beta' = T\overline{\alpha} \Rightarrow T^{-1}\beta' = \overline{\alpha} \Rightarrow \alpha = \overline{T^{-1}\beta'}.$$

We let $a = \beta'$, then

$$(T^*(\alpha))' = T^*(\alpha) = T^*(\overline{T^{-1}\alpha}).$$

We have completed the proof of Lemma.

Next, we give an equivalent characterization of cyclic matrix.

**Lemma 7.50** *Let* $A = (a_{ij})_{N\times N}$, $a = \begin{bmatrix} a_{11} \\ a_{21} \\ \vdots \\ a_{N1} \end{bmatrix} \in \mathbb{R}^N$ *is the first column of A, then $A = T^*(a)$ is a cyclic matrix if and only if for all $1 \le k \le N$, if $1 + i - j \equiv k (\text{mod } N)$, then $a_{ij} = a_{k1}$.*

***Proof*** If $A = T^*(a)$ is a cyclic matrix, by simple observation, there is

$$\begin{cases} a_{11} = a_{22} = \cdots = a_{NN} = a_{11} \\ a_{21} = a_{32} = \cdots = a_{NN-1} = a_{21} \\ \vdots \\ a_{(N-1)1} = a_{N2} \\ a_{N1} = a_{N1} \end{cases}.$$

Thus, $1 + i - j = k$. The same applies to $i < j$. We have

$$k = N + 1 + i - j \Rightarrow 1 + i - j \equiv k (\text{mod } N).$$

So the Lemma holds.

The following lemma characterizes the main properties of cyclic matrices.

**Lemma 7.51** *If* $a = \begin{bmatrix} a_1 \\ \vdots \\ a_N \end{bmatrix}$, $b = \begin{bmatrix} b_1 \\ \vdots \\ b_N \end{bmatrix}$ *are two column vectors, then*

*(i)* $T^*(a) + T^*(b) = T^*(a + b)$.
*(ii)* $T^*(a) \cdot T^*(b) = T^*([T^*a] \cdot b)$, *and* $T^*(a)T^*(b) = T^*(b)T^*(a)$.
*(iii)* $\det(T^*(a)) = \prod_{k=1}^{N}(a_1 + a_2\xi_k + \cdots + a_N\xi_k^{N-1})$. *Where* $\xi_k (1 \le k \le N)$ *is the root of all* $N$*-th units of 1.*
*(iv)* *If the cyclic matrix* $T^*(a)$ *is reversible, the inverse matrix is* $(T^*(a))^{-1} = T^*(b)$, *Where* $b$ *is the first column of* $T^*(a)$.

***Proof*** (i) is trivial, because

$$T^*(a) + T^*(b) = [a + b, T(a + b), \ldots, T^{N-1}(a + b)] = T^*(a + b).$$

To prove (ii), using the row vector block of cyclic matrix (see (7.153)), then

$$[T^*(a)]b = \begin{bmatrix} \overline{a}'T_1 \\ \overline{a}'T_1^2 \\ \vdots \\ \overline{a}'T_1^N \end{bmatrix} b = \begin{bmatrix} \overline{a}'T_1 b \\ \overline{a}'T_1^2 b \\ \vdots \\ \overline{a}'T_1^N b \end{bmatrix},$$

and

$$T^*(a) \cdot T^*(b) = \begin{bmatrix} \overline{a}'T_1 \\ \overline{a}'T_1^2 \\ \vdots \\ \overline{a}'T_1^N \end{bmatrix} [b, Tb, \ldots, T^{N-1}b] = (A_{ij})_{N\times N}.$$

where

$$A_{ij} = \overline{a}'T_1^i \cdot T^{j-1}b = \overline{a}'T_1^{N+i-j+1}b = \overline{a}'T_1^{i+1-j}b.$$

By Lemma 7.50, then $T^*(a) \cdot T^*(b) = T^*([T^*(a)]b)$, so there is the first conclusion of (ii). We notice that

$$A_{ij} = A'_{ij} = b'T_1^{j-1}T^i\overline{a} = b'T_1^{N-i-1+j}\overline{a} = b'T_1^{j-i-1}\overline{a}.$$

It is easy to prove that for any row vector $x$ and column vector $y$, there is $x \cdot \overline{y} = \overline{x} \cdot y$, and

$$\overline{xT_1^k} = \overline{x} \cdot T_1^{N-k}, 1 \le k \le N. \tag{7.154}$$

Thus,

$$A_{ij} = b'T_1^{j-i-1}\overline{a} = \overline{b'}T_1^{N+i+1-j}a = \overline{b'}T_1^{i+1-j}a.$$

This proves that $T^*(a)T^*(b) = T^*(b)T^*(a)$; that is, the multiplication of cyclic matrix to matrix is commutative.

To prove (iii), suppose $(T^*(a))' = A$, but $\det(T^*(a)) = \det((T^*(a))')$, so we just need to calculate $\det(A)$. Make polynomial $f(x) = a_1 + a_2x + \cdots + a_Nx^{N-1}$, and let

$$V = \begin{bmatrix} 1 & 1 & 1 & \cdots & 1 \\ \xi_1 & \xi_2 & \xi_3 & \cdots & \xi_N \\ \xi_1^2 & \xi_2^2 & \xi_3^2 & \cdots & \xi_N^2 \\ \cdots & \cdots & \cdots & \cdots & \cdots \\ \xi_1^{N-1} & \xi_2^{N-1} & \xi_3^{N-1} & \cdots & \xi_N^{N-1} \end{bmatrix}.$$

Then

$$AV = \begin{bmatrix} f(\xi_1) & f(\xi_2) & \cdots & f(\xi_N) \\ \xi_1 f(\xi_1) & \xi_2 f(\xi_2) & \cdots & \xi_N f(\xi_N) \\ \cdots & \cdots & \cdots & \cdots \\ \xi_1^{N-1} f(\xi_1) & \xi_2^{N-1} f(\xi_2) & \cdots & \xi_N^{N-1} f(\xi_N) \end{bmatrix}.$$

So

$$\det(A)\det(V) = \det(AV) = f(\xi_1)f(\xi_2)\cdots f(\xi_N)\det(V).$$

Because $\xi_i$ is different from each other, that is $\det(V) \neq 0$, so

$$\begin{aligned} \det(A) &= f(\xi_1)f(\xi_2)\cdots f(\xi_N) \\ &= \prod_{k=1}^{N} f(\xi_k) \\ &= \prod_{k=1}^{N}(a_1 + a_2\xi_k + \cdots + a_N\xi_k^{N-1}). \end{aligned}$$

Now prove (iv). Let $e = \begin{bmatrix} 1 \\ 0 \\ \vdots \\ 0 \end{bmatrix} \in \mathbb{R}^N$, then

$$T^*(e) = [e, Te, \ldots, T^{N-1}e] = I_N.$$

So take $b \in \mathbb{R}^N$ to satisfy

$$T^*(a) \cdot b = e \Rightarrow b = (T^*(a))^{-1}e.$$

Obviously, $b$ is the first column of $(T^*(a))^{-1}$, by (ii),

$$T^*(a)T^*(b) = T^*([T^*(a)]b) = T^*(e) = I_N.$$

Thus, $(T^*(a))^{-1} = T^*(b)$. In other words, the inverse of a reversible cyclic matrix is also a cyclic matrix.

**Corollary 7.10** *Let $N$ be a prime, $a = \begin{bmatrix} a_1 \\ \vdots \\ a_N \end{bmatrix} \in \mathbb{R}^n$, satisfy $a \neq 1$, and $\sum_{i=1}^{N} a_i \neq 0$, then the cyclic matrix $T^*(a)$ generated by $a$ is a reversible square matrix.*

***Proof*** Under given conditions, we can only prove $\det(T^*(a)) \neq 0$. Let $\varepsilon_k = \exp(\frac{2\pi ik}{N})$, $1 \leq k \leq N-1$, be $N-1$ primitive unit roots of $N$-th( because $N$ is a prime), if $\det(T^*(a)) = 0$, because of $\sum_{i=1}^{N} a_i \neq 0$, there must be a $k$, $1 \leq k \leq N-1$, such that

$$a_1 + \varepsilon_k a_2 + \varepsilon_k^2 a_3 + \cdots + \varepsilon_k^{N-1} a_N = 0.$$

In other words, $\varepsilon_k$ is a root of polynomial $\phi(x) = a_1 + a_2 x + \cdots + a_N x^{N-1}$, so $\phi(x)$ and $1 + x + \cdots + x^{N-1}$ have a common root $\varepsilon_k$, therefore, the greatest common divisor of two polynomials

$$(\phi(x), 1 + x + \cdots + x^{N-1}) > 1.$$

Since $1 + x + \cdots + x^{N-1}$ is a circular polynomial, it is an irreducible polynomial, $a \neq 1$, contradiction shows $\det(T^*(a)) \neq 0$, the Corollary holds.

Next, we give an equivalent description of a lattice of NTRU by using the cyclic matrix. Firstly, we define the linear transformation $\sigma$ in the even dimensional Euclidean space $\mathbb{R}^{2N}$, if $x$ and $y$ are two column vectors, define

$$\sigma \begin{bmatrix} x \\ y \end{bmatrix} = \begin{bmatrix} Tx \\ Ty \end{bmatrix} \in \mathbb{R}^{2N}. \tag{7.155}$$

Equivalently, if $x \in \mathbb{R}^N$, $y \in \mathbb{R}^N$ are two row vectors, define

$$\sigma(x, y) = (xT_1, yT_1) \in \mathbb{R}^{2N}. \tag{7.156}$$

Obviously, $\sigma$ defined above is a linear transformation of $\mathbb{R}^{2N} \to \mathbb{R}^{2N}$.

**Definition 7.12** An entire lattice $L \subset \mathbb{R}^{2N}$ is called a convolution $q$-ary lattice, if

(i) $L$ is $q$-ary lattice, that is $q\mathbb{Z}^{2N} \subset L \subset \mathbb{Z}^{2N}$.
(ii) $L$ is closed under the linear transformation $\sigma$, that is, $x, y \in \mathbb{R}^N$ is the column vector,

$$\begin{bmatrix} x \\ y \end{bmatrix} \in L \Rightarrow \sigma \begin{bmatrix} x \\ y \end{bmatrix} = \begin{bmatrix} Tx \\ Ty \end{bmatrix} \in L.$$

Recall that NTRU's private key is two $N-1$-degree polynomials $f = \sum_{i=0}^{N-1} f_i x^i$, $g = \sum_{i=0}^{N-1} g_i x^i$, and write $f$ and $g$ in column vector form:

$$f = \begin{bmatrix} f_0 \\ \vdots \\ f_{N-1} \end{bmatrix} \in \mathbb{Z}^N,\ f' = (f_0, f_1, \ldots, f_{N-1}) \in \mathbb{Z}^N.$$

And

$$g = \begin{bmatrix} g_0 \\ \vdots \\ g_{N-1} \end{bmatrix} \in \mathbb{Z}^N,\ g' = (g_0, g_1, \ldots, g_{N-1}) \in \mathbb{Z}^N.$$

NTRU's parameter system is $N, q, p$ is two positive integers, $N$ is prime, $p < q$, and defines a polynomial set

$$A_d\{p, 0, -p\} = \{f(x) \in \mathbb{Z}^N | d + 1 \text{ coefficients of } f \text{ are } p,\\ d \text{ coefficients of } f \text{ are } p, \text{ others are } 0\}. \tag{7.157}$$

Select two polynomials $f, g \in \mathbb{Z}^N$ of degree $N - 1$, and parameter $d_f$ are positive integers, which meet the following restrictions.

(A) $N, p, q, d_f$ are positive integers, $N$ is a prime, $1 < p < q, (p, q) = 1$;
(B) $f$ and $g$ are two polynomials of degree $N - 1$, and the constant term of $f$ is 1, and

$$f - 1 \in A_{d_f}\{p, 0, -p\}, g \in A_{d_f}\{p, 0, -p\}.$$

(C) $T^*(f)$ is reversible mod $q$.

The above (A)–(C) are the parameter constraints of NTRU. Obviously, under these conditions, $T^*(f)$ and $T^*(g)$ are reversible matrices, and

$$T^*(f) \equiv I_N (\text{mod } p), T^*(g) \equiv 0 (\text{mod } p). \tag{7.158}$$

After the polynomials $f$ and $g$ satisfying the above conditions are selected as the private key, then $\begin{bmatrix} f \\ g \end{bmatrix} \in \mathbb{Z}^{2N}$, let's construct a minimum convolution $q$-ary lattice containing $\begin{bmatrix} f \\ g \end{bmatrix}$. Suppose

$$A = [T_1^*(f'), T_1^*(g')]_{N \times 2N}, \text{ and } A' = \begin{bmatrix} T^*(f) \\ T^*(g) \end{bmatrix}. \tag{7.159}$$

Consider $A$ as an $N \times 2N$-order matrix on $\mathbb{Z}_q$, that is $A \in \mathbb{Z}_q^{N \times 2N}$, then by (7.45), $A$ defines a $2N$ dimensional $q$-ary lattice $\Lambda_q(A)$, that is

$$\Lambda_q(A) = \{y \in \mathbb{Z}^{2N} | \text{ there is } x \in \mathbb{Z}^N \Rightarrow y \equiv A'x (\text{mod } q)\}. \tag{7.160}$$

We prove that $\Lambda_q(A)$ is a convolution $q$-ary lattice containing $\begin{bmatrix} f \\ g \end{bmatrix}$. First, we prove the following general identity

**Lemma 7.52** *Suppose* $a = \begin{bmatrix} a_1 \\ \vdots \\ a_N \end{bmatrix} \in \mathbb{R}^N$, *then for* $\forall\, x \in \mathbb{R}^N$ *and* $0 \leq k \leq N-1$, *we have*

$$T^k(T^*(a)x) = T^*(a)(T^k x), \text{ where } T^0 = I_N.$$

***Proof*** $k = 0$ is trivial, obviously, we can assume $k = 1$, that is

$$T(T^*(a)x) = T^*(a)(Tx). \tag{7.161}$$

By (7.153),

$$T(T^*(a)x) = T\begin{bmatrix} \overline{a}'T_1x \\ \overline{a}'T_1^2x \\ \vdots \\ \overline{a}'T_1^Nx \end{bmatrix} = \begin{bmatrix} \overline{a}'x \\ \overline{a}'T_1x \\ \vdots \\ \overline{a}'T_1^{N-1}x \end{bmatrix}.$$

Because of $T = T_1^{N-1}$, then the right side of Eq. (7.161) is

$$T^*(a)(Tx) = \begin{bmatrix} \overline{a}'T_1Tx \\ \overline{a}'T_1^2Tx \\ \vdots \\ \overline{a}'T_1^NTx \end{bmatrix} = \begin{bmatrix} \overline{a}'x \\ \overline{a}'T_1x \\ \vdots \\ \overline{a}'T_1^{N-1}x \end{bmatrix}.$$

So (7.161) holds, the Lemma holds.

**Lemma 7.53** $\Lambda_q(A)$ *is a convolution* $q$*-ary lattice, and* $\begin{bmatrix} f \\ g \end{bmatrix} \in \Lambda_q(A)$.

***Proof*** By Lemma 7.27, $\Lambda_q(A)$ is a $q$-ary lattice, that is $q\mathbb{Z}^{2N} \subset \Lambda_q(A) \subset \mathbb{Z}^{2N}$, we only prove $\Lambda_q(A)$ is closed under linear transformation $\sigma$. If $y \in \Lambda_q(A)$, then there is $x \in \mathbb{Z}^N \Rightarrow y \equiv A'x \pmod q$, by the definition of $\sigma$,

$$\sigma(y) \equiv \begin{bmatrix} T(T^*(f)x) \\ T(T^*(g)x) \end{bmatrix} = \begin{bmatrix} T^*(f)Tx \\ T^*(g)Tx \end{bmatrix} \equiv A'Tx \pmod q.$$

Because of $x \in \mathbb{Z}^N \Rightarrow Tx \in \mathbb{Z}^N$, thus $\sigma(y) \in \Lambda_q(A)$. That is, $\Lambda_q(A)$ is a convolution $q$-ary lattice, which is proved $\begin{bmatrix} f \\ g \end{bmatrix} \in \Lambda_q(A)$. Let $e = \begin{bmatrix} 1 \\ 0 \\ \vdots \\ 0 \end{bmatrix} \in \mathbb{Z}^N$, then

$T^*(f) \cdot e$ is the first column of $T^*(f)$, that is

$$T^*(f)e = f, T^*(g)e = g.$$

Thus,

$$A'e = \begin{bmatrix} T^*(f)e \\ T^*(g)e \end{bmatrix} = \begin{bmatrix} f \\ g \end{bmatrix} \in \Lambda_q(A).$$

The Lemma holds.

With the above preparation, we now introduce the equivalent form of NTRU in lattice theory.

Public key generation. After selected private key $\begin{bmatrix} f \\ g \end{bmatrix} \in \mathbb{Z}^{2N}$, NTRU's public key is generated as follows: Because the convolution $q$-ary lattice $\Lambda_q(A)$ containing $\begin{bmatrix} f \\ g \end{bmatrix}$ is an entire lattice, $\Lambda_q(A)$ has a unique HNF basis $H$, where

$$H = \begin{bmatrix} I_N & T^*(h) \\ 0 & qI_N \end{bmatrix}, h \equiv [T^*(f)]^{-1}g (\bmod q). \tag{7.162}$$

By (7.48) of Lemma 7.28, the determinant $d(\Lambda_q(A))$ of $\Lambda_q(A)$ is

$$d(\Lambda_q(A)) = |\det(\Lambda_q(A))| = q^{2N-N} = q^N.$$

So the diagonal elements of $H$ are $I_N$ and $qI_N$. By the assumption $T^*(f) \in \mathbb{Z}^{N \times N}$, and reversible mod $q$, $[T^*(f)]^{-1}$ is the inverse matrix of $T^*(f) \bmod q$, $h \in \mathbb{Z}^N$, its component $h_i$ is selected between $-\frac{q}{2}$ and $\frac{q}{2}$, that is $-\frac{q}{2} \leq h_i < \frac{q}{2}$, such an $h$ is the only one that exists. It is not difficult to verify that $H$ is an HNF matrix and the lattice generated by $H$ is $\Lambda_q(A)$, so $H$ is the HNF basis of $\Lambda_q(A)$. $H$ is published as a public key.

Encryption transformation. The message sender encodes the plaintext as $m \in \mathbb{Z}^N$, and randomly select a vector $r \in \mathbb{Z}^N$ to satisfy

$$m \in A_{d_f}\{1, 0, -1\}, \; r \in A_{d_f}\{1, 0, -1\}. \tag{7.163}$$

That is, $m$ has $d_f + 1$ 1, $d_f$ $-1$, other components are 0. Then, the plaintext $m$ is encrypted with the public key $H$ of the message recipient:

$$c = H \begin{bmatrix} m \\ r \end{bmatrix} \equiv \begin{bmatrix} m + [T^*(h)]r \\ 0 \end{bmatrix} (\bmod q). \tag{7.164}$$

$c$ is called cryptosystem text, the first $N$ components are $m + [T^*(h)]r$, the last $N$ components are 0.

Decryption transformation. If all components of $m + [T^*(h)]r$ are between intervals $[-\frac{q}{2}, \frac{q}{2})$, the message receiver can determine that the cryptosystem text $c$ is

$$c = \begin{bmatrix} m + [T^*(h)]r \\ 0 \end{bmatrix}.$$

Then decrypt it with its own private key $T^*(f)$,

$$\begin{aligned} c &\equiv [T^*(f)]m + [T^*(f)][T^*(h)]r (\bmod q) \\ &\equiv [T^*(f)]m + [T^*(g)]r (\bmod q). \end{aligned} \tag{7.165}$$

By the definition of $h$, there is

$$[T^*(f)]h \equiv g (\bmod q) \Rightarrow T^*([T^*(f)]h) \equiv T^*(g)(\bmod q).$$

And by Lemma 7.51, there is $T^*([T^*(f)]h) \equiv T^*(f) \cdot T^*(h)$, so

$$T^*(f)T^*(h) \equiv T^*(g)(\bmod q).$$

Equation (7.165) holds.

If

$$[T^*(f)]m + [T^*(g)]r \in \left[-\frac{q}{2}, \frac{q}{2}\right]^N. \tag{7.166}$$

So do mod $p$ operation on $[T^*(f)]m + [T^*(g)]r$, and by (7.158), thus

$$([T^*(f)]m + [T^*(g)]r) \bmod p = I_N m + 0 \cdot r = m. \tag{7.167}$$

The correctness of decryption transformation is guaranteed.

In order to ensure that (7.167) holds, it can be seen from the above analysis that the following conditions are necessary.

$$\begin{cases} m + [T^*(h)]r \in [-\frac{q}{2}, \frac{q}{2}]^N \\ [T^*(f)]m + [T^*(g)]r \in [-\frac{q}{2}, \frac{q}{2}]^N. \end{cases} \tag{7.168}$$

Obviously, the first condition can be derived from the second condition; that is, the (7.168) can be derived from the (7.166). We first prove the following Lemma.

**Lemma 7.54** *If the parameter meets* $d_f < \frac{(\frac{q}{4}-1)}{2p}$, *then*

$$[T^*(f)]m + [T^*(g)]r \in \left[-\frac{q}{2}, \frac{q}{2}\right]^N.$$

***Proof*** Because all components of $m$ and $r$ are $\pm 1$ or 0, therefore, we only prove that the absolute value of the row vectors of $[T^*(f)]$ and $[T^*(g)]$ is not greater than $\frac{q}{2}$.

Write $f' = (f_0, f_1, \dots, f_{N-1})$, because of $f_0 = 1$,

$$\left|\sum_{i=0}^{N-1} f_i\right| \le \sum_{i=0}^{N-1} |f_i| = 1 + (2d_f + 1)p < \frac{q}{4}.$$

Similarly,

$$\left|\sum_{i=0}^{N-1} g_i\right| \le \sum_{i=0}^{N-1} |g_i| = (2d_f + 1)p < \frac{q}{4}.$$

Thus

$$[T^*(f)]m + [T^*(g)]r \in \left[-\frac{q}{2}, \frac{q}{2}\right]^N.$$

The Lemma holds.

According to the above lemma, NTRU algorithm needs to add the following additional conditions to ensure the correctness of decryption transformation:

(D)

$$d_f < \frac{(\frac{q}{4} - 1)}{2p}.$$

To sum up, when NTRU cryptosystem satisfies the additional restrictions (A)–(D) on the parameter system, the private key is $\begin{bmatrix} f \\ g \end{bmatrix}$ and the public key is HNF matrix $H$, the encryption and decryption algorithm can be based on the algorithm introduced above.

## 7.8 McEliece/Niederreiter Cryptosystem

McEliece/Niederreiter cryptosystem is a cryptosystem designed based on the asymmetry of coding and decoding of a special class of linear codes (Goppa codes) over a finite field. It was proposed by McEliece and Niederreiter in 1978 and 1985. It is included in the category of postquantum cryptography. We start with cyclic codes. Recall the concept of linear code in Chap. 2, let $\mathbb{F}_q$ be a $q$-element finite field, also known as the alphabet, and the elements in $\mathbb{F}_q$ are called letters or characters. The $N$-dimensional linear space $\mathbb{F}_q^N$ on $\mathbb{F}_q$ is called the codeword space of length $N$. Any a vector $a = (a_0, a_1, \dots, a_{N-1}) \in \mathbb{F}_q^N$, $a$ is called a codeword of length $N$, which is usually written as $a = a_0 a_1 \cdots a_{N-1} \in \mathbb{F}_q^N$, from the previous section, we have

$$aT_1 = (a_0, a_1, \dots, a_{N-1})T_1 = (a_{N-1}, a_0, a_1, \dots, a_{N-2}). \tag{7.169}$$

The reverse codeword $\overline{a}$ of a codeword $a = a_0 a_1 \cdots a_{N-1}$ is defined as

$$\overline{a} = a_{N-1}a_{N-2}\cdots a_1a_0 \in \mathbb{F}_q^N. \tag{7.170}$$

If $C \subset \mathbb{F}_q^N$, and $C$ is a $k$-dimensional linear subspace of $\mathbb{F}_q^N$, which is called a linear code, usually written as $C = [N, k]$, $k = 0$, or $k = N$, $[N, 0]$ and $[N, N]$ is called trivial code, actually,

$$[N, 0] = \{0 = 00\cdots 0\}, [N, N] = \mathbb{F}_q^N.$$

The reverse order code $\overline{C}$ of code $C$ is defined as $\overline{C} = \{\overline{c} | c \in C\}$, obviously, if $C = [N, k]$, then $\overline{C} = [N, k]$.

**Definition 7.13** A linear code $C$ of length $N$ is called a cyclic code, if $\forall\, c \in C \Rightarrow cT_1 \in C$.

Next, we give an algebraic expression of cyclic codes using ideal theory. For this purpose, note that $\mathbb{F}_q[x]$ is a univariate polynomial ring on $\mathbb{F}_q$, and $\langle x^N - 1\rangle$ is the principal ideal generated by polynomial $x^N - 1$. Write $R = \mathbb{F}_q[x]/\langle x^N - 1\rangle$ as quotient ring. If $a = a_0a_1\cdots a_{N-1} \in \mathbb{F}_q^N$, then $a(x) = a_0 + a_1x\cdots + a_{N-1}x^{N-1} \in R$, so $a \to a(x)$ is a 1-1 correspondence of $\mathbb{F}_q^N \to R$ and an isomorphism between additive groups. In this correspondence, we equate codeword $a$ with polynomial $a(x)$. That is $a = a(x) \Rightarrow \mathbb{F}_q^N = R = \mathbb{F}_q[x]/\langle x^N - 1\rangle$, and any code $C \subset \mathbb{F}_q^N$.

$$C = C(x) = \{c(x) | c \in C\} \subset R.$$

That is, a code $C$ is equivalent to a subset of $\mathbb{F}_q[x]/\langle x^N - 1\rangle$. The following lemma reveals the algebraic meaning of a cyclic code.

**Lemma 7.55** *$C \subset \mathbb{F}_q^N$ is a cyclic code $\Leftrightarrow$ $C(x)$ is an ideal in $\mathbb{F}_q[x]/\langle x^N - 1\rangle$.*

***Proof*** If $C(x)$ is an ideal of $\mathbb{F}_q[x]/\langle x^N - 1\rangle$, obviously $C$ is a linear code, for any code $c = c_0c_1\cdots c_{N-1}$, there is $c(x) = c_0 + c_1x + \cdots + c_{N-1}x^{N-1} \in C(x)$, thus $xc(x) = c_{N-1} + c_ox + c_1x^2 + \cdots + c_{N-2}x^{N-1} \in C(x)$. So $cT_1 = c_{N-1}c_0c_1\cdots c_{N-2} \in C$, $C$ is a cyclic code on $\mathbb{F}_q$. Conversely, if $C$ is a cyclic code, then $cT_1 \in C$, thus $cT_1^k \in C$, for all $0 \le k \le N-1$ holds. Where $T_1^0 = I_N$ is the $N$-th order identity matrix. Since the polynomial $cT_1^k(x)$ corresponding to $cT_1^k$ is

$$cT_1^k(x) = x^kc(x). \tag{7.171}$$

So $\forall\, g(x) \in R \Rightarrow g(x)c(x) \in C(x)$. This proves that $C(x)$ is an ideal. The Lemma holds.

Using the homomorphism theorem of rings, we give the mathematical expressions of all ideals in $R$. Let $\pi$ be the natural homomorphism of $\mathbb{F}_q[x] \xrightarrow{\pi} \mathbb{F}_q[x]/\langle x^N - 1\rangle$, then all ideals in $R$ correspond to all ideals containing $\ker\pi = \langle x^N - 1\rangle$ in $\mathbb{F}_q[x]$ one by one, that is

$$\ker\pi = \langle x^N - 1\rangle \subset A \subset \mathbb{F}_q[x] \xrightarrow{\pi} \mathbb{F}_q[x]/\langle x^N - 1\rangle = R.$$

Since $\mathbb{F}_q[x]$ is the principal ideal ring and $A$ is an ideal of $\mathbb{F}_q[x]$, and $\langle x^N - 1\rangle \subset A$, then

$$A = \langle g(x)\rangle, \text{ where } g(x)|x^N - 1. \tag{7.172}$$

Therefore, all ideals in $R$ are finite principal ideals, which can be listed as follows

$$\{\langle g(x)\rangle \bmod N - 1 | g(x) \text{ divide } x^N - 1\}.$$

where $\langle g(x)\rangle \bmod x^N - 1$ represents the principal ideal generated by $g(x)$ in $R$, that is

$$\langle g(x)\rangle \bmod x^N - 1 = \{g(x)f(x) | 0 \leq \deg f(x) \leq N - \deg(g(x)) - 1\}. \tag{7.173}$$

This proves that $\mathbb{F}_q[x]/\langle x^N - 1\rangle$ is a ring of principal ideals, and the number of principal ideals is the number $d + 1$ of positive factors of $x^N - 1$. The so-called positive factor is a polynomial with the first term coefficient of 1. Therefore, the Corollary is as follows:

**Corollary 7.11** *Let $d$ be the number of positive factors of $x^N - 1$, then the number of cyclic codes with length $N$ is $d + 1$.*

A cyclic code $C$ corresponds to an ideal $C(x) = \langle g(x)\rangle \bmod x^N - 1$ in $R$, we define

**Definition 7.14** Let $C$ be a cyclic code, if $C(x) = \langle g(x)\rangle \bmod x^N - 1$, then $g(x)$ is called the generating polynomial of $C$, where $g(x)|x^N - 1$.

If $g(x) = x^N - 1$, then $\langle x^N - 1\rangle \bmod x^N - 1 = 0$, corresponding to zero ideal in $R$. Thus, the corresponding cyclic code $C = \{0 = 00\cdots 0\}$ is called zero code. If $g(x) = 1$, then $\langle g(x)\rangle \bmod x^N - 1 = R$. The corresponding code $C = \mathbb{F}_q^N$. Therefore, there are always two trivial cyclic codes in cyclic codes of length $N$, zero code and $\mathbb{F}_q^N$, which correspond to zero ideal in $R$ and $R$ itself, respectively.

**Lemma 7.56** *Let $g(x)|x^N - 1$, $g(x)$ be the generating polynomial of cyclic code $C$, and* $\deg g(x) = N - k$, *then $C$ is $[N, k]$ linear code, further, let $g(x) = g_0 + g_1 x + \cdots + g_{N-k-1}x^{N-k-1} + g_{N-k}x^{N-k}$, the corresponding codeword $g = (g_0, g_1, \ldots, g_{N-k}, 0, 0, \ldots, 0) \in C$, then the generating matrix $G$ of $C$ is*

$$G = \begin{bmatrix} g \\ gT_1 \\ \vdots \\ gT_1^{k-1} \end{bmatrix}_{k\times N}. \tag{7.174}$$

***Proof*** Let $C$ correspond to ideal $C(x) = \langle g(x)\rangle \bmod x^N - 1$, then $g(x), xg(x), \ldots, x^{k-1}g(x) \in C(x)$, their corresponding codewords are $\{g, gT_1, \ldots, gT_1^{k-1}\} \subset C$, let's prove that $\{g, gT_1, \ldots, gT_1^{k-1}\}$ is a set of bases of $C$. If $\exists\, a_i \in \mathbb{F}_q \Rightarrow \sum_{i=0}^{k-1} a_i gT_1^i = 0$, then its corresponding polynomial is 0, that is

$$\left(\sum_{i=0}^{k-1} a_i gT_1^i\right)(x) = \sum_{i=0}^{k-1} a_i gT_1^i(x) = \sum_{i=0}^{k-1} a_i x^i g(x) = 0.$$

Thus

$$\sum_{i=0}^{k-1} a_i x^i = 0 \Rightarrow \forall\, a_i = 0, 0 \le i \le k-1.$$

That is, $\{g, gT_1, \ldots, gT_1^{k-1}\}$ is a linear independent group in $C$. Further $\forall c \in C$, we can prove that $c$ can be expressed linearly. suppose $c \in C$, then $c(x) \in C(x)$, by (7.174), there is $f(x)$,

$$\begin{aligned} f(x) &= f_0 + f_1 x + \cdots + f_{k-1}x^{k-1} \Rightarrow c(x) = g(x)f(x) \\ &= \sum_{i=0}^{k-1} f_i x^i g(x) \Rightarrow c = \sum_{i=0}^{k-1} f_i gT_1^i. \end{aligned}$$

This proves that the dimension of linear subspace $C$ is $N - \deg g(x) = k$; that is, $C$ is $[N, k]$ linear code. Its generating matrix $G$ is

$$G = \begin{bmatrix} g \\ gT_1 \\ \vdots \\ gT_1^{k-1} \end{bmatrix}_{k\times N}.$$

The Lemma holds.

Next, we discuss the dual code of cyclic code and its check matrix.

**Lemma 7.57** *Let $C \subset \mathbb{F}_q^N$ be a cyclic code and $g(x)$ be the generating polynomial of $g(x)$, $\deg g(x) = N - k$, let $g(x)h(x) = x^N - 1$, $h(x) = h_0 + h_1 x + \cdots + h_k x^k$, $h = (h_0, h_1, \ldots, h_k, 0, 0, \cdots, 0) \in \mathbb{F}_q^N$ is the corresponding codeword. $\overline{h}$ is the reverse order codeword, then the check matrix of $C$ is*

$$H = \begin{bmatrix} \overline{h} \\ \overline{h}T_1 \\ \vdots \\ \overline{h}T_1^{N-k-1} \end{bmatrix}_{(N-k)\times N}. \qquad (7.175)$$

*The dual code $C^{\perp}$ of $C$ is $[N, N-k]$ linear code, and*

$$C^{\perp} = \{aH | a \in \mathbb{F}_q^{N-k}\},$$

*$h(x)$ is called the check polynomial of cyclic code $C$.*

***Proof*** By Lemma 7.56, $C$ is a $k$-dimensional linear subspace, and the generating matrix $G$ is given by (7.175). Because of $g(x)h(x) = x^N - 1$, then there is $g(x)h(x) = 0$ in ring $R$. Equivalently,

$$g_0 h_i + g_1 h_{i-1} + \cdots + g_{N-k} h_{i-N+k} = 0, \forall\, 0 \leq i \leq N-1.$$

The matrix of the above formula is expressed as $GH' = 0$, so $H$ is the generation matrix of dual code of $C$, and we have Lemma holds.

***Remark 7.5*** The polynomial $\overline{h(x)}$ corresponding to the reverse codeword $\overline{h}$ is

$$\overline{h(x)} = h_0 x^{N-1} + h_1 x^{N-2} + \cdots + h_k x^{N-k-1}.$$

In general, when $h(x)|x^N - 1$, $\overline{h(x)} \nmid x^N - 1$, therefore, the dual code of cyclic code is not necessarily cyclic code.

**Definition 7.15** Let $x^N - 1 = g_1(x)g_2(x)\cdots g_t(x)$ be the irreducible decomposition of $x^N - 1$ on $\mathbb{F}_q$, where $g_i(x)(1 \leq i \leq t)$ is the irreducible polynomial with the first term coefficient of 1 in $\mathbb{F}_q[x]$. Then the cyclic code generated by $g_i(x)$ is called the $i$-th maximal cyclic code in $\mathbb{F}_q^N$, denote as $M_i^+$. The cyclic code generated by $\frac{x^N-1}{g_i(x)}$ is called the $i$-th minimal cyclic code, denote as $M_i^-$.

Minimal cyclic codes are also called irreducible cyclic codes because they no longer contain the nontrivial cyclic codes of $\mathbb{F}_q^N$ in $M_i^-$. The irreducibility of minimal cyclic codes can be derived from the fact that the ideal $M_i^-(x)$ in $R$ corresponding to $M_i^-$ is a field. We can give a proof of pure algebra.

**Corollary 7.12** *Let $M_i^-$ be the $i$-th minimal cyclic code of $\mathbb{F}_q^N (1 \leq i \leq t)$, $M_i^-(x)$ is the ideal corresponding to $M_i^-$ in $R$, then $M_i^-(x)$ is a field, thus, $M_i^-$ no longer contains any nontrivial cyclic code of $\mathbb{F}_q^N$.*

***Proof*** Let $g(x) = (x^N - 1)/g_i(x)$, $g_i(x)$ be an irreducible polynomial in $\mathbb{F}_q[x]$, by (7.175),

$$M_i^-(x) = g(x)\mathbb{F}_q[x]/(x^N - 1)\mathbb{F}_q[x] \cong \mathbb{F}_q[x]/g_i(x)\mathbb{F}_q[x],$$

where $g(x)\mathbb{F}_q[x]$ is the principal ideal generated by $g(x)$ in $\mathbb{F}_q[x]$. Since $g_i(x)$ is an irreducible polynomial, so $M_i^-(x)$ is a field.

***Example 7.1*** All cyclic codes with length of 7 are determined on binary finite field $\mathbb{F}_2$.

**Solve:** Polynomial $x^7 - 1$ has the following irreducible decomposition on $\mathbb{F}_2$

$$x^7 - 1 = (x-1)(x^3+x+1)(x^3+x^2+1).$$

Therefore, $x^7 - 1$ has 7 positive factors on $\mathbb{F}_2$, by Corollary 7.11, there are 8 cyclic codes with length of 7 on $\mathbb{F}_2$. Where 0 and $\mathbb{F}_2^7$ are two trivial cyclic codes. There are three maximal cyclic codes generated by $g(x) = x - 1$, $g(x) = x^3 + x + 1$ and $g(x) = x^3 + x^2 + 1$, respectively. The dimensions of the corresponding cyclic codes are 6 dimension, 4 dimension and 4 dimension. Similarly, there are three minimal cyclic codes, corresponding to the dimension of one and two three-dimensional cyclic codes.

Another characterization of cyclic codes is zeroing polynomials, if $x^N - 1 = g_1(x) \cdots g_t(x)$, the ideal $M_i^+(x)$ in $R$ corresponding to the maximum cyclic code $M_i^+ (1 \le i \le t)$ generated by $g_i(x)$ is

$$M_i^+(x) = \{g_i(x)f(x) | 0 \le \deg f(x) \le N - \deg g_i(x) - 1\}.$$

Let $\beta$ be a root of $g_i(x)$ in the split field. Then $g_i(x)$ is the minimal polynomial of $\beta$ in $\mathbb{F}_q[x]$, all $c(x) \in M_i^+(x) \Rightarrow c(\beta) = 0$. Therefore,

$$M_i^+(x) = \{c(x) | c(x) \in R, \text{ and } c(\beta) = 0\}.$$

***Example 7.2*** Suppose $N = (q^m - 1)/q - 1$, $(m, q-1) = 1$, $\beta$ is an $N$-th primitive unit root in $\mathbb{F}_{q^m}$, then the cyclic code

$$C = \{c(x) | c(\beta) = 0, c(x) \in R\}$$

is equivalent to Hamming code $[N, N-m]$.

***Proof*** Because $(m, q-1) = 1$, and

$$N = q^{m-1} + q^{m-2} + \cdots + q + 1 = (q-1)(q^{m-2} + 2q^{m-3} + \cdots + (m-1)) + m.$$

So $(N, q-1) = 1$. Therefore, $\beta^{i(q-1)} \neq 1$, for $1 \le i \le N-1$, in other words, $\beta^i \notin \mathbb{F}_q$ for $\forall\, 1 \le i \le N-1$ holds. In $\mathbb{F}_{q^m}$, any two elements of $\{1, \beta, \beta^2, \ldots, \beta^{N-1}\}$ are linearly independent on $\mathbb{F}_q$. If each element is regarded as an $m$-dimensional column vector on $\mathbb{F}_q$, then the $m \times N$-order matrix

$$H = [1, \beta, \beta^2, \ldots, \beta^{N-1}]_{m\times N}$$

constitutes the check matrix of cyclic code $C$, and any two rows of $H$ are linearly independent on $\mathbb{F}_q$, by the definition, $C$ is $[N, N-m]$ Hamming code.

**Lemma 7.58** *Let $C \subset \mathbb{F}_q^N$ be a cyclic code, $C(x) \subset F_q[x]/\langle x^N - 1\rangle$ be an ideal, $(N, q) = 1$, then $C(x)$ contains a multiplication unit element $c(x) \in C(x) \Rightarrow$*

$$c(x)d(x) \equiv d(x) (\mathrm{mod}\, x^N - 1), \forall\, d(x) \in C(x).$$

*The unit element $c(x)$ in $C(x)$ is unique.*

***Proof*** Because $(N, q) = 1 \Rightarrow x^N - 1$ has no double root in $\mathbb{F}_q$, let $g(x)$ be the generating polynomial of $C$ and $h(x)$ be the checking polynomial of $C$, that is $g(x)h(x) = x^N - 1$. Therefore, $(g(x), h(x)) = 1$, and there is $a(x), h(x) \in \mathbb{F}_q[x], \Rightarrow$

$$a(x)g(x) + b(x)h(x) = 1.$$

Let $c(x) = a(x)g(x) = 1 - b(x)h(x) \in C(x)$, so for $\forall\, d(x) \in C(x)$, write $d(x) = g(x)f(x)$, thus

$$\begin{aligned} c(x)d(x) &= a(x)g(x)g(x)f(x) \\ &= (1 - b(x)h(x))g(x)f(x) \\ &= g(x)f(x) - b(x)h(x)g(x)f(x). \end{aligned}$$

Therefore

$$c(x)d(x) \equiv d(x)(\text{mod } x^N - 1).$$

There is $c(x)d(x) = d(x)$ in $R = \mathbb{F}_q[x]/\langle x^N - 1\rangle$. That is, $c(x)$ is the multiplication unit element of $C(x)$. obviously, $c(x)$ exists only. The Lemma holds.

**Definition 7.16** $C \subset \mathbb{F}_q^N$ is a cyclic code, and the multiplication unit element $c(x)$ in $C(x)$ is called the idempotent element of $C$. If $C = M_i^-$ is the $i$-th minimal cyclic code, the idempotent element of $C$ is called the primitive idempotent element, denote as $\theta_i(x)$.

**Lemma 7.59** *Let $C_1 \subset \mathbb{F}_q^N$, $C_2 \subset \mathbb{F}_q^N$ are two cyclic codes, $(N, q) = 1$, Idempotent elements are $c_1(x)$ $c_2(x)$, respectively, then*

*(i) $C_1 \bigcap C_2$ is also the cyclic code of $\mathbb{F}_q^N$, idempotent element is $c_1(x)c_2(x)$.*
*(ii) $C_1 + C_2$ is also the cyclic code of $\mathbb{F}_q^N$, idempotent element is $c_1(x) + c_2(x) + c_1(x)c_2(x)$.*

***Proof*** It is obvious that $C_1 \bigcap C_2$ and $C_1 + C_2$ are cyclic codes in $\mathbb{F}_q^N$, because they correspond to ideal $C_1(x)$ and $C_2(x)$ in $R$, we have

$$C_1(x) \cap C_2(x) \text{ and } C_1(x) + C_2(x)$$

is still the ideal in $R$. Therefore, the corresponding codes $C_1 \cap C_2$ and $C_1 + C_2$ are still cyclic codes, and the conclusion on idempotents is not difficult to verify. The Lemma holds.

In 1959, A. Hocquenghem and 1960, R. Bose and D. Chaudhuri independently proposed a special class of cyclic codes, which required minimal distance. At present, it is generally called BCH codes in academic circles.

**Definition 7.17** A cyclic code $C \subset \mathbb{F}_q^N$ with length $N$ is called a $\delta$-BCH code. If its generating polynomial is the least common multiple of the minimal polynomial of $\beta, \beta^2, \ldots, \beta^{\delta-1}$, where $\delta$ is a positive integer, $\beta$ is a primitive $N$-th unit root. $\delta$-BCH code is also called BCH code with design distance of $\delta$. If $\beta \in \mathbb{F}_{q^m}$, $N = q^m - 1$, such BCH codes are called primitive.

**Lemma 7.60** *Let $d$ be the minimal distance of a $\delta$-BCH code, then we have $d \geq \delta$.*

***Proof*** Suppose $x^N - 1 = (x-1)g_1(x)g_2(x)\cdots g_t(x)$, $\beta$ is a primitive $N$-th unit root on $\mathbb{F}_q$, then $\beta$ is the root of a $g_i(x)$. Let $\deg g_i(x) = m \Rightarrow \beta \in \mathbb{F}_{q^m}$. Because of $[\mathbb{F}_{q^m} : \mathbb{F}_q] = m$, we can think of $\beta, \beta^2, \ldots, \beta^{\delta-1}$ as an $m$-dimensional column vector. Let $H$ be the following $m(\delta-1) \times N$-order matrix.

$$H = \begin{bmatrix} 1 & \beta & \beta^2 & \cdots & \beta^{N-1} \\ 1 & \beta^2 & \beta^4 & \cdots & \beta^{2(N-1)} \\ \vdots & \vdots & \vdots & \ddots & \vdots \\ 1 & \beta^{\delta-1} & \beta^{2(\delta-1)} & \cdots & \beta^{(N-1)(\delta-1)} \end{bmatrix}_{m(\delta-1)\times N}.$$

In fact, $H$ is the check matrix of $\delta$-BCH code $C$, that is

$$c \in C \Longleftrightarrow cH' = 0.$$

We prove that any $(\delta - 1)$ column vectors of $H$ are linear independent vectors. Let the first component of these $(\delta - 1)$ column vectors be $\beta^{i_1}, \beta^{i_2}, \ldots, \beta^{i_{\delta-1}}$, where $i_j \geq 0$, the corresponding determinant is Vandermonde determinant $\triangle$, and

$$\triangle = \beta^{i_1+i_2+\cdots+i_{\delta-1}} \prod_{r>s}(\beta^{i_r} - \beta^{i_s}) \neq 0.$$

Therefore, any $(\delta - 1)$ column vectors of $H$ are linearly independent. Thus, the minimum distance of $C$ is $d \geq \delta$.

Now, we can introduce the design principle of McEliece/Niederreiter cryptosystem. Its basic mathematical idea is based on the decoding principle of error correction code. Recall the concept of error correction code in Chap. 2, a code $C \subset \mathbb{F}_q^N$ is called $t$-error correction code ($t \geq 1$ is a positive integer). If for $\forall\, y \in \mathbb{F}_q^N$, there is at most one codeword $c \in C \Rightarrow d(c, y) \leq t$, $d(c, y)$ is the Hamming distance between $c$ and $y$. We know that if the minimum distance of a code $C$ is $d$, then $C$ is a $t$-error correction code, where $t = \lceil \frac{d-1}{2} \rceil$ is the smallest integer not less than $\frac{d-1}{2}$. Lemma 7.60 proves the existence of $t$-error correction codes for any positive integer $t$, i.e., $2t+1$-BCH code ($\delta = 2t + 1$), this kind of code is called Goppa code (see the next section), which provides a theoretical basis for McEliece/Niederreiter cryptosystem. Next, we will introduce the working mechanism of this kind of cryptosystem in detail. First, let's look at the generation of key.

Private key: Select a $t$-error correction code $C \subset \mathbb{F}_q^N$, $C = [N, k]$, $H$ is the check matrix of $C$, $H$ is an $(N - k) \times N$-dimensional matrix. For $\forall\, x \in \mathbb{F}_q^N$, $x \to xH' \in$

$\mathbb{F}_q^{N-k}$ is a correspondence of Spaces $\mathbb{F}_q^N$ to $\mathbb{F}_q^{N-k}$, let's prove that this correspondence is a single shot on a special codeword whose weight is not greater than $t$.

**Lemma 7.61** *$\forall\, x, y \in \mathbb{F}_q^N$, if $xH' = yH'$, and $w(x) \leq t$, $w(y) \leq t$, then $x = y$.*

***Proof*** By hypothesis,

$$xH' = yH' \Rightarrow (x-y)H' = 0 \Rightarrow x - y \in C.$$

Obviously, the Hamming distance $d(0, x) = w(x) \leq t$ between $x$ and 0, and the Hamming distance $d(x, x-y)$ between $x$ and $x-y$ is

$$d(x, x-y) = w(x-(x-y)) = w(y) \leq t.$$

Because $C$ is $t$-error correction code, then $x - y = 0$, the Lemma holds.

We use $t$-error correction code $C$ and check matrix $H$ as the private key.

Public key: In order to generate the public key, we randomly select a permutation matrix $P_{N\times N}$ so that $I_N$ is an $N$-order identity matrix, $I_N = [e_1, e_2, \ldots, e_N]$, $\sigma \in S_N$ is an $N$-ary substitution, then

$$P = \sigma(I_N) = [e_{\sigma(1)}, e_{\sigma(2)}, \ldots, e_{\sigma(N)}].$$

This kind of matrix is also called Wyel matrix. A nonsingular diagonal matrix $\text{diag}\{\lambda_1, \lambda_2, \ldots, \lambda_N\}(\lambda_i \in \mathbb{F}_q, \lambda_i \neq 0)$ can also be randomly selected, and suppose

$$P = \sigma(\text{diag}\{\lambda_1, \lambda_2, \ldots, \lambda_N\}) = \text{diag}\{\lambda_{\sigma_1}, \lambda_{\sigma_2}, \ldots, \lambda_{\sigma_N}\}.$$

Let $M$ be an $(N-k)\times(N-k)$-order invertible matrix. The public key is the $(N-k)\times N$-order matrix $K$ generated as follows,

$$K = PH'M, \text{ this is } N \times (N-k) \text{ ordermatrix.}$$

We take $K$ as the public key and $H$, $P$ and $M$ as the private key.

Encryption: Let $m \in \mathbb{F}_q^N$ be a codeword, $w(m) \leq t$, encrypt $m$ as plaintext as follows.

$$c = mK \in \mathbb{F}_q^{N-k}, c \text{ is cryptosystem text.}$$

In fact, a plaintext with length $N$ and weight no greater than $t$ on $\mathbb{F}_q$ is encrypted into a cryptosystem text with length $(N-k)$ on $\mathbb{F}_q$ through public key $K$.

Decrypt: After receiving cryptosystem text $c$, decrypt it through private keys $H$, $P$ and $M$.

$$c \cdot M^{-1} = mKM^{-1} = mPH'MM^{-1} = mPH'.$$

Since $mP \in \mathbb{F}_q^N$ and $m$ have the same root, that is

$$w(m) = w(mP) \leq t.$$

Using the decoding principle of error correction code: all codewords $xH' = mPH'$ satisfying $x \in \mathbb{F}_q^N$ actually constitute an additive coset of code $C$, as the leader vector of this additive coset, $mP$ can be obtained accurately. That is

$$mPH' \overset{\text{decode}}{\longrightarrow} mP.$$

Finally, we have $m = (mP) \cdot P^{-1}$, and get plaintext.

## 7.9 Ajtai/Dwork Cryptosystem

By choosing an appropriate $n \times m$-order matrix $A \in \mathbb{Z}_q^{n\times m}$, two $m$-dimensional $q$-element lattices $\Lambda_q(A)$ and $\Lambda_q^{\perp}(A)$ are defined (see (7.45) and (7.46)),

$$\Lambda_q(A) = \{y \in \mathbb{Z}^m | \exists\, x \in \mathbb{Z}^n \Rightarrow y \equiv A^{'}x (\mathrm{mod}\, q)\}$$

and

$$\Lambda_q^{\perp}(A) = \{y \in \mathbb{Z}^m | Ay \equiv 0 (\mathrm{mod}\, q)\}.$$

Using matrix $A$, an anti-collision hash function can be defined:

$$f_A : \{0, 1, \ldots, d-1\}^m \to \mathbb{Z}_q^n, \tag{7.176}$$

where for any $y \in \{0, 1, \ldots, d-1\}^m$, define $f_A(y)$ as

$$f_A(y) = Ay \mod q, \tag{7.177}$$

If parameter $d, q, n, m$ is satisfied

$$n \log q < m \log d \Rightarrow \frac{n \log q}{\log d} < m. \tag{7.178}$$

Then Hash function $f_A$ will produce collision, that is there is $y, y^{'} \in \{0, 1, \ldots, d-1\}^m$, $y \neq y^{'}$, and $f_A(y) = f_A(y^{'})$. By (7.177), we have it directly

$$A(y - y^{'}) \equiv 0 (\mathrm{mod}\, q) \Rightarrow y - y^{'} \in \Lambda_q^{\perp}(A),$$

this shows that the collision points $y$ and $y^{'}$ of Hash function $f_A$ directly lead to a shortest vector $y - y^{'}$ on $q$-element lattice $\Lambda_q^{\perp}(A)$.

In order to obtain the anti-collision Hash function, the selection of $n \times m$-order matrix $A$ is very important. First, we can select the parameter system: let $d = 2$,

$q = n^2$, $n|m$, and $m \log 2 > n \log q$, where $n$ is a positive integer. In Ajtai/Dwork cryptographic algorithm, there are two choices of parameter matrix $A$, one is cyclic matrix and the other is more general ideal matrix. Their corresponding $q$-element lattice $\Lambda_q^{\perp}(A)$ are cyclic lattice and ideal lattice, respectively.

Cyclic lattice

Because $n|m$, $A$ can be divided into $\frac{m}{n}$ $n \times n$-order cyclic matrices, that is

$$A = [A^{(1)}, A^{(2)}, \ldots, A^{(\frac{m}{n})}], \tag{7.179}$$

where $\alpha^{(i)} \in \mathbb{Z}_q^n$ is the $n$-dimensional column vector and $A^{(i)}$ is the cyclic matrix generated by $\alpha^{(i)}$ (see (7.149)), that is

$$A^{(i)} = T^*(\alpha^{(i)}) = [\alpha^{(i)}, T\alpha^{(i)}, \ldots, T^{n-1}\alpha^{(i)}], 1 \leq i \leq \frac{m}{n}.$$

$A$ is called an $n \times m$-dimensional generalized cyclic matrix, and the $q$-element lattice in $\mathbb{R}^m$ defined by $A$,

$$\Lambda_q^{\perp}(A) = \{y \in \mathbb{Z}^m | Ay \equiv 0 (\bmod q)\}$$

is called a cyclic lattice. The Ajtai/Dwork cryptosystem based on cyclic lattice can be stated as follows:

Algorithm 1: Hash function based on cyclic lattice.

Parameter: $q, n, m, d$ is a positive integer, $n \mid m$, $m \log d > n \log q$.

Secret key: $\frac{m}{n}$ column vectors $\alpha^{(i)} \in \mathbb{Z}_q^n$, $1 \leq i \leq \frac{m}{n}$.

Hash function $f_A : \{0, 1, \ldots, d-1\}^m \longrightarrow \mathbb{Z}_q^n$ define as

$$f_A(y) \equiv Ay (\bmod q),$$

the cyclic matrix $A \in \mathbb{Z}_q^{n \times m}$ is given by (7.179).

We can extend the above concepts of cyclic matrix and cyclic lattice to more general cases and obtain the concepts of ideal matrix and ideal lattice. Let $h(x)$ be the first integer coefficient polynomial of $n$ degree, $h(x) = x^n + a_{n-1}x^{n-1} + \cdots + a_1 x + a_0 \in \mathbb{Z}[x]$, define the rotation matrix $T_h$ as

$$T_h = \begin{pmatrix} 0 & \cdots & 0 & -a_0 \\ & & & -a_1 \\ & I_{n-1} & & \vdots \\ & & & -a_{n-1} \end{pmatrix}, \tag{7.180}$$

if $h(x) = x^n - 1$ is a special polynomial, then $T_h = T$. $T$ is highlighted in Sect. 7.7 of this chapter. Here, we discuss the more general $T_h$. Obviously, when the constant term $a_0 \neq 0$, $T_h$ is a reversible $n$-order square matrix, and $T_h = \det(T_h) = (-1)^n a_0$.

**Lemma 7.62** *The characteristic polynomial of rotation matrix $T_h$ is $f(\lambda) = h(\lambda)$.*

***Proof*** By the definition, the characteristic polynomial $f(\lambda)$ of $T_h$ is

$$
\begin{aligned}
f(\lambda) &= \det(\lambda I_n - T_h) \\
&= \begin{vmatrix} \lambda & 0 & \cdots & 0 & a_0 \\ -1 & \lambda & \cdots & \vdots & \vdots \\ \cdots & \cdots & \cdots & \cdots & \cdots \\ 0 & \cdots & \cdots & \lambda & a_{n-2} \\ 0 & \cdots & \cdots & -1 & a_{n-1} \end{vmatrix} \\
&= \frac{1}{\lambda}\frac{1}{\lambda^2}\cdots\frac{1}{\lambda^{n-1}} \begin{vmatrix} \lambda & 0 & \cdots & 0 & a_0 \\ 0 & \lambda & \cdots & \vdots & a_1\lambda + a_0 \\ \cdots & \cdots & \cdots & \cdots & \cdots \\ 0 & \cdots & \cdots & \cdots & \lambda^n + a_{n-1}\lambda^{n-1} + \cdots + a_1\lambda + a_0 \end{vmatrix} \\
&= \lambda^n + a_{n-1}\lambda^{n-1} + \cdots + a_1\lambda + a_0 = h(\lambda).
\end{aligned}
$$

**Lemma 7.63** *Let $h(x) = x^n + a_{n-1}x^{n-1} + \cdots + a_1 x + a_0 \in \mathbb{Z}[x]$, if $a_0 \neq 0$, then the rotation matrix $T_h$ is a reversible $n$-order square matrix, and*

$$
T_h^{-1} = \begin{bmatrix} -a_0^{-1}\alpha & I_{n-1} \\ -a_0^{-1} & 0 \end{bmatrix}, \quad \alpha = \begin{bmatrix} a_1 \\ a_2 \\ \vdots \\ a_{n-1} \end{bmatrix} \in \mathbb{Z}^{n-1}.
$$

***Proof*** By the definition of $T_h$,

$$
\begin{aligned}
T_h \cdot \begin{bmatrix} -a_0^{-1}\alpha & I_{n-1} \\ -a_0^{-1} & 0 \end{bmatrix} &= \begin{bmatrix} 0 & -a_0 \\ I_{n-1} & -\alpha \end{bmatrix} \begin{bmatrix} a_0^{-1}\alpha & I_{n-1} \\ -a_0^{-1} & 0 \end{bmatrix} \\
&= \begin{bmatrix} 1 & 0 \\ 0 & I_{n-1} \end{bmatrix} = I_n.
\end{aligned}
$$

So

$$
T_h^{-1} = \begin{bmatrix} -a_0^{-1}\alpha & I_{n-1} \\ -a_0^{-1} & 0 \end{bmatrix}.
$$

For a given first polynomial $h(x) = x^n + a_{n-1}x^{n-1} + \cdots + a_1 x + a_0 \in \mathbb{Z}[x]$ of degree $n$, let $R$ be a residue class ring of module $h(x)$ in $\mathbb{Z}[x]$, i.e.,

$$
R = \mathbb{Z}[x]/\langle h(x) \rangle, \tag{7.181}
$$

where $\langle h(x) \rangle$ is the ideal generated by $h(x)$ in $\mathbb{Z}[x]$. Because of $\deg h(x) = n$, then polynomial $g(x) \in R$ in $R$ has a unique expression: $g(x) = g_{n-1}x^{n-1} + g_{n-2}x^{n-2} + \cdots + g_1 x + g_0 \in R$, define mapping $\sigma : R \longrightarrow \mathbb{Z}^n$ as

$$\sigma(g(x)) = \begin{bmatrix} g_0 \\ g_1 \\ \vdots \\ g_{n-1} \end{bmatrix} \in \mathbb{Z}^n. \tag{7.182}$$

Obviously, $\sigma$ is an Abel group isomorphism of $R \longrightarrow \mathbb{Z}^n$. Therefore, any polynomial $g(x)$ in $R$ can be regarded as an $n$-dimensional integer column vector.

**Definition 7.18** For any $n$-dimensional column vector $g = \sigma(g(x)) = \begin{bmatrix} g_0 \\ g_1 \\ \vdots \\ g_{n-1} \end{bmatrix} \in \mathbb{Z}^n$ in $\mathbb{Z}^n$, define

$$T_h^*(g) = [g, T_h(g), T_h^2(g), \ldots, T_h^{n-1}(g)]_{n \times n}, \tag{7.183}$$

the $n$-order square matrix $T_h^*(g)$ is called an ideal matrix generated by vector $g$.

Ideal matrix is a more general generalization of cyclic matrix. The former corresponds to a first $n$-degree polynomial $h(x)$, and the latter corresponds to a special polynomial $x^n - 1$. We first prove that the ideal matrix $T_h^*(g)$ and the rotation matrix $T_h$ generated by any vector $g \in \mathbb{Z}^n$ are commutative under matrix multiplication.

**Lemma 7.64** *For any given first n-degree polynomial* $h(x) \in \mathbb{Z}[x]$, *and n-dimensional column vector* $g \in \mathbb{Z}^n$, *we have*

$$T_h \cdot T_h^*(g) = T_h^*(g) \cdot T_h.$$

***Proof*** Let $h(x) = x^n + a_{n-1}x^{n-1} + \cdots + a_1 x + a_0 \in \mathbb{Z}[x]$, by Lemma 7.62, the characteristic polynomial of rotation matrix $T_h$ is $h(\lambda)$, then by Hamilton–Cayley theorem, we have

$$T_h^n + a_{n-1}T_h^{n-1} + \cdots + a_1 T_h + a_0 = 0, \tag{7.184}$$

there is

$$\begin{aligned} T_h^*(g)T_h &= [g, T_h g, T_h^2 g, \ldots, T_h^{n-1} g] \begin{bmatrix} 0 & -a_0 \\ I_{n-1} & -\alpha \end{bmatrix} \\ &= [T_h g, T_h^2 g, \ldots, -a_0 g - a_1 T_h g - \cdots - a_{n-1} T_h^{n-1} g] \\ &= [T_h g, T_h^2 g, \ldots, (-a_0 - a_1 T_h - \cdots - a_{n-1} T_h^{n-1}) g] \\ &= [T_h g, T_h^2 g, \ldots, T_h^n g] \\ &= T_h [g, T_h g, \ldots, T_h^{n-1} g] \\ &= T_h \cdot T_h^*(g). \end{aligned}$$

When the monic $n$-degree integer coefficient polynomial $h$ is selected, we want to establish the corresponding relationship between the ideal and the integer lattice $L \subset \mathbb{Z}^n$ in the quotient ring $R = \mathbb{Z}[x]/\langle h(x) \rangle$. First, we define the concept of an ideal lattice. In short, an ideal lattice is an integer lattice generated by the ideal matrix.

**Definition 7.19** Let $g = (g_0, g_1, \ldots, g_{n-1})^T \in \mathbb{Z}^n$ be a given column vector and $T_h^*(g)$ be the ideal matrix generated by $g$, and call the integer lattice $L = L(T_h^*(g))$ an ideal lattice.

Our main result is the 1-1 correspondence between ideal and ideal lattice in $R = \mathbb{Z}[x]/\langle h(x) \rangle$. This also explains the reason why $L(T_h^*(g))$ is called ideal lattice.

**Theorem 7.10** *The principal ideal in $R = \mathbb{Z}[x]/\langle h(x) \rangle$ 1-1 corresponds to the ideal lattice in $\mathbb{Z}^n$. Specifically,*

*(i) If $N = \langle g(x) \rangle$ is any principal ideal in R, then*

$$\sigma(N) = \{\sigma(f) | f \in N\} = L(T_h^*(\sigma(g(x)))) = L(T_h^*(g)).$$

*(ii) If $g = (g_0, g_1, \ldots, g_{n-1})^T \in \mathbb{Z}^n$, $T_h^*(g) \subset \mathbb{Z}^n$ is any ideal lattice, then*

$$\sigma^{-1}(T_h^*(g)) = \{\sigma^{-1}(b) | b \in T_h^*(g)\} = \langle g(x) \rangle \subset R,$$

*where $g(x) = g_0 + g_1 x + \cdots + g_{n-1}x^{n-1} = \sigma^{-1}(g)$.*

***Proof*** We first prove (i). Let $g(x) = g_0 + g_1 x + \cdots + g_{n-1}x^{n-1} \in R$ be a given polynomial, $N = \langle g(x) \rangle \subset R$ is a principal ideal generated by $g(x)$ in $R$, by (7.182),

$$\sigma(g(x)) = (g_0, g_1, \ldots, g_{n-1})^T = T_h^*(g) \cdot \begin{bmatrix} 1 \\ 0 \\ \vdots \\ 0 \end{bmatrix} \in L(T_h^*(g)).$$

And because

$$\begin{aligned} xg(x) &= g_{n-1}x^n + g_{n-2}x^{n-1} + \cdots + g_1 x^2 + g_0 x \\ &= (g_{n-2} - g_{n-1}a_{n-1})x^{n-1} + (g_{n-3} - g_{n-1}a_{n-2})x^{n-2} + \cdots \\ &\quad + (g_0 - g_{n-1}a_1)x - g_{n-1}a_0, \end{aligned}$$

so

$$\sigma(xg(x)) = \begin{bmatrix} -g_{n-1}a_0 \\ g_0 - g_{n-1}a_1 \\ \vdots \\ g_{n-2} - g_{n-1}a_{n-1} \end{bmatrix} = T_h \cdot g = T_h^*(g) \begin{bmatrix} 0 \\ 1 \\ 0 \\ \vdots \\ 0 \end{bmatrix} \in L(T_h^*(g)).$$

For the same reason, for $0 \le k \le n-1$, we have

$$\sigma(x^k g(x)) = T_h^k \cdot g = T_h^*(g) \cdot \begin{bmatrix} 0 \\ 0 \\ \vdots \\ 1 \\ 0 \\ 0 \end{bmatrix} \in L(T_h^*(g)).$$

Suppose $f(x) \in N = \langle g(x) \rangle$, then $f(x) = b(x) \cdot g(x)$, where $b(x) = b_0 + b_1 x + \cdots + b_{n-1}x^{n-1}$, then we have

$$\begin{aligned} \sigma(f(x)) &= \sigma(b(x)g(x)) \\ &= \sum_{k=0}^{n-1} b_k \sigma(x^k g(x)) \\ &= T_h^*(g) \begin{bmatrix} b_0 \\ b_1 \\ \vdots \\ b_{n-1} \end{bmatrix} \in L(T_h^*(g)). \end{aligned} \tag{7.185}$$

That proves

$$\sigma(x) = \sigma(\langle g(x) \rangle) \subset L(T_h^*(g)).$$

Conversely, for any lattice point $\alpha \in L(T_h^*(g))$, then

$$\alpha = T_h^*(g)b = T_h^*(g) \begin{bmatrix} b_0 \\ b_1 \\ \vdots \\ b_{n-1} \end{bmatrix},$$

since $\sigma$ is 1-1 corresponds, by (7.185), then

$$f(x) = \sigma^{-1}(f(x)) = \sigma^{-1}(T_h^*(g)b) \in N = \langle g(x) \rangle.$$

So we have

$$\sigma(N) = \sigma(\langle g(x) \rangle) = L(T_h^*(g)).$$

(i) holds. Again, $\sigma$ is 1-1 corresponds, so (ii) can be derived directly. We complete the proof of Theorem 7.10.

The above discussion on ideal matrix and ideal lattice can be extended to a finite field $\mathbb{Z}_q$, because any quotient ring $\mathbb{Z}_q[x]/\langle h(x) \rangle$ on polynomial ring $\mathbb{Z}_q[x]$ in finite field is a principal ideal ring. Therefore, we can establish the 1-1 correspondence between all ideals in $R = \mathbb{Z}_q[x]/\langle h(x)\rangle$ and linear codes on $\mathbb{Z}_q$.

Back to the Ajtai/Dwork cryptosystem, let $h(x) \in \mathbb{Z}_q[x]$ be a given polynomial, and select an $n \times m$-dimensional matrix $A \in \mathbb{Z}_q^{n\times m}$ as the generalized ideal matrix, i.e.,

$$A = [A_1, A_2, \ldots, A_{\frac{m}{n}}], \tag{7.186}$$

where $A_i (1 \le i \le \frac{m}{n})$ is the ideal matrix generated by $g^{(i)} \in \mathbb{Z}_q^n$, that is

$$A_i = T_h^*(g^{(i)}) = [g^{(i)}, T_h g^{(i)}, \ldots, T_h^{n-1} g^{(i)}], \tag{7.187}$$

we get the second algorithm of Ajtai/Dwork cryptosystem:

Algorithm 2: Hash function based on ideal lattice.

Parameter: $q, n, m, d$ are positive integers, $n|m$, $m \log d > n \log q$.

Secret key: $\frac{m}{n}$ column vectors $g^{(i)} \in \mathbb{Z}_q^n (1 \le i \le \frac{m}{n})$, polynomial $h(x) = x^n + a_{n-1}x^{n-1} + \cdots + a_1 x + a_0 \in \mathbb{Z}_q[x]$.

Hash function $f_A : \{0, 1, \ldots, d-1\}^m \longrightarrow \mathbb{Z}_q^n$ defined as

$$f_A(y) \equiv Ay (\mathrm{mod}\, q),$$

The ideal matrix $A \in \mathbb{Z}_q^{n\times m}$ is given by Eq. (7.186).

We will not introduce the anti-collision performance of hash functions constructed by cyclic lattices and ideal lattices here. Interested students can refer to the reference Micciancio and Regev (2009) in this chapter.

## Exercise 7

1. $L \subset \mathbb{R}^n$ is a lattice (full rank lattice), if $L^*$ is a dual lattice of $L$, then the integer lattice $L = \mathbb{Z}^n$ is a self-dual lattice, that is $(\mathbb{Z}^n)^* = \mathbb{Z}^n$. Let $L = 2\mathbb{Z}^n$, find $L^* =$?
2. Is it correct that $L$ is a self-dual lattice if and only if $L = \mathbb{Z}^n$? Why?
3. Under the assumption of exercise 1, let $\lambda_1(L)$ be the shortest vector length of $L$ and $\lambda_1(L^*)$ be the shortest vector length of dual lattice $L^*$. Then

$$\lambda_1(L) \cdot \lambda_1(L^*) \le n.$$

4. Let $\lambda_1(L), \lambda_2(L), \ldots, \lambda_n(L)$ be the length of the Successive Shortest vector of lattice $L$, prove

$$\lambda_1(L) \cdot \lambda_n(L^*) \ge 1.$$

5*. Let $L$ be a lattice, $B = [\beta_1, \beta_2, \ldots, \beta_n]$ is the generating matrix of $L$, $B^* = [\beta_1^*, \beta_2^*, \ldots, \beta_n^*]$ is the corresponding orthogonal matrix. Prove: any lattice $L$ has a set of bases $\{\beta_1, \beta_2, \ldots, \beta_n\}$, such that

$$\frac{1}{n}\lambda_1(L) \le \min\{|\beta_1^*|, |\beta_2^*|, \ldots, |\beta_n^*|\} \le \lambda_1(L).$$

(Hint: use KZ basis on lattice $L$).

6. Under the assumption of exercise 5, let $\lambda_1(L), \lambda_2(L), \ldots, \lambda_n(L)$ be the continuous minimum of lattice $L$, prove:

$$\lambda_j(L) \geq \min_{j \leq i \leq n} |\beta_i^*|, \quad 1 \leq j \leq n.$$

7. For a full rank lattice $L \subset \mathbb{R}^n$, define its coverage radius $\mu(L)$ as

$$\mu(L) = \max_{x \in \mathbb{R}^n} |x - L|.$$

Prove: the covering radius of any lattice $L$ exists.

8. Prove: $\mu(\mathbb{Z}^n) = \frac{1}{2}\sqrt{n}$.
9. For any lattice $L \subset \mathbb{R}^n$, prove: $\mu(L) \geq \frac{1}{2}\lambda_n(L)$.
10. For any lattice $L \subset \mathbb{R}^n$, prove the following theorem:

$$\lambda_1(L) \cdot \mu(L^*) \leq n.$$

## References

Ajtai, M. (2004). Generating hard instances of lattice problems. In *Quad. Mat.: Vol. 13. Complexity of computations and proofs* (pp. 1–32). Dept. Math., Seconda Univ. Napoli. Preliminary version in STOC 1996.

Ajtai, M., & Dwork, C. (1997). A public-key cryptosystem with worst-case/average-case equivalence. In *Proceedings of 29th Annual ACM Symposium on Theory of Computing (STOC)* (pp. 284–293).

Babai, L. (1986). On Loväasz lattice reduction and the nearest lattice point problem. *Combinatorica, 6*, 1–13.

Cassels, J. W. S. (1963). *Introduction to diophantine approximation*. Cambridge University Press.

Cassels, J. W. S. (1971). *An introduction to the geometry of numbers*. Springer.

Gama, N., & Nguyen, P. Q. (2008a). Finding short lattice vectors within Mordell's inequality. In *Proceedings of 40th ACM Symposium on Theory of Computing (STOC)* (pp. 207–216).

Gama, N., & Nguyen, P. Q. (2008b). Predicting lattice reduction. In *Lecture Notes in Computer Science: Advances in cryptology. Proceedings of Eurocrypt'08*. Springer

Goldreich, O., Goldwasser, S., & Halevi, S. (1997). Public-key cryptosystems from lattice reduction problems. In *Lecture Notes in Computer Science: Vol. 1294. Advances in cryptology* (pp. 112–131). Springer.

Hoffstein, J., Pipher, J., & Silverman, J. H. (1998). NTRU: A ring based public key cryptosystem. In *LNCS: Vol. 1423. Proceedings of ANTS-III* (pp. 267–288). Springer.

Klein, P. (2000). Finding the closest lattice vector when it's unusually close. In *Proceedings of 11th Annual ACM-SIAM Symposium on Discrete Algorithms* (pp. 937–941).

Lenstra, A. K., Lenstra, H. W., Jr., & Lovasz, L. (1982). Factoring polynomials with rational coefficients. *Mathematische Annalen, 261*(4), 515–534.

McEliece, R. (1978). *A public-key cryptosystem based on algebraic number theory*. Technical Report, Jet Propulsion Laboratory. DSN Progress Report 42-44.

Micciancio, D. (2001). Improving lattice based cryptosystems using the Hermite normal form. In J. Silverman (Ed.), *Lecture Notes in Computer Science: Vol. 2146. Cryptography and lattices conference—CaLC 2001* (pp. 126–145). Springer.

Micciancio, D., & Regev, O. (2009). *Lattice-based cryptography*. Springer.

Niederreiter, H. (1986). Knapsack-type cryptosystems and algebraic coding theory. *Problems of Control and Information Theory/Problemy Upravlen. Teor. Inform., 15*(2), 159–166.

Peikert, C. (2016). *A decade of lattice cryptography*. Foundations & Trends in Theoretical Computer Science.

Regev, O. (2004). *Lattices in computer science* (Lecture 1–Lecture 7). Tel Aviv University, Fall.

# References

1. Apostol, T. M. (1976). *Introduction to analytic number theory*. Springer.
2. Hardy, G. H., & Wright, E. M. (1979). *An introduction to the theory of number*. Oxford University Press.
3. Leveque, W. J. (1977). *Fundamentals of number theory*. Addison-Wesley.
4. Lidl, R., & Niederreiter, H. (1983). *Finite fields*. Addison-Wesley.
5. VanLint, J. H. (1991). *Introduction to coding theory*. Springer.
6. Rosen, K. H. (1984). *Elementary number theory and its applications*. Addison-Wesley.
7. Rosen, M. H. (2002). *Number theory in function fields*. Springer.
8. Spencer, D. (1982). *Computers in number theory*. Computer Science Press.
9. Rényi, A. (1970). *Probability theory*. North-Holland.
10. Vander Walden, B. L. (1963). *Algebra (I)* (S. Ding, K. Zeng & F. Hao, Trans.). Science Press (in Chinese).
11. Vander Walden, B. L. (1976). *Algebra (II)* (X. Cao, K. Zeng & F. Hao, Trans.). Science Press (in Chinese).
12. Jacobson, N. (1989). *Basic algebra (I)* Translated by the Department of Algebra, Department of Mathematics, Shanghai Normal University, Beijing. Higher Education Press (in Chinese).
13. Nie, L., & Ding, S. (2000). *Introduction to algebra*. Higher Education Press (in Chinese).
14. Li, X. (2010). *Basic probability theory*. Higher Education Press (in Chinese).
15. Long, Y. (2020). *Probability theory and mathematical statistics*. Higher Education Press (in Chinese).
16. Berlekamp, E. R. (1968). *Algebraic coding theory*. McGraw-Hill.
17. Berlekamp, E. R. (1972). *Decoding the Golay code*. JPL Technical Report 32-1256 (Vol. IX, pp. 81–85). Jet Propulsion Laboratory.
18. Best, M. R. (1978). *On the existence of perfect codes*. Report ZN 82/78. Mathematical Centre.
19. Best, M. R. (1980). Binary cades with a minimum distance of four. *IEEE Transactions on Information Theory, 26*, 738–742.
20. Bussey, W. H. (1905). Galois field tables for $p^n$ 169. *Bulletin of the American Mathematical Society, 12*, 22–38.
21. Bussey, W. H. (1910). Tables of Galois fields of order less than 1000. *Bulletin of the American Mathematical Society, 16*, 188–206.
22. Cameron, P. J., & van Lint, J. H. (1991). *London Mathematical Society Student Texts: Vol. 22. Designs, graphs, codes and their links*. Cambridge University Press.
23. Curtis, C. W., & Reiner, I. (1962). *Representation theory of finite groups and associative algebras*. Interscience.



Z. Zheng, *Modern Cryptography Volume 1*, Financial Mathematics and Fintech,
https://doi.org/10.1007/978-981-19-0920-7

24. Delsarte, P., & Goethals, J. M. (1975). Unrestricted codes with the Golay parameters are unique. *Discrete Mathematics, 12*, 211–224.
25. Elias, P. Coding for noisy channels. IRE Conv, Record, Part 4, pp. 37–46.
26. Feller, W. (1950). *An introduction to probability theory and its applications* (Vol. I). Wiley.
27. Forney, G. D. (1970). Convolutional codes I: algebraic structure. *IEEE Transactions on Information Theory, 16*, 720–738. Ibid., 17, 360 (1971).
28. Gallagher, R. G. (1968). *Information theory and reliable communication*. Wiley.
29. Goethals, J. M. (1977). The extended Nadler code is unique. *IEEE Transactions on Information Theory, 23*, 132–135.
30. Goppa, V. D. (1970). A new class of linear error-correcting codes. *Problems of Information Tansmission, 6*, 207–212.
31. Goto, M. (1975). A note on perfect decimal AN codes. *Information and Control, 29*, 385–387.
32. Goto, M., & Fukumara, T. (1975). Perfect nonbinary AN codes with distance three. *Information and Control, 27*, 336–348.
33. Graham, R. L., & Sloane, N. J. A. (1980). Lower bounds for constant weight codes. *IEEE Transactions on Information Theory, 26*, 37–40.
34. Gritsenko, V. M. (1969). Nonbinary arithmetic correcting codes. *Problems of Information and Transmission, 5*, 15–22.
35. Helgert, H. J., & Stinaff, R. D. (1973). Minimum distance bounds for binary linear codes. *IEEE Transactions on Information Theory, 19*, 344–356.
36. Justesen, J. (1975). An algebraic construction of rate $\frac{1}{v}$ convolutional codes. *IEEE Transactions on Information Theory, 21*, 577–580.
37. Kasami, T. (1969). An upper bound on k/n for affine invariant codes with fixed d/n. *IEEE Transactions on Information Theory, 15*, 171–176.
38. Levenshtein, V. I. (1975). Minimum redundancy of binary error-correcting codes. *Information and Control, 28*, 268–291.
39. van Lint, J. H. (1971). Nonexistence theorems for perfect error-correcting codes. In *Computers in algerbra and theory. SIAM-AMS Proceedings* (Vol. IV).
40. van Lint, J. H. (1999). *Introduction to coding theory, GTM86*. Springer.
41. van Lint, J. H. (1972). A new description of the Nadler code. *IEEE Transactions on Information Theory, 18*, 825–826.
42. van Lint, J. H. (1975). A survey of perfect codes. *Rocky Mountain Journal of Mathematics, 5*, 199–224.
43. van Lint, J. H., & Macwilliams, F. J. (1978). Generalized quadratic residue codes. *IEEE Transactions on Information Theory, 24*, 730–737.
44. Macwilliams, F. J., & Sloane, N. J. A. (1977). *The theory of error-correcting codes*. North Holland.
45. Massey, J. L., & Garcia, O. N. (1972). Error-correcting codes in computer arithmetic. In: J. T. Ton (Ed.) *Advances in information systems science* (Vol. 4, Chap. 5). Plenum Press.
46. Massey, J. L., Costello, D. J., & Justesen, J. (1973). Polynomial weights and code construction. *IEEE Transactions on Information Theory, 19*, 101–110.
47. McEliece, R. J., Rodemich, E. R., Rumsey, H. C., & Welch, L. R. (1977). New upper bounds on the rate of a code via the Delsarte-MacWilliams inequalities. *IEEE Transactions on Information Theory, 23*, 157–166.
48. McEliece, R. J. (1977). The theory of information and coding. In *Encyclopedia of mathematics and its applications* (Vol. 3). Addison-Wesley.
49. McEliece, R. J. (1979). The bounds of Delsarte and Lovasz and their applications to coding theory. In: G. Longo (Ed.) *CISM Courses and Lectures: Vol. 258. Algebraic coding theory and applications*. Springer.
50. Peterson, W. W., & Weldon, E. J. (1972). *Error-correcting codes* (2nd Ed.). MIT Press.
51. Piret, Ph. (1977). *Algebraic properties of convolutional codes with automorphisms* (Ph.D. Dissertation, Université Catholique de Louvain).
52. Posner, E. C. (1968). Combinatorial structures in planetary reconnaissance. In H. B. Mann (Ed.), *Error correcting codes* (pp. 15–46). Wiley.

53. Rao, T. R. N. (1974). *Error coding for arithmetic processors*. Academic Press.
54. Roos, C. (1979). On the structure of convolutional and cyclic convolutional codes. *IEEE Transactions on Information Theory, 25*, 676–683.
55. Shannon, C. E. (1948). A mathematical theory of communication. *Bell Labs Technical Journal, 27*, 379–423, 623–656.
56. Sloane, N. J. A., Reddy, S. M., & Chen, C. L. (1972). New binary codes. *IEEE Transactions on Information Theory, 18*, 503–510.
57. Solomon, G., & van Tilborg, H. C. A. (1979). A connection between block and convolutional codes. *SIAM Journal on Applied Mathematics, 37*, 358–369.
58. Tietäváinen, A. (1973). On the nonexistence of perfect codes over finite fields. *SIAM Journal on Applied Mathematics, 24*, 88–96.
59. van der Geer, G., & van Lint, J. H. (1988). *Introduction to coding theory and algebraic geometry*. Birkhäuser.
60. Hong, Y. (1984). On the nonexistence of unknown perfect 6- and 8-codes in Hamming schemes $H(n, q)$ with $q$ arbitrary. *Osaka Journal of Mathematics, 21*, 687–700.
61. Kerdock, A. M. (1972). A class of low-rate nonlinear codes. *Information and Control, 20*, 182–187.
62. van Oorschot, P. C., & Vanstone, S. A. (1989). *An introduction to error correcting codes with applications*. Kluwer.
63. Peek, J. H. (1985). Communications aspects of the compact disc digital audio system. *IEEE Communications Magazine, 23*(2), 7–15.
64. Piret, Ph. (1988). *Convolutional codes, an algebraic approach*. The MIT Press.
65. Barg, A. M., Katsman, S. L., & Tsfasman, M. A. (1987). Algebraic geometric codes from curves of small genus. *Problems of Information Transmission, 23*, 34–38.
66. Conway, J. H., & Sloane, N. J. A. (1994). Quaternary constructions for the binary single-error-correcting codes of Julin, Best, and others. *Designs, Codes and Cryptography, 41*, 31–42.
67. Feng, G.-L., & Rao, T. R. N. (1994). A simple approach for construction of algebraic-geometric codes from affine plane curves. *IEEE Transactions on Information Theory, 40*, 1003–1012.
68. Høholdt, T., & Pellikaan, R. (1995). On the decoding of algebraic-geometric codes. *IEEE Transactions on Information Theory, 41*, 1589–1614.
69. Høholdt, T., van Lint, J. H., & Pellikaan, R. (1998). Algebraic geometry codes. In V. S. Pless, W. C. Huffman& R. A. Brualdi (Eds.), *Hand-book of coding theory*. Elsevier Science Publishers.
70. Justesen, J., Larsen, K. J., Jensen, E. H., & Havemose, A., & Høholdt, T. (1989). Construction and decoding of a class of algebraic geometry codes. *IEEE Transactions on Information Theory, 35*, 811–821.
71. van Lint, J. H. (1990). Algebraic geometric codes. In D. Ray-Chaudhuri (Ed.), *Coding theory and design theory I. The IMA Volumes in Mathematics and its Applications* (Vol. 20). Springer.
72. van Lint, J. H., & Wilson, R. M. (1992). *A course in combinatorics*. Cambridge University Press.
73. Stichtenoth, H. (1993). *Algebraic function fields and codes. Universitext*. Springer.
74. Bassoli, R., Marques, H., & Rodriguez, J. (2013). Network coding theory, a survey. *IEEE Communications Surveys & Tutorials, 15*(4), 1950–1978.
75. Berger, T. (1971). *Rate distortion theory: a mathematical basis for data compression*. Prentice-Hall.
76. Billingsley, P. (1965). *Ergodic theory and information*. Wiley.
77. Chung, K. L. (1961). A note on the ergodic theorem of information theory. *Annals of Mathematical Statistics, 32*, 612–614.
78. Cover, T. M., & Thomas, J. A. (1991). *Elements of information theory*. Wiley.
79. Csiszár, I., & Körner, J. (1981). *Information theory: Coding theorems for discrete memoryless systems*. Academic Press.
80. El Gamal, A., & Kim, Y. H. (2011). *Network information theory*. Cambridge University Press.

81. Fragouli, C., Le Boudec, J. Y., & Widmer, J. (2006). Network coding: an instant primer. *ACMSIGCOMM Computer Communication Review, 36*, 63–68.
82. Gallager, R. G. (1968). *Information theory and reliable communication*. Wiley.
83. Gray, R. M. (1990). *Entropy and information theory*. Springer.
84. Guiasu, S. (1977). *Inormation theory with applications*. McGraw-Hill.
85. Ho, T., & Lun, D. (2008). Network coding: An introduction. *Computer Journal*.
86. Hu, X. H., & Ye, Z. X. (2006). Generalized quantum entropy. *Journal of Mathematical Physics, 47*(2), 1–7.
87. Ihara, S. (1993). *Information theory for continuous systems*. World Scientific.
88. Kakihara, Y. (1999). *Abstact methods in information theory*. World Scientific.
89. McMillan, B. (1953). The basic theorems of information theory. *Annals of Mathematical Statistics, 24*(2), 196–219.
90. Moy, S. C. (1961). A note on generalizations of Shannon-McMilllan theorem. *Pacific Journal of Mathematics, 11*, 705–714.
91. Nielsen, M. A., & Chuang, I. L. (2000). *Quantum computation and quantum information*. Cambridge University Press.
92. Shannon, C. E. (1948). A mathematical theory of communication. *The Bell System Technical Journal, 27*(4), 379–423, 623–656.
93. Shannon, C. E. (1959). Coding theorem for a discrete source with a fidelity criterion. *IRE National Convention Record, 4*, 142–163.
94. Shannon, C. E. (1958). Channels with side information at the transmitter. *IBM Journal of Research and Development, 2*(4), 189–193.
95. Shannon, C. E. (1961). Two-way communication channels. In *Proceedings of the Fourth Berkeley Symposium on Mathematical Statistics and Probability* (Vol. 1, pp. 611–644).
96. Thomasian, A. J. (1960). An elementary proof of the AEP of information theory. *Annals of Mathematical Statistics, 31*(2), 452–456.
97. Wolfowitz, J. (1978). *Coding theorems of information theory* (3rd Ed.). Springer.
98. Ye, Z. X., & Berger, T. (1998). *Information measures for discrete random fields*. Science Press.
99. Yeung, R. W. (2002). *A first course in information theory*. Kluwer Academic.
100. Qiu, P. (2003). *Information theory and coding*. Higher Education Press (in Chinese).
101. Qiu, P., Zhang, C., Yang, S., et al. (2012). *Multi user information theory*. Science Press (in Chinese).
102. Ye, Z. (2003). *Fundamentals of information theory*. Higher Education Press (in Chinese).
103. Zhang, Z., & Lin, X. (1993). *Information theory and optimal coding*. Shanghai Science and Technology Press (in Chinese).
104. Adleman, L. M. (1979). A subexponential algorithm for the discrete logarithm problem with application to cryptography. In *Proceedings of the 20th Annual Symposium on the Foundations of Computer Science* (pp. 55–60).
105. Adelman, L. M., Rivest, R. L., & Shamir, A. (1978). A method for obtaining digital signatures and public-key crypto system. *Communications of ACM, 21*, 120–126.
106. Blum, M. Coin-flipping by telephone—A protocol for Solving impossible problems. *IEEE Proceeding*, 133–137.
107. Diffie, W., & Hellman, M. E. (1976). New direction in crytography. *IEEE Transactions in Information Theory, IT-22*, 644–654.
108. Hellman, M. E. (1979). The mathematics of public-key cryptography. *Scientific America, 241*, 146–157.
109. Hellman, M. E., & Merkle, R. C. (1978). Hiding information and signatures in trap door knapsacks. *IEEE Transactions in Information Theory, IT-24*, 525–530.
110. Garey, M. R., & Johnson, D. S. (1979). *Computers and intractability. A Guide to the Theory of NP-Completeness*. W.H. Freeman.
111. Coppersmith, D. (1984). Fast evaluation of logarithms in fields of characteristic two. *IEEE Transactions in Information Theory, IT-30*, 587–594.

112. El Gamal, T. (1985). A public key cryptosystem and a signature scheme based on discrete logarithms. *IEEE Transactions in Information Theory, IT-314*, 469–472.
113. Gordon, J. A. (1985). Strong prime are easy to find. In *Advance in cryptology. Proceedings of Euro Crypt84* (pp. 216–223). Springer.
114. Fait, A., & Shamir, A. (1986). How to prove yourself: Practical solutions to identifications and signature problems. In *A advance in cryptology-CRYPTO'86* (Lvcs 263, pp. 186–194). Springer.
115. Goldreich, O. (2001). *Foundation of cryptography*. Cambridge University Press.
116. Hill, L. S. (1931). Concerning certain linear transformation apparatus of cryptography. *The American Mathematical Monthly, 38*, 135–154.
117. Knuth, D. E. (1973). *The art of computer programming*. Addison-Wesley.
118. Kranakis, E. (1986). *Primality and cryptography*. Wiley.
119. Kahn, D. (1967). *The codebreakers: The story of secret writing*. Macmillan.
120. Massey, J. L. (1983). Logarithms in finite cyclic group-cryptographic issues. In *Proceedings of the 4th Benelux Symposium on Informations Theory* (pp. 17–25).
121. Koblitz, N. (1994). *A course in number theory and cryptograph*. Springer.
122. Ruggiu, G. (1985). Cryptology and complexity theories. In *Advances in cryptology. Proceedings of Eurocrypt 84* (pp. 3–9). Springer.
123. Rivest, R. L. (1985). RSA chips (past, present, and future). In *Advances in cryptology. Proceedings of Eurocrypt 84* (pp. 159–165).
124. Schneier, B. (1996). *Applied cryptography*. Wiley.
125. Shannon, C. E. (1949). Communication theory of secrecy system. *The Bell System Technical Journal, 28*, 656–715.
126. Odlyzko, A. M. (1985). Discrete logarithms in finite fields and their cryptographic significance. In *Advance in cryptology. Proceedings of Eurocrypt 84* (pp. 224–314). Springer.
127. Wah, P., & Wang, M. Z. (1984). Realization and application of Massey-Omura lock. In *Proceedings of the International, Zürich Seminar* (pp. 175–182).
128. Shamir, A. (1982). A polynomial time algorithm for breaking the basic Markle-Hellman cryptosystem. In *Proceedings of the 23rd Annual Symposium on the Foundations of Computer Science* (pp. 145–152).
129. Stinson, D. R. (2003). *Principles and practice of cryptography* (G. Feng, Trans.). Electronic Industry Press (in Chinese).
130. Cover, T. M. (2003). *Fundamentals of information theory*. Tsinghua University Press (in Chinese).
131. Trappe, W., & Washington, L. C. (2008). *Cryptography and coding theory* (Q. Wang et al., Trans.). People's Posts and Telecommunications Publishing House (in Chinese).
132. Adelman, L. M., Pomerance, C., & Rumely, R. S. (1983). On distinguishing prime number from composite numbers. *Annuals of Mathematics, 117*, 173–206.
133. Blair, W. D., Lacampague, C. B., & Selfridge, J. L. (1986). Factoring large numbers on a Pocket calculator. *The American Mathematical Monthly, 93*, 802–808.
134. Brent, R. P. (1980). An improved Monte Carlo factorization algorithm. *BIT, 20*, 176–184.
135. Berent, R. P., & Pollared, J. M. (1981). Factorization of the eighth Fermat number. *Mathematics of Computation, 36*, 627–630.
136. Cohen, H., & Lenstra, H. W. (1984). Primality testing and Jacobi sums. *Mathematics of Computation, 142*, 297–330.
137. Davenport, H. (1982). *The higher arithemetic*. Cambridge University Press.
138. Dickson, L. E. (1952). *History of the theory of number* (Vol. 1). Chelsea.
139. Dixon, J. D. (1984). Factorization and primality tests. *The American Mathematical Monthly, 91*, 333–352.
140. Guy, R. K. (1975). How to factor a number. In *Proceedings of the 5th Manitoba Conference on Numerical Mathematics* (pp. 49–89).
141. Kranakis, E. (1986). *Primality and cryptography*. Wiley.
142. Lehmer, D. H., & Powers, R. E. (1931). On factoring large number. *Bulletin of the American Mathematical Society, 37*, 770–776.

143. Lehman, R. S. (1974). Factoring large number. *Mathematics of Computation, 28*, 637–646.
144. Miller, G. L. Riemann's hypothesis and tests for primality. In *Proceedings of the 7th Annual ACM Symposium on the Theory of Computing* (pp. 234–239).
145. Morrison, M. A., & Brillhart, J. (1975). A method of factoring and the factorization of $\mathbb{F}_7$. *Mathematics of Computation, 29*, 183–205.
146. Pomerance, C. (1981). Recent development in primality testing. *The Mathematical Intelligencer, 3*, 97–105.
147. Pomerance, C. (1982). The search for prime number. *Scientific American, 427*, 136–147.
148. Pomerance, C. (1982). Analysis and comparison of some integer factoring algorithms. In *Computation Methods in Number Theory, Part 1*. Mathematics Chentrum.
149. Pomerance, C., & Wagstaff, S. S. (1983). Implementation of the continued fraction integer factor in algorithm. In *Proceedings of the 12th Winnipeg Conference on Numerical Methods and Computing*.
150. Rabin, M. O. (1980). Probabilistic algorithms for testing Primality. *Journal of Number Theory, 12*, 128–138.
151. Pollard, J. M. (1975). A Monte Carlo method for factorization. *BIT, 15*, 331–334.
152. Solovag, R., & Strassen, V. (1977). A fast Munte Carlo test for primality. *SIAM Journal for Computing, 6*, 84–85.
153. Wagon, S. (1986). Primality testing. *The Mathematical Intelligence, 8*, 58–61.
154. Wunderlich, M. C. (1979). A running time and analysis of Brillhart's continued fraction factoring method. *Number Theory, Carbondale, Springer Lecture Notes, 175*, 328–342.
155. Wunderlich, M. C. (1985). Implementing the continued fraction factoring algorithm on parallel machines. *Mathematics of Computation, 44*, 251–260.
156. Fulton, W. (1969). *Algebraic curves*. Benjamin.
157. Koblitz, N. (1984). *Introduction to elliptic curves and modular forms*. Springer.
158. Koblitz, N. (1982). Why study equations over finite fields. *Mathematics Magazine, 55*, 144–149.
159. Koblitz, N. (1987). Elliptic curves cryptosystems. *Mathematics of Computation, 48*.
160. Koblitz, N. Primality of the number of points on an elliptic curve over finite field.
161. Gupta, R., & Murty, M. R. (1986). Primitive points on elliptic curves. *Compositio Mathematica, 58*, 13–44.
162. Lenstra, H. W., Jr. (1986). *Factoring integers with elliptic curves*. Report 86-18. Mathematic Institute University of Van Amsterdam.
163. Lenstra, H. W., Jr. (1986). *Elliptic curves and number-theoretic algorithms*. Report 86-19. Mathematics Institute University of Van Amsterdam.
164. Lang, S. (1978). *Elliptic curves: diophantine analysis*. Springer.
165. Miller, V. (1985). Use of elliptic curves in cryptography. In *Abstracts for Crypto'85*.
166. Odlyzko, A. M. (1985). Discrete logarithms in finite fields and their cryptographic significance. In *Advance in cryptography. Proceedings of Eurocrypt 84* (pp. 224–314). Springer.
167. Schoof, H. (1985). Elliptic curves over finite fields and the computation of square roots mod $p$. *Mathematics of Computation, 44*, 483–494.
168. Silverman, J. (1986). *The arithmatic of elliptic curves*. Springer.
169. Pollard, J. M. (1974). Theorems on factorization and primality testing. *Mathematical Proceedings of the Cambridge Philosophical Society, 76*, 521–528.

The manufacturer's authorised representative in the EU is Springer Nature Customer Service Centre GmbH, Europaplatz 3, 69115 Heidelberg, Germany. If you have any concerns regarding our products, please contact ProductSafety@springernature.com

Printed and bound by CPI Group (UK) Ltd, Croydon, CR0 4YY
30/09/2026
02225246-0017